Recognized as an
American National Standard (ANSI)

IEEE Std 739-1995
(Revision of IEEE Std 739-1984)

IEEE Recommended Practice for Energy Management in Industrial and Commercial Facilities

Sponsor

**Energy Systems Committee
of the
Industrial and Commercial Power Systems Department
of the
IEEE Industry Applications Society**

Approved 12 December 1995

IEEE Standards Board

Approved 16 July 1996

American National Standards Institute

Abstract: This recommended practice serves as an engineering guide for use in electrical design for energy conservation. It provides a standard design practice to assist engineers in evaluating electrical options from an energy standpoint. It establishes engineering techniques and procedures to allow efficiency optimization in the design and operation of an electrical system considering all aspects (safety, costs, environment, those occupying the facility, management needs, etc.).
Keywords: break-even analysis; cogeneration; demand control; electrical energy; electric rate structure; energy audit; energy balance; energy conservation program; energy monitoring; energy-rate method; energy savings; heating, ventilating, and air conditioning (HVAC); levelized cost analysis; life cycle costing (LCC); lighting; load management; load type; loss evaluation; marginal cost analysis; metering; power bill; process energy; process modification; product energy rate; space conditioning; utility rate structure

Grateful acknowledgment is made to the following organizations for having granted permission to reprint material in this document as listed below:

ABB Power T&D Company, Inc., Bland, VA for annex 5I.

Acme Electric Co., Lumberton, NC, for data used in 5.4.3, including tables 5-19, 5-20, 5-21, and 5-22.

American Society of Heating, Refrigerating, and Air-Conditioning Engineers (ASHRAE), Atlanta, GA, for table 5-6 from the 1992 ASHRAE Handbook—Systems and Equipment; and for table 5-18 from the 1993 ASHRAE Handbook—Fundamentals.

Butterworth Heinemann Publishers, Oxford, England, for data used in 5.4.5 from the *J&P Transformer Book*, 11th Edition, by A. C. Franklin and D. P. Franklin, 1983.

Calmac Manufacturing Corp., Englewood, NJ, for data used in 5.2.12.

Commonwealth Sprague Capacitors, Inc., North Adams, MA, for tables 5-24, 5-25, and 5-26 from Publication PF-2000B, 1987.

Eaton Corporation, Kenosha, WI, for data used in 5.3.8.3.2 from Technical Bulletin No. 700-4006.

Electric Power Research Institute (EPRI), Palo Alto, CA, for table 5-7 from EPRI RP-2918-15, "Research and Development Plan for Advanced Motors & Drives," 1993; for tables 5-8 and 5-9 from EPRI TR-102639, "Drivers of Electricity Growth and the Role of Utility Demand-Side Management," 1993; and for annex 5A from EPRI TR-101021, "Electrotechnology Reference Guide," Revision 2, Aug. 1992.

National Electrical Contractors Association (NECA), Inc., Bethesda, MD, for data from Electrical Design Library publications used in 5.2.1.

General Electric Co., Fort Wayne, IN, for figure 5-7 from "GE Motors: AC Motor Selection and Application Guide," GET-6812B; for data used in 5.3.8.12 from "How to Maximize the Return on Energy Efficient Motors," GEA-10951C; and for data used in table 5-23 from GEP-500J, "GE Motors Stock Catalog."

Ingersoll-Rand Company, Washington, NJ, for figure 5-1 from *Compressed Air and Gas Handbook*, 3rd edition, 1969.

Kato Engineering, Mankato, MN, Division of Reliance Electric Company, for tables 5-27 and data used in 5.5.5 from AC Synchronous Condenser Bulletin SC 8-92; and for table 5-28 and data used in 5.5.5 from AC Synchronous Motor Bulletins SM 1-94 and SM 2-93.

National Electrical Manufacturers Association (NEMA), Washington, DC, for tables 5-10, 5-11, 5-12, and figure 5-9 from NEMA MG 1-1993, Motors and Generators; for table 5-13, figures 5-5, 5-10, 5-11, 5-12, and annex 5F from NEMA MG 10-1994, Energy Management Guide for Selection and Use of Fixed-Frequency Medium AC Squirrel-Cage Polyphase Induction Motors; and for annex 5E from NEMA MG 11-1992, Energy Management Guide for Selection and Use of Single-Phase Motors.

Reliance Electric Co., Cleveland, OH, for figure 5-3, table 5-14, and data used in 5.3.7.6.1 from "Adjustable Speed Drives as Applied to Centrifugal Pumps," Technical Paper D-7100-1, Oct. 1981; for figure 5-4 and data used in 5.3.7.6.2 from "Fan Control for the Glass Industry Using Static Induction Motor Drives," Technical Paper D-7102, Oct. 1981; for figure 5-8 from "Motor Application," Bulletin B-2615; and for annexes 5G and 5H.

Schindler Elevator Corp., Morristown, NJ, for data used in 5.2.2.2.

SEW-Eurodrive, Inc., Troy, OH, for figure 5-6 and data used in 5.3.8.3.1.

Square D Company, Lexington, KY, for material reprinted in 5.5.3 from Bulletin D-412D, "Power Factor Correction Capacitor—Applications."

State Electricity Commission (SEC) of Victoria, Energy Business Centre, Victoria, Australia, for data used in 5.2.3 from "Electricity for Materials Handling"; for data used in 5.2.4 from "Ultra-Violet Heating at Work in Industry"; for data used in 5.2.5 from "Infrared Heating at Work in Industry"; for data used in 5.2.6 from "Resistance Heating at Work in Industry"; for data used in 5.2.7 from "Radio Frequency Heating of Dielectric Materials"; for data used in 5.2.8 from "Tungsten Halogen Heating"; for data used in 5.2.9 from "Induction Heating and Melting for Industry"; for data used in 5.2.10 from "Induction Metal Joining, Productivity and Energy Efficiency"; for data used in 5.2.11.1 from "Electric Steam Raising"; for data used in 5.2.14, for figure 2, and for annex 5C from the "Compressed Air Savings Manual."

Trans-Coil, Inc., Milwaukee, WI, for the table in 5.5.1.2 from a "Guide to Power Quality."

Most of Chapter 6, including figures and tables, was originally published in a different form in *The Dranetz Field Handbook for Electrical Energy Management,* copyright 1992, Dranetz Technologies, Inc., and is reprinted with permission from Dranetz Technologies. All rights reserved.

First Printing
November 1996
SH94387

The Institute of Electrical and Electronics Engineers, Inc.
345 East 47th Street, New York, NY 10017-2394, USA

ISBN 1-55937-696-1

Introduction

(This introduction is not a part of IEEE Std 739-1995, IEEE Recommended Practice for Energy Management in Industrial and Commercial Facilities.)

IEEE Std 739-1984, IEEE Recommended Practice for Energy Conservation and Cost-Effective Planning in Industrial Facilities, was the precursor to this revision. That publication was born out of a need to convey conservation techniques to electrical engineers, designers, and operators. Much had been written for mechanical and architectural engineers at that point in time, but little had been written and disseminated to electrical engineers.

This new version has changed in several ways and has added new material that is the result of a decade of research and innovation by IEEE and others. The most obvious change is the title of this standard, "IEEE Recommended Practice for Energy Management in Industrial and Commercial Facilities." The title shows a recognition of the need to manage a valuable resource—electrical energy. The title change also shows an expansion of scope. The scope expansion resulted when the parent committee sponsoring the work changed to the Energy Systems Committee. The Energy Systems Committee is one of five main technical committees in the Industrial and Commercial Power Systems Department of the IEEE Industry Applications Society. This new sponsorship expanded the focus of this recommended practice to include commercial facilities.

We are thankful to those who have given time and effort to the birthing of this recommended practice and who are no longer members of this committee. In particular, John Linders, Mel Chiogioji, Art Killin, H. L. (Sonny) Harkins, and Terry McGowan should be remembered as pioneers in the establishment of this recommended practice.

This IEEE recommended practice continues to serve as a companion publication to the following other recommended practices prepared by the IEEE Industrial and Commercial Power Systems Department:

— IEEE Std 141-1993, IEEE Recommended Practice for Electric Power Distribution for Industrial Plants (IEEE Red Book).

— IEEE Std 142-1991, IEEE Recommended Practice for Grounding of Industrial and Commercial Power Systems (IEEE Green Book).

— IEEE Std 241-1990, IEEE Recommended Practice for Electric Power Systems in Commercial Buildings (IEEE Gray Book).

— IEEE Std 242-1986 (Reaff 1991), IEEE Recommended Practice for Protection and Coordination of Industrial and Commercial Power Systems (IEEE Buff Book).

— IEEE Std 399-1990, IEEE Recommended Practice for Industrial and Commercial Power Systems Analysis (IEEE Brown Book).

— IEEE Std 446-1995, IEEE Recommended Practice for Emergency and Standby Power Systems for Industrial and Commercial Applications (IEEE Orange Book).

— IEEE Std 493-1990, IEEE Recommended Practice for the Design of Reliable Industrial and Commercial Power Systems (IEEE Gold Book).

- IEEE Std 602-1996, IEEE Recommended Practice for Electric Systems in Health Care Facilities (IEEE White Book).

- IEEE Std 1100-1992, IEEE Recommended Practice for Powering and Grounding Sensitive Electronic Equipment (IEEE Emerald Book).

Participants

The Bronze Book Working Group for the 1995 edition had the following membership:

Carl E. Becker, *Chair*

Chapter 1: Overview—**Carl E. Becker**, *Chair;*
Charles N. Claar; Daniel L. Goldberg; Lawrence G. Spielvogel

Chapter 2: Organizing for energy management—**Carl E. Becker**, *Chair;*
Wayne L. Stebbins

Chapter 3: Translating energy into cost—**Carl E. Becker**, *Chair;*
Barry N. Hornberger; Douglas Cato; Joseph Eto

Chapter 4: Load management—**Richard C. Lennig,** *Chair;*
Kao Chen; Daniel L. Goldberg; R. Gerald Irvine; Sukanta Sengupta;
Wei-Jen Lee

Chapter 5: Energy management for motors, systems, and electrical equipment—
R. Gerald Irvine, *Chair;* Joseph Eto; Douglas Cato; Wei-Jen Lee; Chris Duff;
Pat O'Neal; Walter Rusuck; Paul Moser

Chapter 6: Metering for energy management—**Wayne L. Stebbins**, *Chair;*
Carl E. Becker

Chapter 7: Energy management for lighting—**Kao Chen**, *Chair;*
Daniel L. Goldberg; R. Gerald Irvine; John Stolshek; John Verderber

Chapter 8: Cogeneration—**Barry N. Hornberger**, *Chair;*
Richard S. Bono; C. Grant Keough

The following persons were on the balloting committee:

James Beall	Daniel L. Goldberg	Richard C. Lennig
Carl E. Becker	Erling Hesla	Dan Love
Douglas Cato	Barry N. Hornberger	Gregory Nolan
Charles N. Claar	James Jones	James Pfafflin
Bruce Douglas	C. Grant Keough	Wayne L. Stebbins
	Wei-Jen Lee	

When the IEEE Standards Board approved this recommended practice on 12 December 1995, it had the following membership:

Contents

IEEE Recommended Practice for Energy Management in Industrial and Commercial Facilities

Chapter 1
Overview

1.1 Scope

This IEEE recommended practice was conceived in the early 1970s shortly after the oil embargo. The purpose was to publish an engineering guide for use in electrical design for energy conservation. The purpose of this recommended practice continues to be one of providing a standard design practice to assist engineers in evaluating electrical options from an energy standpoint. Hence, it is a recommended practice for energy management in design and operation of an electrical system.

This recommended practice is not intended to be one to set minimum values for regulatory (law making) purposes. The intent is rather to establish engineering techniques and procedures to allow efficiency optimization in the design and operation of an electrical system considering all aspects (safety, costs, environment, those occupying the facility, management needs, etc.). Other national standards are mentioned where applicable for reference by the reader. State and local governments usually adopt some or all of these national standards, which makes them law, and on occasion, the governing body prepares its own standard(s).

1.2 General discussion

IEEE Std 739-1995, IEEE Recommended Practice for Energy Management in Industrial and Commercial Facilities (commonly known as the IEEE Bronze Book) is published by the Institute of Electrical and Electronics Engineers (IEEE) to provide a recommended practice for electrical energy management in industrial and commercial facilities. It has been prepared by engineers and designers on the Energy Systems Committee of the IEEE Industrial and Commercial Power Systems Department (ICPS) with the assistance of the Production and Application of Light Committee (PAL).

This recommended practice will probably be of greatest value to the power-oriented engineer with some design or operation experience with industrial and commercial facilities. It can be an aid, however, to engineers and designers at all levels of experience. It should be considered a guide and reference rather than a detailed manual, to be supplemented by the many excellent publications available.

There will be an overlap with other fields of engineering, particularly mechanical and architectural, in the area of building systems, which are covered in this recommended practice.

When references are made to codes, standards, laws, and regulations, it is essential that the document referred to or current authentic interpretation be consulted. Such material changes frequently in contrast to conventional engineering information, and it is impractical to include complete current regulations or detailed interpretations in a text of this size.

1.3 Management

Energy management embodies engineering, design, applications, utilization, and to some extent the operation and maintenance of electric power systems to provide for the optimal use of electrical energy. *Optimal* in this case refers to the design or modification of a system to use minimum overall energy where the potential or real energy savings are justified on an economic or cost benefit basis. Optimization also involves factors such as comfort, healthful working conditions, the practical aspects of productivity, aesthetic acceptability of the space, and public relations.

Energy management involves the following professions and fields:

— Engineering
— Management
— Economics
— Financial analysis
— Operations research (system analysis)
— Public relations (selling conservation)
— Environmental engineering

Some of the tools that are utilized include the following:

— Meters and measurement
— Demand and energy limiters
— Highly efficient energy devices
— Control systems (e.g., building management systems)

While this recommended practice emphasizes electrical energy management, other forms of energy are also discussed. Chapter 8 covers the combined efficient use of electrical and thermal energy (steam) in a highly efficient and cost-effective manner. Chapter 2 points out that the reduction of energy in one area can have a negative effect on energy needs in another area. For example, the reduction of lighting levels in large buildings in winter can cause increased operation of the heating system because of the loss of heat from the lamps and ballasts.

1.4 Fuel cost effects on electrical energy

Since fuel cost impacts the cost of electrical energy, this subclause discusses the factors that affect the costs of today's major fuels.

All known fuels or energy sources have a life cycle that begins with the development of a means to effectively manufacture and utilize the fuel itself. The energy source then grows in usage to a point of maximum production at which the usage begins to decline. This decline is caused by many factors, including obsolescence and depletion of supply. Today's energy sources can be divided into three major categories: fossil, solar, and nuclear fuels.

1.4.1 Fossil fuel

The three most common fossil fuels are coal, natural gas, and oil. Their cleanliness, heat content, and availability will affect their use and electrical costs in the future, so these aspects are covered in this subclause.

a) Coal is generally acknowledged as the most plentiful fossil fuel resource in the U.S. At current consumption rates, the supply should last at least 100 years. The Btu content and quality of coal varies from region to region, and it is not uniformly available throughout the country. Problems affecting the use and availability of coal are: mining, including site restoration and water pollution; transportation; air pollution; and lack of a coherent governmental coal-burning policy that encourages the use of this resource. However, industry and government continue to fund programs for clean coal burning research. Some new techniques are being tested for the use of coal, and new coal burning technologies should be available in the coming years.

   One advantage of coal as an energy source is that the industry is unregulated. This fact makes coal costs subject, to some extent, to the laws of supply and demand. However, government health and safety mining regulations, in addition to inflationary pressures and the energy situation in general, have caused coal prices to soar. Coal prices should track inflation under the present conditions.

b) Natural gas is the cleanest burning fossil fuel, but it has the least proven reserves. Natural gas presents few air pollution problems and is probably the least capital intensive of the fossil fuels. Questionable availability remains its chief disadvantage, although deregulation has encouraged exploration and the use of wells that were previously thought to be uneconomical to operate.

   In summary, gas prices should rise at the rate of inflation, but deregulation could adversely affect the price if, at some point, the supply fails to keep pace with demand.

c) Oil is the second cleanest burning fuel, and it is not very capital intensive when compared to coal. Because of these two factors, it has distinct advantages as an industrial and commercial fuel. Availability could be a future problem. As long as the U.S. imports oil there should be few problems with availability. The price of domestic oil was regulated by the Federal government and is now in the process of deregulation, while the world price of oil is unregulated.

The problems caused by the oil embargo of 1973 and the Iraq conflict are well documented. In summary, the price of foreign oil has no relationship to the laws of supply and demand. The U.S. demand for oil has outstripped our domestic supply. Without foreign oil, we would deplete our reserves in approximately two decades. The U.S. is confronted, therefore, with a problem concerning the use of oil now and in the immediate future. Oil prices will continue to fluctuate due to world market changes.

1.4.2 Solar fuel

Solar energy encompasses the use of the sun's radiation to produce electricity or as a direct source for heating and cooling. Solar energy has gained attention because of governmental and general public pressure due to its non-polluting characteristic.

Five primary types of solar energy are: photovoltaic, thermal, wind, hydropower, and biomass. Photovoltaic involves the direct conversion of the sun's rays into electricity. Thermal conversion uses the sun's rays to create heat, which is either used directly or to create steam for electric generation. Wind is generally used to provide shaft horsepower to a generator. Biomass primarily involves trash as a heat source for a steam-driven turbine generator.

a) Photovoltaics are probably the most interesting means of converting solar energy into electricity. They have relatively poor efficiencies and cost several times more per kilowatthour than other methods. They can only generate electricity on sunny days. However, the strides in development have been great, and they will probably be a viable alternate in future years.

b) Thermal conversion has not proven economical for generating electricity. However, heating and cooling systems are now marketed that compete favorably with other energy sources on a five- to ten-year payback basis. A central collector has been proposed for use in generating steam for a turbine-driven generator. However, the land requirement is large at over 150 acres per electric megawatt output, and the output is restricted to daylight hours.

c) Wind conversion has been used successfully to generate electrical energy. Some disadvantages of wind generation include noise, radio and TV interference, very fast tip speeds (60–80 ft/s), and size limitation. However, there are areas of the country with typical wind speeds of 7–30 mi/h where windmill farms successfully generate bulk power.

d) Most electricity generated today by hydropower is by high head dams, large scale operations that only large investors can afford to finance. However, low head dams are becoming more attractive as a means to generate electricity. The use of low head dams could improve the availability of electricity on a local basis.

e) Biomass is being used in conjunction with other fuels, due to low heat content. This source may begin to supply small amounts of power for municipalities and industrial complexes. As with coal, there are many pollution and handling problems.

1.4.3 Nuclear fuel

There are three nuclear conversion techniques: fission, breeder, and fusion reactors. While this is the cleanest and most economical source of power, regulations and political pressures make future dependence on this source questionable. Fission conversion has proven economical for three to four decades. However, the disposal of nuclear waste and eventual plant decommissioning methods have not received universal concurrence. Furthermore, the known uranium reserve would only supply energy for 100 years to a nation totally reliant on fission power.

The breeder reactor can multiply the nuclear supply by a factor of ten by recycling nuclear waste. This concept has been demonstrated on a large scale, but development in the U.S. has been virtually at a standstill.

The fusion concept has been sustained on a laboratory level, but full development is not anticipated for decades. Fusion could ultimately provide a virtually unlimited supply of energy at a very low cost without pollution. However, the infancy of fusion technology makes this source inappropriate for consideration at this time.

From all indications, the cost of electricity will increase at a rate at least equal to that of annual inflation. Coal will probably be the primary source of fuel in the U.S. Petroleum and gas will probably increase in price at a rate faster than electricity, but the relative costs and rates will differ in various areas.

1.5 Periodicals

The energy engineer needs to keep up-to-date on new technologies in this rapidly expanding field. The following periodicals offer new engineering information on energy conservation.

a) *Spectrum,* the monthly magazine of the IEEE, covers all aspects of electrical and electronics engineering, including conservation. It contains references to IEEE books and other publications; technical meetings and conferences; IEEE group, society, and committee activities; abstracts of papers and publications of the IEEE and other organizations; and other material essential to the professional advancement of the electrical engineer.

b) *IEEE Transactions on Industry Applications* contains papers presented to conferences, a number of which include the subject of energy management. All members of the Industry Applications Society (IAS) receive this publication.

c) Publications that are specifically oriented toward energy conservation include:

    1) *Energy User News*[1] is a monthly news publication in tabloid format that covers the entire field of energy conservation with emphasis on keeping the reader

---

[1]Radnor, PA.

informed of new technologies. It is quite open in discussing the effectiveness or failures of systems. A legislative and regulatory scoreboard is included.

2) *NTIS,* the National Technical Information Service of the U.S. Department of Commerce,[2] is an abstract newsletter published weekly. It contains abstracts of hundreds of articles relating to energy, listed by subject. An order form for purchasing many of the articles is included with each issue.

d) The Association of Energy Engineers[3] publishes several documents including the bimonthly *Energy Engineering* and the quarterly *Strategic Planning for Energy and the Environment.* In addition, the Association sponsors conferences.

e) Some other periodicals of a more general nature that are heavily involved with electrical energy conservation are as follows:

1) *Electrical Construction and Maintenance,* 1221 Avenue of the Americas, New York, NY 10020.

2) *Electrical Consultant,* One River Road, Cos Cob, CT 06807.

3) *LD & A,* Illuminating Engineering Society, 345 East 47th Street, New York, NY 10017.

4) *Plant Engineering,* 1350 Touhy Avenue, P. O. Box 5080, Des Plaines, IL 60017-5080.

5) *Power,* 1221 Avenue of the Americas, New York, NY 10020.

6) *Power Engineering,* 1301 South Grove Avenue, Barrington, IL 60010.

7) *Electric Light and Power,* 1301 South Grove Avenue, Barrington, IL 60010.

8) *Consulting-Specifying Engineer,* 44 Cook Street, Denver, CO 80206-5800.

f) Two technical periodicals that contain material of interest are as follows:

1) *Energy Management* is directed at management in business and industry. It is published bimonthly by Penton, 614 Superior Avenue West, Cleveland, OH 44113.

2) *Plant Energy Management* is published bimonthly by Walker-Davis Publication, Inc., 2500 Office Center, Willow Grove, PA 19090.

It should be noted that new periodicals can be expected in the expanding field of energy. Some of the publications are available at no cost to qualified individuals engaged in energy management work.

---

[2]5285 Port Royal Road, Springfield, VA 22161.
[3]4025 Pleasantdale Road, Suite 420, Atlanta, GA 30340.

1.6 Standards and Recommended Practices

A number of organizations in addition to the National Fire Protection Association (NFPA)[4] publish documents that affect electrical design. Required adherence to these documents can be written into design specifications.

a) The American National Standards Institute (ANSI)[5] coordinates the review of proposed standards among all interested affiliated societies and organizations to ensure a consensus approval. It is in effect the *clearing house* in the U.S. for technical standards of all types.

b) Underwriters' Laboratories, Inc. (UL)[6] is a nonprofit organization, operating laboratories for investigation of materials and products, especially electrical appliances and equipment with respect to hazards affecting life and property.

c) The Edison Electric Institute (EEI) and the Electric Power Research Institute (EPRI) represent the investor-owned utilities and publish extensively. The following handbooks are available through EEI:[7]

1) *Electrical Heating and Cooling Handbook*
2) *A Planning Guide for Architects and Engineers*
3) *A Planning Guide for Hotels and Motels*
4) *Electric Space-Conditioning*
5) *Industrial and Commercial Power Distribution*
6) *Industrial and Commercial Lighting*

d) The National Electrical Manufacturer's Association (NEMA)[8] represents equipment manufacturers. Their publications standardize the manufacture of, and provide testing and operating standards for, electrical equipment. The design engineer should be aware of any NEMA standard that might affect the application of any equipment that he or she specifies.

e) The IEEE publishes several hundred electrical standards relating to safety, measurements, equipment testing, application, and maintenance. The following three publications are general in nature and are important for the preparation of plans:

1) IEEE Std 100-1992, The New IEEE Standard Dictionary of Electrical and Electronics Terms (ANSI).

2) IEEE Std 315-1975 (Reaff 1993), IEEE Standard Graphic Symbols for Electrical and Electronics Diagrams, and IEEE Std 315A-1986 (Reaff 1993), Supplement to IEEE Std 315-1975 (ANSI).

---

[4] 1 Batterymarch Park, P.O. Box 9101, Quincy, MA 02269-9101.
[5] 11 West 42nd Street, 13th Floor, New York, NY 10036.
[6] 333 Pfingsten Road, Northbrook, IL 60062-2096.
[7] 701 Pennsylvania Ave. NW, Washington, DC 10004-2696.
[8] 1300 N. 17th St., Ste. 1847, Rosslyn, VA 22209.

3) IEEE Std Y32.9-1972 (Reaff 1989), American National Standard Graphic Symbols for Electrical Wiring and Layout Diagrams Used in Architecture and Building Construction.

f) The Building Officials and Code Administrations International, Inc. (BOCA®)[9] is an organization that promulgates the BOCA building construction model codes. It also provides educational services in code administration and enforcement. Many governmental bodies have mandated its codes as the governing construction code. They published the BOCA National Energy Conservation Code of 1993.

g) The Council of American Building Officials (CABO)[10] published the "Model Energy Code" in 1992 with input from several organizations: Building Code Officials and Code Administrators International, Inc. (BOCA); International Conference of Building Officials (ICBO); National Conference of States on Building Codes and Standards (NCSBCS); and Southern Building Code Congress International, Inc. (SBCCI).

1.7 Industry Applications Society (IAS)

The IAS is one of 37 IEEE groups and societies that specialize in various technical areas of electrical and electronics engineering. Each group or society conducts meetings and publishes papers on developments within its specialized area. Some of the more relevant technical committees include the Petroleum and Chemical Industry, the Cement Industry, the Glass Industry, and the Industrial and Commercial Power Systems Committees. Papers of interest to electrical engineers and designers involved in the fields covered by the IEEE Bronze Book are, for the most part, contained in the *IEEE Transactions on Industry Applications*.

1.8 IEEE publications

The IEEE Bronze Book is one of a series of standards that are published by IEEE and are known as the IEEE Color Books. These standards are prepared by the Industrial and Commercial Power Systems Department of the IEEE Industry Applications Society.[11] They are as follows:

— IEEE Std 141-1993, IEEE Recommended Practice for Electric Power Distribution for Industrial Plants (IEEE Red Book).

— IEEE Std 142-1991, IEEE Recommended Practice for Grounding of Industrial and Commercial Power Systems (IEEE Green Book).

---

[9]4051 West Flossmoor Road, Country Club Hills, IL 60478-5795.

[10]4205 Leesburg Pike, Falls Church, VA.

[11]IEEE publications are available from the Institute of Electrical and Electronics Engineers, 445 Hoes Lane, P.O. Box 1331, Piscataway, NJ 08855-1331.

— IEEE Std 241-1990, IEEE Recommended Practice for Electrical Power Systems in Commercial Buildings (IEEE Gray Book).

— IEEE Std 242-1986 (Reaff 1991), IEEE Recommended Practice for Protection and Coordination of Industrial and Commercial Power Systems (IEEE Buff Book).

— IEEE Std 399-1990, IEEE Recommended Practice for Industrial and Commercial Power System Analysis (IEEE Brown Book).

— IEEE Std 446-1987, IEEE Recommended Practice for Emergency and Standby Power Systems for Industrial and Commercial Applications (IEEE Orange Book).

— IEEE Std 493-1990, IEEE Recommended Practice for the Design of Reliable Industrial and Commercial Power Systems (IEEE Gold Book).

— IEEE Std 602-1996, IEEE Recommended Practice for Electric Systems in Health Care Facilities (IEEE White Book).[12]

— IEEE Std 739-1995, IEEE Recommended Practice for Energy Management in Industrial and Commercial Facilities (IEEE Bronze Book).

— IEEE Std 1100-1992, IEEE Recommended Practice for Powering and Grounding Sensitive Electronic Equipment (IEEE Emerald Book).

1.9 Governmental regulatory agencies

a) Actions of the U.S. government regulatory agencies are embodied in the *Federal Register.*[13] Often the rules have sufficient legal complexity that interpretations are best left to experts in the field. The agencies making such rules shall operate within the scope of laws enacted by Congress.

b) The Department of Energy (DOE) sponsors experimental or demonstration projects through grants and other incentives. Reports are issued on programs such as the fluorescent lamp solid-state ballast under grant to Lawrence Berkeley Laboratories of the University of California.

c) Some individual states have their own Departments of Energy. Agencies such as the DOE are able to exert a powerful influence on these local departments through law or by the ability to withhold funds for areas deemed not in compliance with federal regulations. At the present time, ASHRAE/IESNA 90.1-1989[14] is accepted as a basis for design of new and existing commercial buildings. While some states have accepted these standards almost verbatim, a few other states have issued their own equivalents. It is important to recognize which rulings are based on law in the locality under consideration.

---

[12]As this standard goes to press, IEEE Std 602-1996 is approved but not yet published. The draft standard is, however, available from the IEEE. Anticipated publication date is March 1997. Contact the IEEE Standards Department at 1 (908) 562-3800 for status information.

[13]This daily publication is available from the U.S. Government Printing Office, Washington DC 20402.

[14]ASHRAE publications are available from the Customer Service Dept., American Society of Heating, Refrigerating and Air-Conditioning Engineers, 1791 Tullie Circle, NE, Atlanta, GA 30329.

d) ASHRAE/IESNA 90.1-1989, Energy Efficient Design of New Buildings Except Low-Rise Residential Buildings, was developed by the American Society of Heating, Refrigerating and Air-Conditioning Engineers (ASHRAE) in conjunction with members of the Illuminating Engineering Society of North America (IESNA). The production of this recommended practice was encouraged by various governmental agencies to fill a need for a standard developed on a consensus basis by voluntary professional organizations knowledgeable in energy conservation. However, this and other standard documents have legal status only where local authorities have incorporated their recommendations in law.

e) Where generation (except for emergency and standby purposes) is involved and especially where the resale and redistribution of energy for use by others is concerned, regulations of the Federal Energy Regulatory Commission (FERC) may apply. PURPA, the Public Utility Regulatory Policies Act of 1978, provides guidelines for the sale and resale of energy and for exemption from FERC rulings, which are applicable to utilities. Industries employing cogeneration or generation not requiring fossil fuels fall under PURPA jurisdiction. These industries are exempt from certain FERC regulations. They also have certain advantages in their relations with utility companies.

f) Public Service Control Boards or Utility Regulatory Commissions govern the actions of the local utilities. They enact rules and establish rates, after public hearings, responsive to the needs of the utilities and their customers and to the pressures of governmental bodies as described above. During recent years, these Commissions have recognized energy conservation needs by establishing time of day, incremental, cogeneration, and similar rates intended to minimize energy and peak demand power usage. The effects of such actions are expressed in the rates and rules of the local utilities.

g) Utilities usually provide energy conservation services in addition to their more conventional advice on rates and conditions of service. Some utilities will provide energy audits and some may be required to check the energy budget before providing service. However, the advice they provide is usually more general than that of the consultant or in-house engineer.

1.10 Keeping informed

The following suggestions offer ways to remain current amid the abundance of information being generated in the energy field:

a) Read the aforementioned periodicals. Some are dedicated to energy only, while others can be scanned for energy articles.

b) Attend energy-oriented conferences and courses, hundreds of which are given throughout the country. The courses are often conducted for one or two days and may cover the subject in general or concentrate on special aspects.

c) Become active in the energy committees of societies such as the IEEE, IESNA, AIA (American Institute of Architects),[15] and ASHRAE on a local or national basis.

d) Read professional books and handbooks dealing with conservation. These will often be advertised or listed in the periodicals referred to above.

e) Read advertising literature, which, although often biased, can provide an excellent guide to new equipment, its application, and the basis of justification.

f) If you are a manager, establish the programs described in the following chapters, utilize the services of consultants, and attend energy seminars specifically designed for the manager, executive architect, financial planner, etc.

1.11 Professional activities

If an engineer is to practice publicly, professional registration is often required. Many organizations require such registration for certain levels of engineering. Regulatory agencies require that designs for public and commercial buildings be prepared under the jurisdiction of state-licensed professional architects or engineers. Information on such registration may be obtained from the appropriate state agency or from the local chapter of the National Society of Professional Engineers.

To facilitate obtaining registration in different states under the reciprocity rule, a National Professional Certificate is issued by the Records Department of the National Council of Examiners for Engineering and Surveying (NCEE)[16] to engineers who obtained their home-state license by examination. All engineering graduates are encouraged to start on the path to full registration by taking the engineer-in-training examination as soon after graduation as possible. The final written examination in the field of specialization is usually conducted after four years of progressive professional experience.

1.12 Coordination with other disciplines

During the course of work, the energy engineer will often span several of the conventional disciplines and will often work with engineers from other disciplines.

The energy engineer is concerned with professional associates such as the architect, the mechanical engineer, the structural engineer, and, where underground services are involved, the civil engineer. The energy engineer is also concerned with the builder and the building owner or operator who, as clients, may take an active interest in the design or redesign. The energy engineer will work directly with the architect and will cooperate with the safety engineer, fire-protection engineer, environmental engineer, and a host of other concerned people, such as interior decorators, all of whom may have a say in the ultimate design or modification for conservation.

In performing conservation design, it is essential, at the outset, to prepare a checklist of all the design stages that have to be considered. Major items include staging, clearances, access

[15]1735 New York Ave. NW, Washington, DC 20006.
[16]P.O. Box 1686, Clemson, SC 29633-1686.

to the site, and review by others. It is important to note that certain conservation work may appear in several chapters of the general construction specifications. Building control systems, even if electronically based, may appear in the mechanical section or in a separate section. Many organizations utilize the specification format of the Construction Specifications Institute (CSI)[17] where the energy conservation work of the electrical-mechanical engineering will usually fall in Chapter 15, Chapter 16 (electrical), and (if included) Chapter 17 (special building control systems). For example, furnishing and connecting of chiller electric motors may be covered in the mechanical section of the specifications. For administrative purposes, the work may be divided into a number of contracts, some of which may be awarded to contractors of different disciplines. The designer will be concerned with items such as preliminary estimates, final cost estimates, plans or drawings, specifications (which are the written presentation of the work), materials, manuals, factory inspections, laboratory tests, and temporary power. The designer may also be involved in providing information on how the proposed conservation items affect financial justification of the project in terms of owning and operating costs, amortization, return on investment, and related items.

The energy engineer is encouraged to proceed with care in making changes that affect people. A test installation is often justified in determining the acceptability of an installation. Human reactions to changed environment often cannot be modified without considering the other physical factors. Some typical pitfalls are listed below.

— Are lighting color changes acceptable and compatible?

— Are workers uncomfortable in the changed ventilation conditions?

— Is task lighting producing a gloomy environment?

— Is machinery operating at reduced power because of voltage reduction in a brownout?

— Is the area uncomfortably humid because of loss of reheat?

— Is the building computer properly programmed and debugged?

— Is an otherwise perfectly good lighting redesign producing a rash of headache and eye complaints?

— Are the building tenants sharing in energy savings?

The experienced energy engineer will have encountered most of these conditions, which are covered in this text, and be able to avoid the possible negative side of energy conservation. A host of tools described in this recommended practice are available to ease the work of the engineer in preparing a well conceived, energy-conserving design.

---

[17]601 Madison Avenue, Industrial Park, Alexandria, VA 22314.

1.13 Text organization

Subsequent chapters cover energy management in the following sequence of subjects:

Chapter 2: Organizing for energy management.
The proper organization and management commitment are prerequisites to a successful energy conservation effort.

Chapter 3: Translating energy into cost.
Several models for evaluating energy alternatives are provided, as well as a method for motor and transformer loss evaluation. Electric rate structures are explained.

Chapter 4: Load management.
The use of demand control can reduce electric bills but not necessarily energy use. Demand control techniques and, more importantly, energy control techniques, are discussed.

Chapter 5: Energy management for motors, systems, and electrical equipment.
Proper application of equipment along with correct design and maintenance of the electrical system and its components are key factors in energy conservation and system efficiency.

Chapter 6: Metering for energy management.
Proper measurement techniques ensure a viable and continually effective energy conservation program.

Chapter 7: Energy management for lighting systems.
New devices and proper application of existing equipment can provide significant energy savings.

Chapter 8: Cogeneration.
The concurrent use of steam (or other heat sources) for process (and comfort) heating, and electric generation provides a very efficient process, provided that specific requirements are met.

Design details are too intricate for discussion in this text. The reader is therefore advised to make extensive use of design manuals and other sources.

Chapter 2
Organizing for energy management

2.1 Introduction

To understand the energy consumption patterns in a facility, it is important to understand the applications of energy processes. Energy applications are grouped into six major types.

2.1.1 Types of energy applications

a) *Space conditioning.* Energy used directly for heating or cooling an area for comfort conditioning.

b) *Boiler fuel.* This is subdivided into space conditioning and process energy, depending on how the steam or hot water from the boiler is utilized.

c) *Direct process heat.* Energy used to heat the product being processed, e.g., for kilns, reheat furnaces, etc., excluding energy used in boiler steam or hot water.

d) *Feedstock.* Fuel used as an ingredient in the process, e.g., electroplating and sodium production.

e) *Lighting.*

f) *Mechanical drive.* Motors used for ventilation systems, pumps, crushers, grinders, production lines, etc. possess this mechanical drive.

Energy savings can result from improvements in the efficiency of the energy conversion processes, either by recycling waste energy or by reusing waste materials.

Any process requires a certain minimum consumption of energy. Energy (or equipment) additions beyond this minimum require an evaluation of the incremental cost of more efficient equipment or techniques versus the resulting energy savings or costs. Some of the more intensive users of industrial energy, including chemicals, paper, and petroleum refining, have long found it competitively advantageous to design for energy conservation. Virtually all new commercial/industrial facility designs consider energy conservation in some form.

Two economic incentives exist for the development of an energy management program on a facility-by-facility basis: savings realized by reducing energy use, and preventing economic losses by minimizing the probability of fuel supply curtailment.

With increasing costs of energy, every facility can benefit by seeking to reduce energy consumption.

2.1.2 Energy saving methods

The four general categories in which energy savings can be grouped are as follows:

a) *Housekeeping measures.* Energy savings can result from better maintenance and operation. Such measures include shutting off unused equipment; improving electricity demand management; reducing winter temperature settings; turning off lights; and eliminating steam, compressed air, and heat leaks. Proper lubrication of equipment, proper cleaning and replacement of filters in equipment, and periodic cleaning and lamp replacement in lighting systems will result in optimal energy use in existing facilities.

b) *Equipment and process modifications.* These can be either applied to existing equipment (retrofitting) or incorporated in the design of new equipment. Examples include the use of more durable or more efficient components; the implementation of novel, more efficient design concepts; or the replacement of an existing process with one using less energy.

c) *Better utilization of equipment.* This can be achieved by carefully examining the production processes, schedules, and operating practices. Typically, industrial plants are multiunit, multiproduct installations that evolved as a series of independent operations with minimum consideration of overall plant energy efficiency. Improvements in plant efficiency can be achieved through proper sequencing of process operations, rearranging schedules to utilize process equipment for continuous periods of operation to minimize losses associated with start-up; scheduling process operations during off-peak periods to level electrical energy demand; and conserving the use of energy during peak demand periods. Commercial facilities will typically achieve energy savings by relamping, installing adjustable speed drives in ventilation systems, and considering solar effects.

d) *Reduction of losses in the building shell.* Reduction in heat loss is achieved by adding insulation, closing doors, reducing exhaust, utilizing process heat, etc.

Management should provide effective personnel motivation, planning, and administration to achieve meaningful energy savings. The establishment of a formalized energy management responsibility is highly desirable to give the effort both the focus and the direction required. An energy management function gives the line manager the tools to get the job done. Line managers need to know their energy use and costs, the future energy supply availability and its cost, problems or opportunities of energy situations, and those alternative solutions worth pursuing.

The following clauses attempt to identify and describe the steps needed to implement an effective energy management program. The engineer or plant manager can be a catalyst in developing a solid, functional energy management program.

2.2 Organizing the program

At any level of the corporate structure, an individual should understand the incentives and motivations of top management. An alert engineer will become aware of the hidden lines of authority and the key persons who make decisions. These key persons should be convinced of the positive merits of an energy management program before it can be successful. Therefore, it is the engineer's job to see that the proper facts are presented through the proper channels to convince top management that they should make the energy commitment. Sometimes the informal organizational structure is an effective tool.

The five critical factors in organizing an effective energy management program are as follows:

a) *Obtain top management energy commitment.* This is a formally communicated, financially supported dedication to reducing energy consumption while maintaining or improving the functioning of a facility. This commitment shall be active and clearly and visibly communicated to all levels of the organization in terms of words and actions by top management.

b) *Obtain people commitment.* People at all levels of the organization should be involved in the program. Ideas should be encouraged with rewards for significant contributions to the energy management program. People should be shown why their help is needed, and a team approach should evolve. The most successfully planned program can be devastated by a single person trying to subvert the program.

c) *Set up a communication channel.* The purpose of this channel is to report to the organization the results of your efforts, to recognize high achievers, and to identify reward recipients. Use the channel to advertise the program and to encourage cooperation.

d) *Change or modify the organization to give authority and commensurate responsibility for the conservation effort and develop an energy management program.*

e) *Set up a means to monitor and control the program.*

An organizational plan, using the above criteria, should then be developed for both implementing and monitoring specific energy management programs. This plan should also define the responsibilities of the energy management coordinator or committee; describe an effective communication system between coordinator and major divisions, departments, and employees; establish an energy accounting and monitoring system; and provide the means for educating and motivating employees. Realistic goals should be established that are specific in both amount and time. The goals should take into consideration the effect of diminishing returns.

The energy team should at least consist of representatives from each major facility, engineering, operations, and the union or labor representative. Other members might include personnel from purchasing, accounting, and finance. The team should have authority adequate to investigate prevailing energy supply and demand situations and to implement policy recommendations. Their tasks should be assigned a priority consistent with the current or potential importance of energy problems in the company operation. They should utilize those talents within the firm in making the analysis and implementing the programs.

2.3 Surveying energy uses and losses

The first step for the energy team is to use an energy audit to determine the amount of energy that enters and leaves a plant. This determination will probably be an approximation at first, but the accuracy should improve with experience. The audit consists of a survey and appraisal of energy and utility systems at various levels of detail.

There is a direct relationship between the extent of data collection and consolidation, and the subsequent evaluation of energy conservation opportunities. While an insufficient database may prevent the identification of several energy-saving opportunities, too extensive a baseline survey may prove unnecessary and wasteful by diverting funds and time from more rewarding conservation opportunities. While the best results can be obtained by conducting a thorough, comprehensive survey and analyzing all site energy and utility systems, time and budgetary constraints may impose limitations on the extent of survey and appraisal for various site energy systems.

Certain items should be covered. Power bills should be reviewed considering the quantities noted in 2.3.1. At least a "walk through" audit should be conducted and preferably a detailed audit. An audit will uncover such items as unnecessary operation of equipment, unnecessary high levels of lighting, no one assigned to turn off unnecessary lights and equipment, improper thermostat setting, unnecessary or excess ventilation, and obvious waste. A comprehensive tour will also involve acquisition of nameplate data, motor and lighting loads, and other system details.

2.3.1 Establishing the energy relationship and the power bill

It is important, from several standpoints, to establish the existing pattern of electrical usage and to identify those areas where energy consumption could be reduced. A history on a month-by-month basis of electric usage is available from the electric bills, and this usage should be carefully recorded in a format (possibly graphic) that will facilitate future reference, evaluation, and analysis.

The following list of items (where appropriate) should be recorded in the electric usage history:

— Billing month
— Reading date
— Days in billing cycle
— Kilowatthours (or kilovoltampere hours if billed on this basis)
— Billing kilowatt demand (or kilovoltampere demand if billed on this basis)
— Actual kilowatt demand (or kilovoltampere demand if billed on this basis)
— Kilovar demand (actual and billed)
— Kilovar hours (actual and billed)
— Power factor
— Power bill (broken down into the above categories plus fuel cost)
— Production level

- Occupancy level
- Heating or cooling degree days
- Additional column(s) for remarks (such as vacation or high vacancy periods)

2.3.2 Project lists

This method utilizes a list of specific projects to reduce energy and costs. The engineer chooses a project from the list without study and evaluation. The project is implemented without an auditing of the electrical system to uncover and categorize areas of needed improvement. Shopping lists of projects are obtained from associates, newspapers, U.S. Department of Energy (DOE) magazines, the local utility, or trade groups. Local utilities often have demand-side management programs that offer financial incentives if certain conservation measures are pursued. These lists include readjustment of thermostats for heating, cooling, and hot water; removing lamps in lighting fixtures; installing storm windows and doors; caulking; additional insulation; etc. An excellent reference is a do-it-yourself guide called *Identifying Retrofit Projects for Buildings* (Report 116)[1] by the Federal Energy Management Program (FEMP) [B4].[2]

Nationally oriented standards can be helpful. Some have been issued by the American Society of Heating, Refrigerating, and Air-Conditioning Engineers, Inc. (ASHRAE) as part of ASHRAE/IEEE 90A-1-1988 [B1]. Other information sources include the Illuminating Engineering Society of North America (IESNA) documents, Building Energy Performance Standards (BEPS) by the National Institute of Standards and Technology (NIST), and *Total Energy Management* [B8], developed jointly by the National Electrical Contractors Association and National Electrical Manufacturers Association in cooperation with the DOE. Energy codes that specify permissible usage levels on various bases such as watts per square foot, or allowable illuminance levels, have been published by a number of states. Such information is likely to be available from a state energy office.

2.3.3 The energy audit

The amount of attention given will depend on the type of audit. A walk-through audit will require only a general notation of the performance, while a comprehensive audit will require specific information.

A comprehensive audit requires the development of a complete energy audit listing major energy-using equipment with nameplate data relating to energy, any efficiency tests, and estimates or measured hours of operation per month. This will include monthly utility data, amounts, and total costs. National Weather Service monthly degree days for heating and cooling must be included. Generally, a one- or two-year compilation of data is used. The crucial part is an intelligent appraisal of energy usage, what is done with energy as it flows through the processes and facility, and how this compares with accepted known standards. This study can lead to the development of a prioritized list of projects with a high rate of return.

---

[1]This publication is available from the U.S. Department of Commerce, Technology Administration, National Technical Information Service (NTIS), Springfield, VA 22161.

[2]The numbers in brackets correspond to those of the bibliography in 2.9.

A walk-through audit consists of a careful observation of operations while walking through the facility with the express purpose of uncovering energy conservation opportunities (ECOs). These ECOs are noted on a clipboard or other means during the tour of the facility. Six categories for energy use should be analyzed as the tour is conducted. These six categories are as follows:

a) Lighting: interior, exterior, natural, and artificial

b) The heating, ventilating, and air conditioning (HVAC) system and the heating and cooling effects of conduction, convection, and radiation

c) Motors and drives

d) Processes

e) Other electrical equipment (transformers, contactors, conductors, switchgear, etc.)

f) Building shell (thermal infiltration, insulation, and transmission)

Within each of these categories, the four basic energy saving methods (see 2.1.2) shall be considered.

2.3.4 Establishing load type for demand control

Chapter 4 covers the topic of demand control in detail. However, this subject is discussed now, because the foundation of demand control is the identification of those loads that can be controlled. The second part, of course, is controlling those loads to reduce demand. An important part of the audit process is, therefore, the identification of the curtailability of each process or piece of equipment. Each piece of equipment or process shall be identified by at least the following four "control" categories:

a) *Critical.* Critical equipment/processes are those whose energy curtailment will be detrimental to the operation of the facility. Equipment that is required for safety reasons and/or for "high worth" to the facility is in this category and must be left energized. These loads shall not be made a part of a demand control scheme nor be turned off indiscriminately.

b) *Necessary.* Necessary equipment/processes are those whose continual energization is important to the operation of the facility. However, these items can be deenergized at some specific loss to the facility. (Safety systems are never categorized as "Necessary." They are always categorized as "Critical.") Economic models can be developed to determine the benefit versus loss upon deenergizing "Necessary" loads. Disconnecting this equipment must not violate health, safety, or environmental needs of the facility and its operation.

c) *Deferrable.* Deferrable load is that equipment/process that can be deenergized for a given period of time without any financial, production, or other loss. However, caution shall be taken to ensure that the equipment/process 1) can be safely disconnected and reconnected, 2) can withstand the frequency and duration of the anticipated starts and stops, and 3) does not indirectly involve the health environment or safety and/or production level of the facility.

d) *Unnecessary.* This equipment should not be in operation; so, it shall be shut off as soon as possible. In addition, means should be installed to ensure that this load is deenergized when not required.

2.4 The six equipment audit categories

The ensuing subclauses propose questions that direct attention to possible energy savings. This is not a comprehensive list, but it will be helpful in elevating an awareness of energy waste and conservation opportunities.

2.4.1 Lighting

The first item that gets attention is lighting because of its visibility. Before making changes, read Chapter 7. Key efforts in analyzing and considering lighting from a conservation standpoint are noted below. It is wise to make checks both day and night.

— Is the light intensity sufficient for the task? [15–40 fc (160–450 lx) for halls; 70–100 fc (750–1100 lx) for detailed work.]

— Is the luminaire proper for directing the light where it is needed?

— Is the reflection good?

— Is the color right for the task?

— Is the luminaire too high or too low?

— Can task lighting be used effectively?

— Is good use being made of available natural light?

— Can desks or machines be grouped by task light required?

— Are lights and luminaires cleaned periodically?

— Are lamps turned off when not in use?

— How many luminaires can be turned off by a single switch?

— Who turns the lights off?

— Who uses the space and how often?

— Can a different, lower wattage lamp be used in the fixture?

— Do the surfaces reflect or absorb light?

— Are the luminaires strategically located?

— Does the luminaire location cause glare?

— Can lighting be used to heat?

— Can more efficient light sources be used?

— Can timers or photocells be effectively used?

2.4.2 Heating, ventilating, and air conditioning (HVAC)

Key factors in evaluating and better utilizing the HVAC system are as follows:

a)   Are there obstructions in the ventilating system?

    1)   Do filters, radiator fins, or coils need cleaning?

    2)   Are ducts, dampers, or passages and screens clogged?

b)   Is the wrong amount of air being supplied at various times?

    1)   Are dampers stuck?

    2)   Is exhaust or intake volume too high or too low?

    3)   Are all dampers functioning in the most efficient manner?

c)   Can the system exhaust only the area needing ventilation?

d)   Can the system intake only the amount required?

e)   Can air be recycled rather than exhausted?

f)   Can the intake or exhaust be closed when the facility is unoccupied?

g)   Can the system be turned off at night?

h)   Is the temperature right for the area's use? [A 40–50°F temperature may be acceptable for storage.]

i)   Can temperature setback be used effectively?

j)   Will an adjustable speed drive be more efficient?

k)   How many fixtures can be turned off by a single switch?

l)   Is solar energy being effectively utilized?

    1)   Light, but minimum heat in summer

    2)   Light and heat in winter

m)   Can waste heat be used?

n)   Are belts properly tensioned?

o)   Are pulleys and drives properly maintained and lubricated?

p)   Is the refrigerant proper?

q)   Can heat be redirected?

r)   Is the proper system being used?

s)   Is there too much or too little ventilation?

t)   Can the natural environment be used more effectively?

u)   Are doors, windows, or other openings letting out valuable heat?

v)   Can weather-strip, caulking, or other leaks be repaired?

w)   Can additional insulation be justified?

x) Can all hooded exhaust systems have their own air supply or can they be used as part of the exhaust requirement for the building?

y) Is the blower cycled or run continuously?

2.4.3 Motors and drives

Since motors use nearly 70% of the electrical energy consumed in the U.S., they provide great opportunity for reducing energy waste. The following questions point to very common kinds of waste:

a) Does the motor match the load?

b) Can the motor be stopped and then restarted rather than idled?

c) Is the motorized process needed at all? Can it be done manually?

d) Who lubricates the motor and associated drives? Is this done at the proper intervals?

e) Can motor heat be recycled?

f) What type of drive is used? Is it the most efficient?

g) What is the voltage and is it balanced?

h) Can the motor be cleaned to lower heat buildup?

i) How is load adjusted?

j) Will two (or more) motors in tandem work better?

k) Is the motor well maintained and in good condition? Are there any electrical leaks to ground? Is the motor in a wet environment?

l) Who turns the motor off and on? How often?

m) How efficient is the motor?

2.4.4 Processes

Normally, processes rely heavily on motors, but there are other electrical parts. Process heating is probably the most common non-motor electric process load. The following questions point to areas where improved efficiencies can be made:

a) Can equipment or processes be grouped together to eliminate the transportation of the equipment or material in process?

b) Is the temperature too high?

c) Does heat escape? Can insulation be used effectively?

d) Can the heated energy be recirculated for comfort, process heat, or cogeneration? Can it be exhausted for summer comfort?

e) Is preheat required?

f) Can the process be staged or interlocked?

g)   Is the product heated, cooled, and then reheated again? If so, a continuous process may be appropriate.

h)   Can the processes be lined up for more effective use of equipment?

i)   Are the drives, bearings, etc. correctly lubricated?

j)   Can the conveyer system be eliminated or modified?

k)   Can hot areas be isolated from cold areas?

l)   Is one large motor or many small motors better?

m)   What equipment can be switched off at night?

n)   Would two or three shifts be more efficient?

o)   Is any equipment kept idling rather than switched off when in hold or waiting?

p)   Are screens cleaned and dampers checked for proper operation of pollution controls, and are they maintained at proper intervals, etc.?

q)   Is compressed air made in two or three stages? Is a storage tank being used and is pressure too high?

r)   Is process water too hot?

s)   Can fluid be recirculated?

t)   Is the fluid cooled too much?

u)   Is the hot water heater close to where the hot water is needed?

v)   Is the process exhaust higher than required for safety or quality, or both?

w)   Is hot and cold piping insulated where appropriate?

x)   Is temperature controlled so that only necessary heat is added?

y)   Is heat supplied or added at the point of use or is it transmitted some distance?

2.4.5 Other electrical equipment

There is a significant amount of electrical equipment that is taken for granted or rarely noticed. The list below contains some key questions regarding efficiency.

a)   Are the transformers required?

b)   Is the transformer too hot?

c)   Can the transformer be turned off when not in use?

d)   Are wiring connections tight? (Improper voltage, unbalanced voltage, and excess heat can result from a bad connection.)

e)   Can heat from the switchgear room be utilized? (Remove this heat during the summer.)

f)   Are voltage taps in the proper position?

g)   Are heaters applied properly?

h) Can heaters be switched off at times?

i) Are contactors in good working order?

j) Is equipment properly bonded and grounded?

k) Are conductors sized properly for the load?

l) Is the power factor too low?

2.4.6 Building environmental shell

The list below is applicable to electrically heated buildings and also where other energy sources are used.

a) Is allowance made for transition from cold to hot areas and vice versa with an air curtain or vestibule?

b) Can a wind screen keep air infiltration down?

c) Is the proper level of insulation applied?

d) Can full advantage of solar heat be taken? (Remove solar heating effect in summer.)

e) Is automatic door closing appropriate?

f) Can covered loading and unloading areas be utilized to keep heat in?

g) It is possible to caulk, weather-strip, glaze, or close-off windows?

h) Can double- or triple-pane glass be used?

i) Is a small positive pressure used to keep out drafts?

j) Can areas be staged in progressively cooler or warmer requirements?

k) Would an air screen, radiant heater, etc. be more effective?

l) Are dock seals used on overhead doors?

2.4.7 Overall considerations

All conservation changes shall be made while considering the overall plant. An electrically heated building will use the heat from lighting fixtures in the winter, so reduced lighting may just transfer the heat requirement to baseboard heaters. While this chapter provides many areas for improved efficiency, a detailed analysis and engineering design by a competent consultant will be required. The final design shall include the consideration of proper codes (National Electrical Code®, Occupational Safety and Health Administration Codes, National Fire Protection Association Codes, and Environmental Protection Agency Codes, etc.).

2.4.8 Energy balance

An important step in evaluating opportunities is the development and analysis of major energy balances. An energy balance is a little more complicated than an energy survey and is based on measurement and calculation.

A basic energy balance is shown in figure 2-1. Energy is input to a system that produces a product. There is a certain amount of energy absorbed in the process, which can be called energy output. The difference between the energy input and the energy output is waste. Energy balances should be developed on each process to define, in detail, energy input, raw materials, utilities, energy consumed in waste disposal, energy credit for by-products, net energy charges to the product, and energy dissipated or wasted.

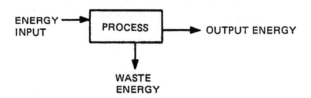

Figure 2-1—Energy balance

All process energy balances should be analyzed in depth. Various questions should be asked and answered, including:

a)   Can waste heat be recovered to generate steam or to heat water or a raw material?

b)   Can a process step be eliminated or modified in some way to reduce energy use?

c)   Can an alternate raw material with lower process energy requirements be used?

d)   Is there a way to improve yield?

e)   Is there justification for replacing old equipment with new equipment requiring less energy or replacing an obsolete, inefficient process with a new and different process using less energy?

The energy survey and energy balances identify energy wasting situations and differentiate between those that can be corrected by maintenance and operations actions and those that require capital expenditures. The former can be corrected in a short time and the results are almost immediate—savings with little effort or delay. The latter will require some investment dollars and delivery time for materials and equipment.

The survey can be conducted by an individual, a team, or an outside consultant. The work can be segregated according to plant areas or departments. However, those conducting the survey shall be aware of the importance of finding waste, determining the cost involved, and reducing the energy costs. The initial survey shall be followed up periodically with additional surveys to ensure that waste is curbed and new problems are avoided.

2.5 Energy conservation opportunities

A key element of the energy management process is the identification and analysis of energy conservation opportunities (ECOs). These opportunities involve the previously listed spectrum of activities, from simply changing procedures or switching off lights to new long-range technologies.

The equipment used in operating a facility or in manufacturing a product plays another key role in the energy conservation effort. By understanding the relative energy consumption by equipment type, one can determine the opportunities that exist for decreasing this consumption.

After developing the energy balance and listing all of the energy conservation projects, each project should be evaluated for implementation using the following procedure:

a)   Calculate annual energy savings for each project.

b)   Project future energy costs and calculate annual dollar savings.

c)   Estimate the cost of the project, including both capital and expense items.

d)   Evaluate investment merit of projects using measures such as return on investment, etc.

e)   Assign priorities to projects.

f)   Select conservation projects for implementation and request authorization to proceed.

g)   Implement authorized projects. Examples of electrical energy conservation opportunities are described in 2.4. Application of a specific ECO requires careful evaluation to determine if its use is appropriate because there is the possibility that an ECO could be counterproductive. For instance, the proposed change may cause existing equipment to exceed operating limits.

2.6 Energy monitoring and forecasting

Managing energy is more than the implementation of energy conservation opportunities. Since energy management is a continuous process, it is extremely important to monitor energy usage and to use the results to gauge future actions. In some energy applications, such as in combustion processes, measurement or control (or both) at each point of application may be necessary to ensure efficient use of energy. Measuring methods that best describe the process are thus necessary to control, evaluate, and manage efforts to conserve energy. It is also important to budget the anticipated energy usage so that performance can be evaluated and plans can be made. A measure against budget also provides input to determine if additional action is required.

Several types of reports can be prepared to show energy usage over time. Summary reports for upper management typically note monthly or quarterly performance by facility or division. The type of report will vary with each specific facility. Some typical reports are the percent energy reduction method, the product energy-rate method, and the activity method. These reports are detailed in 2.6.1. Quick response to energy wastes and evaluation of the energy efficiency of a facility is better gauged using energy forecasts and tracking. This type of activity is described in 2.6.2.

2.6.1 Energy management reports

A number of very useful methods to maintain energy conservation performance in plants and offices worldwide are described in [B2]. Other established effective methods are given in [B5], [B6], [B7], and [B8]. In engineering new facilities, potential conservation is monitored using either the percent reduction energy-rate method or the design energy-savings report. For existing facilities, [B2] states that energy conservation is monitored using the activity method, the energy-rate method, the variable energy-rate method, or tracking charts.

2.6.1.1 Percent reduction energy rate method

This reduction is determined by comparing the project's product energy rate, expressed in Btu/lb (kJ/kg) with the existing plant's product energy rate. An example is shown in table 2-1.

The product energy requirements consist of all energy supplied minus credit for all usable energy exported. The requirements include purchased energy such as gas, oil, and electricity, plus generated in-plant utilities such as steam, refrigeration, compressed air, and cooling water. All energy, purchased or generated, has to be expressed in the same units (e.g., Btu). Likewise, the manufacturing department is given × credit for usable energy that is exported such as condensate and steam.

Table 2-1—Product energy rate

Current production rate	= 1950 Btu/lb
Projected production rate	= 1480 Btu/lb

$$\text{Percent reduction in product energy rate} = \frac{1950 - 1480}{1950} \times 100 = 24\%$$

2.6.1.2 Design energy-savings report

The design energy-savings report covers an energy-savings idea that is incorporated in the design of a project. The report serves three purposes:

a) It is a means of information exchange on energy-saving ideas.

b) It provides an opportunity to monitor energy conservation capital requirements versus energy savings on projects.

c) It assists in an energy-awareness program in the engineering department.

This report is prepared by a project participant when that participant incorporates an energy-saving idea or innovation that will reduce energy requirements. The guideline for an energy-saving idea is that it deviates from present design or plant practices for the product line. If it is a new product, the reference point is the pilot plant or project definition report.

A summary report is issued quarterly on the number of energy-saving ideas submitted by each design group. This quarterly report encourages competition between design groups.

2.6.1.3 Activity method

The activity method compares the anticipated energy saved with energy purchased. The percent energy savings is based on the annual energy savings, expressed in Btu, compared with the plant's total purchased energy.

The activity method gives a rapid response to conservation results. Equally important, savings are not affected by changes in product energy efficiency as the production rate varies. For these reasons, it is an excellent method for monitoring performance.

A typical quarterly report for a company using the activity method is that shown in table 2-2. A similar report can also be prepared expressing the savings in Btu and costs; these statistics are most useful for internal communications and external publicity.

Table 2-2—Activity method report

XYZ Manufacturing Co. Energy-conservation results Activity method (% energy savings)					
Plant	3 Quarters	4th Quarter	Year 0 Total	Planned activities	
				Year 1	Year 2
Fulton	2.0	0.2	2.2	2.1	1.0
Grace	1.8	0.1	1.9	0.8	0.4
Nixon	1.5	0.2	1.7	2.3	0.3
St. James	0.8	0.4	1.2	0.1	0
Company TOTAL	1.7	0.3	2.0	1.8	0.4

The activity method yields a good visual report on activities that have been completed and planned to save energy. However, it does not reflect the true reduction in energy rate (Btu/lb of products) because the savings are annualized.

2.6.1.4 Energy-rate method

The Chemical Manufacturers Association (CMA) developed the chemical-industry energy-rate method for reporting energy conservation results. An example of the calculations is shown in table 2-3. The Year 0 base period rate is calculated using the total "pounds of product" manufactured in Year 0 and the equivalent purchased energy consumed by the department manufacturing the product. The equivalent energy includes energy consumed by the specific line plus a proportionate share of energy that cannot be assigned (or measured) to any one area.

For any reporting year, the percent reduction in energy consumption rate is calculated by comparing the base period energy to the total energy purchased by the plant, excluding feed-stock energy. The comparison base period energy for each product is calculated using the base year product energy rate times the weight of the product manufactured in the current year.

The CMA energy-rate method of reporting conservation results compensates for product mix, the addition and deletion of products, and for OSHA and environmental energy requirements. Changes in production rate have a major effect on conservation results because the product energy-rate requirements are made up of fixed and variable energies.

2.6.1.5 Variable energy-rate method

Many variables need to be considered in a conservation monitoring method, including the following:

a) Changes in energy rate as production rate varies (see figures 2-2 and 2-3)

b) Product energy requirement changes resulting from ambient temperature changes during the year

c) Changes in raw material quality

d) Changes in the plant heat balance that may cause the process to operate turbine drives instead of motor drives, and vice versa

e) Minor changes in product quality or specifications for different customers

f) A change in the equipment or process to increase output

The product energy-rate versus production-rate curve in figure 2-2 was developed for each product considering only production rate and its effect on the energy rate as variables. To calculate the conservation performance for any year, the product production rate for that year is used to establish the base period energy rate from the energy-rate curve of figure 2-4. For example, using the CMA energy-rate method for Year 0, the product energy rate is 10 000 Btu/lb. The actual base energy rate is 11 500 Btu/lb.

The 11 500 Btu/lb base energy rate instead of 10 000 is used in calculating the comparison base-period energy rate for product G. This same procedure is then duplicated for each product. This method, although not 100% accurate in compensating for all variables, does compensate for the major variable, production rate, and does prevent wide swings in conservation results as the production rate varies.

2.6.2 Energy management and forecasting methodology

Tracking of energy on a quarterly or annual basis does not afford the engineer with an opportunity to quickly react to energy problems. Also, it is important to utilize audit information to establish an energy budget. The use of a budget, and/or a projected usage allows evaluation of energy efforts and properly gauges the energy portion of the cost of doing business.

Table 2-3—Energy rate method report

Fulton plant Products manufactured	Year 0 (Base period)			Total production (×10⁶ lb) Column 4	Total energy (×10⁶ Btu) Column 5	Year 3	
	Total production (×10⁶ lb) Column 1	Total energy (×10⁶ Btu) Column 2	Base energy rate (Btu/lb) Column 3			Comparison base period energy (×10⁶ Btu) Column 6 (Column 3 × Column 4)	% Reduction energy consumption rate Column 7
A	200	10 000	50	300		15 000	
B	10 000	30 000	3	12 000		36 000	
C	2 000	20 000	10	3 000		30 000	
D	3 000	60 000	20	6 000		120 000	
	15 200	120 000		21 300		201 000	
Adjustments to base: New products (+) or discontinued products (−) after Year 0 (base).							
E (Year 1)….	1 000	10 000	10	2 000		20 000	
F (Year 2)….	1 000	5 000	5	1 000		5 000	
	2 000	15 000		3 000		25 000	
Total products	17 200	135 000		24 300	208 000	226 000	7.5*
Adjustments for environmental and OSHA					(1000)		
Adjusted grand total	Base 17 200	135 000		24 300	207 000	226 000	8.0*

$$* \text{ \% Reduction energy consumption rate} = \frac{(\text{Column } 6) - (\text{Column } 5)}{(\text{Column } 6)} \times 100 = (\text{Column } 7)$$

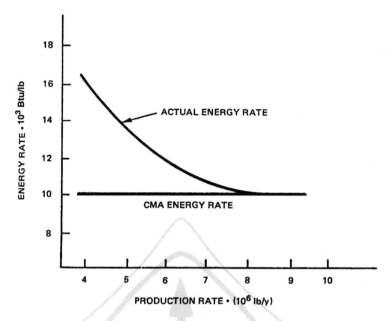

Figure 2-2—Change in production

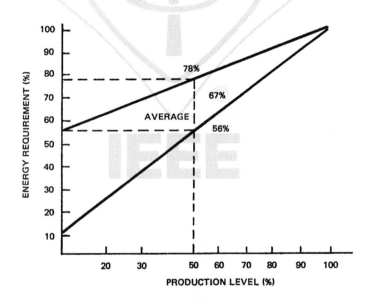

Figure 2-3—Energy versus production

From a forecast viewpoint, energy usage shall be classified into two groups. *Baseline energy* is the energy component that is constant over the measurement period and does not vary with production, weather, etc. *Variable energy* is the component that changes with production,

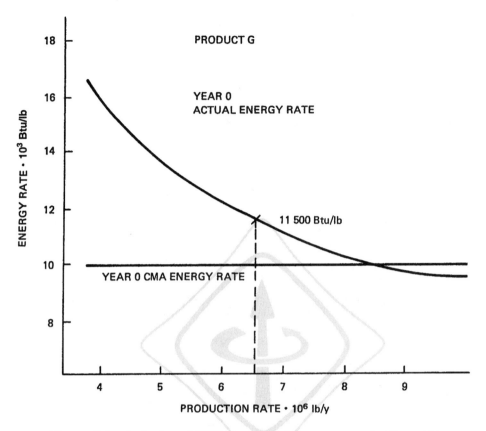

Figure 2-4—Actual and CMA energy rate versus production rate

weather, occupancy, etc. These two components should be determined for the facility in order to properly gauge and monitor energy performance. Although this standard discusses electrical energy, it is recommended that the same technique be applied to other energy source inputs such as steam, chilled water, and direct combustion of other fuels.

This subclause will review an industrial situation to guide the reader through the energy monitoring and budgeting process. The same concept can be applied to commercial buildings by replacing "product" with "occupancy," "weather," etc.

The first step in energy budgeting is the performance of a usage audit. It is preferable to review results of the previous 12 months (see 2.3, 2.4, and all of Chapter 8 for metering and auditing information). Both usage and production should be reviewed. With properly installed meters whose readings were diligently tabulated, results are very easy to obtain. The engineer can use these results in working with production personnel to obtain their forecast of production and the associated projected energy use.

Installation of meters as shown in figure 2-5 allows for the calculation of process efficiency for various input and consumption variables.

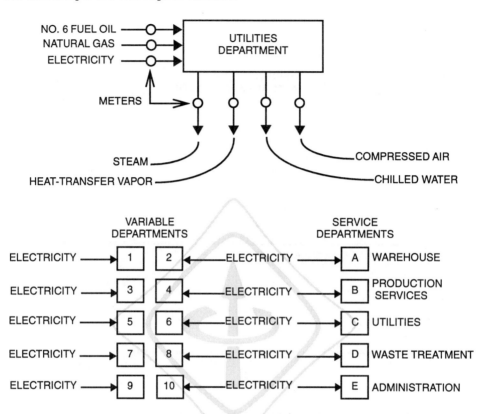

Figure 2-5—Typical plant energy distribution

This subclause will discuss the electric portion only. The production estimate is received from the production people in several forms (e.g., figure 2-6). The anticipated monthly production run can be used in total and then an average daily figure can be calculated using average kWh/lb of daily product. The forecast can also be calculated by using the anticipated hours of machine use for the month and/or day. The calculation can be done on a form similar to that shown in figure 2-6. The resulting information, as well as the base values, can be put on a summary form, an example of which is shown in figure 2-7. Both the estimate and the actual numbers can be conveniently noted. If available, both the production estimate and usage estimate on machines should be used to determine the variable portion of the electrical usage. The results from using the two estimates should converge as time hours progresses, unless one or the other is not a good gauge of kWh usage.

Total actual electrical consumption is then plotted on a graph. Two types of graphs are shown in figures 2-8 and 2-9. Although these graphs are by months, it is important to keep an energy track for shorter periods—even in shifts—by department and/or activity. A review of the information frequently will reveal a more efficient way to produce a product or service. For

```
┌─────────────────────────────────────────────────────────┐
│  _____ Production Forecast    │
│                                                          │
│                          For                             │
│                                                          │
│   Department _____ is _____ MGP         │
│                                                          │
│                                                          │
│   I.   _____ ÷ _____ = _____  │
│            MGP           Days in Month      MGP/Day       │
│                                                          │
│                                                          │
│   II.  _____ ÷ _____ = _____  │
│          MGP/Day           MGP/           Average        │
│                          Machine-Day    Machine Level     │
│                                                          │
│                                                          │
│   III. _____ × _____ = _____  │
│          Average        Days in Month   Machine-Days      │
│       Machine Level                      In Month         │
│                                                          │
└─────────────────────────────────────────────────────────┘
```

NOTE—MGP is thousands of gallons of product per minute.

Figure 2-6—Production forecast form

example, energy information was tracked by pound of product produced for each shift in a cement plant. An analysis of the results over time showed that the most energy-efficient shift kept the machines running the entire shift rather than shutting down frequently to make changes in the production run. The remaining shift supervisors were asked to keep machines running and the resultant savings were over a quarter million dollars in energy costs. Figure 2-10 shows one way of reporting energy usage on a monthly basis.

A plot of production versus energy usage can be used to develop a formula for predicting energy usage. The regression analysis of "least squares" (software is available) or "eye balling" can be used to establish a formula that fits the data points. Figure 2-9 shows a "curve" that "fits" the data. The curve is a straight line with a slope of 1705 and an intercept at 139 500 kWh per month. The equation would be as follows:

$$1705 \times M + 139\,500 = \text{kWh per month}$$

where

M = thousands of pounds of net good product per day, where the plant produces product 24 hours per day for 31 days

35

Energy Analysis for Department: _____ Machine type: _____

	Std	Jan	Feb	Mar	Apr	May	June	July	Aug	Sept	Oct	Nov	Dec	Year Avg
Machine days														
Thousands of net good pounds														
Thousands of kWh per machine day														
kWh per net good pounds														
MSCF air—Machine day														
SCF air—NGP														

NOTE—MSCF = Thousands of standard cubic feet; SCF = Standard cubic feet; NGP = Net good pound of product.

Figure 2-7—Energy usage analysis form

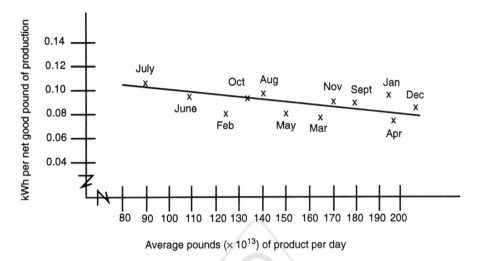

Figure 2-8—Per unit kWh consumption

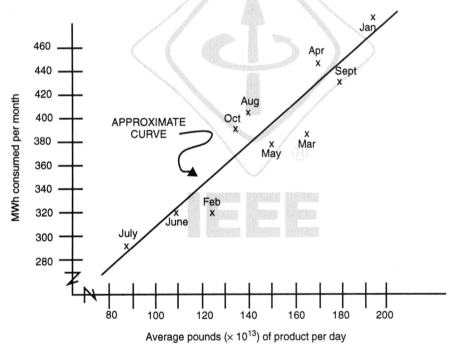

Figure 2-9—Monthly MWh consumption

This analysis shows that the baseline portion of the energy bill is 139 500 kWh per month. The baseline is essentially fixed, but that does not mean that an energy audit of the baseline portion is not a place to reduce electric energy usage. In this example, the baseline energy represents 31% of the energy usage at full production (180 000 lb of good product per day).

Variable and Service Energy Forecast Summary Sheet

_____ , _____
Month   Days

Dept. no.	Machine electricity (kWh × 10³)	Air (SCF × 10³)	Steam (Btu × 10⁶)	Heat transfer vapor (Btu × 10⁶)
1				
2				
3A				
3B				
3C				
3D				
4				
5				
6				
7				
8				
9				
10				
Total variable				
Warehouse				
Utilities				
Waste treatment				
Production services				
Administration				
Construction				
Chillers				
Losses				
—Service air				
—Steam				
Total services				

Figure 2-10—Variable and service energy forecast summary sheet

2.7 Employee participation

An energy conservation program can be successful only if it arouses and maintains the participative interest of the employees. Employees who participate and who feel themselves partners in the planning and implementation of the program will be more inclined to share pride in the results.

Communicating with employees on the subject of energy can be accomplished in many different ways: face to face discussion, seminars and workshops, distribution of informative and descriptive literature, slide presentations and moving pictures, and most important of all, sincere practice of conservation on the part of management at all times.

The use of company newsletters, bulletin boards, or posters for pictorializing energy conservation objectives and accomplishments will help impress employees with the importance of such matters. Employee participation can be increased by communicating examples of energy conservation ideas being implemented, photographs of persons who submitted the ideas, and information on the savings realized.

Competition between departments, sections, or groups within the company in pursuing conservation of energy can also generate enthusiasm among employees. Competitive programs can be initiated among the employees and should be encouraged. Acknowledgment of good ideas and positive reinforcement are keys to this approach. Employee education can take many diverse forms: workshops and training courses for supervisory personnel, articles in the company newsletter, and energy conservation checklists given to each employee.

A clear, concise list of firm do's and don't's to guide employees in performance of their work can be helpful in achieving energy conservation practices. Such lists should be distributed to all employees whose jobs involve the use or control of energy. The list should be updated as often as necessary. Supervisors should be responsible for seeking adherence to all items on the list.

2.8 Summary

Energy management is a broader concept than energy conservation. In energy management, one is attempting to achieve the same level of productivity with a lower expenditure of energy, an adequate energy supply, and the lowest possible cost.

Energy management responsibility requires a significant reorientation of managers toward energy conservation. Corporate individualism operating under the profit motive in a free marketplace has been a major factor contributing to U.S. economic success. These achievements have also been due, in part, to the industrial manager's dedication to the substitution of inexpensive electric and thermal energy for human labor in the production of goods and services.

We now find that it is necessary for management to accept the reality that the future requires a major reorientation because energy input is no longer going to be inexpensive. To secure energy supplies and to minimize energy costs, industry will have to act in conjunction with its own members, government, the public, and all segments of the energy supply industry.

Achievement of meaningful energy savings in existing processes is a function of management committing itself to do the job by effective personnel motivation, planning, and administration. The establishment of a formalized energy management responsibility is highly desirable to give the effort both the focus and the direction required. An energy management function gives the line manager the tools to get the job done. Engineers should inform upper management regarding energy use and costs, future energy supply cost and availability, and the level of savings possible. The engineer can then convince management of the need for alternative solutions that should be pursued.

Obtaining participation from key local plant operators is a prime ingredient to energy conservation success since they know their electrical system and generally are in the best position to take corrective action. However, top management must give them the monetary and technical resources required.

2.9 Bibliography

Additional information may be found in the following sources:

[B1] ASHRAE/IEEE 90A-1-1988, Energy Conservation in New Building Design (Sections 1–9).

[B2] Doerr, R. E., "Six Ways to Keep Score on Energy Savings," *The Oil and Gas Journal,* pp. 130–145, May 17, 1976.

[B3] *Energy Efficiency and Electric Motors.* Springfield, Va.: Arthur D. Little, Inc., National Technical Information Service (Pub PB-259129), 1976.

[B4] FEMP, *Identifying Retrofit Projects for Buildings* (Report 116).

[B5] Gatts, Robert; Massey, Robert; and Robertson, John, *Energy Conservation Program Guide for Industry and Commerce* (NBS Handbook 11.5). Washington, DC: U.S. Department of Commerce, Sept. 1974.

[B6] Gibson, A. A., et al., *Energy Management for Industrial Plants.* British Columbia, Canada, Dec. 1978.

[B7] Myers, John, et al., *Energy Consumption in Manufacturing.* Massachusetts: Barlinger Publishing Co., 1974.

[B8] NEMA and NECA, *Total Energy Management, A Practical Handbook on Energy Conservation and Management,* 2d ed. National Electrical Contractors Association, 1979.

[B9] *Site Energy Handbook.* Energy Research and Development Administration, Washington, DC, Oct. 1976.

[B10] Stebbins, Wayne, "Implementing An Energy Management Program," *Fiber Producer,* Atlanta, Ga.: W. R. C. Smith Publishing Company, vol. 8, no. 5, pp. 44–57, Oct. 1980.

Chapter 3
Translating energy into cost

3.1 Introduction

The use of economic analysis is critical to the conservation program because monetary savings can significantly influence management decisions. The engineer shall, therefore, be able to translate a proposal into monetary value, i.e., expenditures versus savings. This chapter covers basic economic concepts and utility rate structures, and then addresses the subject of loss evaluation.

The engineer can use the economic basics to develop an energy program and to determine the best choice among alternatives. This determination should include understanding payback periods, understanding the time value of money, and weighing other pertinent costs.

An understanding of electric rate structures is essential in an economic analysis because the monetary savings will accrue from lower electric bills. Rate structures are sufficiently complex to warrant careful consideration. The electric rate is particularly important in demand control (load management) projects.

All electrical equipment has losses. The savings from loss reduction is calculated by evaluating the reduced cost of energy over the life of the equipment.

3.2 Important concepts in an economic analysis

Two or more alternatives are considered in an economic analysis (one may be to do nothing). In any case, an investment is usually evaluated over a period of time. The initial investment is called the capital cost. Since an item is usually worth less as it ages, the value is depreciated over its life. A piece of equipment normally has a salvage value (which could be negative if its removal or disposal cost is higher than its resale value) when it is retired and sold. In addition to the initial investment, an alternative usually has recurring costs such as maintenance and energy usage. These costs are grouped as annual costs and are then put in a form that can be added directly to the capital cost.

The capital cost represents the total expenditure for a physical plant or facility. The capital cost is comprised of two components: direct costs and indirect costs. Direct costs are monetary expenditures that can be directly assigned to the project such as material, labor for design and construction, and start-up costs. Indirect costs or overheads are expenditures that cannot be directly assigned to a project (e.g., taxes, rent, employee benefits, management, corporate offices, etc.).

Depreciation is the distribution of a capital cost over the anticipated life of the process or equipment. These depreciation amounts are then used to reduce the value of the capital investment. There are several means of distributing these costs; the simplest method is straight-line depreciation. In straight-line depreciation, the annual depreciation is merely the capital cost divided by the estimated life.

Equipment, material, and buildings have two values used for life expectancy: the book life and the expected or useful life. The book life is the number of years used to financially depreciate the investment. The expected life is the anticipated length of time that the investment will be utilized. Computers tend to become obsolete in only a few years due to rapid advances in technology today. However, computers can, and do, continue to compute long after their technological obsolescence; so the expected life in the case of a computer could be the time until obsolescence.

The salvage value is the amount of money that can be returned to the company at the end of the expected life of the capital investment. This salvage value is equal to the anticipated resale value minus any cost associated with 1) the sale of the equipment and 2) its physical disassembly and subsequent removal.

The annual costs include such items as fuel or energy, operation/maintenance, labor, taxes, and other recurring costs. The fuel cost depends on the amount, quality, rate of use, and the rate schedule. The maintenance cost includes both routine work and purchase of parts, such as the periodic relining of boilers.

3.3 Economic models—their applications and limitations

There are two general means of evaluating energy options: simple break-even analysis and a more complex method called life cycle costing. When there is a large energy savings for a small investment, simple payback may be the best evaluating tool. Furthermore, housekeeping projects with minimal or no cost may not need evaluating at all. For example, the installation of a timer on an exterior light circuit that will significantly reduce "on" time or a decision to switch off unused equipment may not need an economic evaluation. However, the subsequent effect of these actions should be shown to encourage management and support the energy conservation effort. Life cycle costing is most likely needed when the project costs are large compared to the energy savings or when there are significant future costs. The ensuing subclauses detail the various modeling methods.

3.3.1 Break-even analysis

The break-even methods do not use the time value of money, and they all answer the same question: At what point will I get my money back? Common terms for this model are simple payback analysis, break-even point analysis, and minimum payback analysis. All of these methods are essentially the same. They all relate the capital investment to the savings. Some methods chart results while others use a not-to-exceed value. The basic mathematical description is

$$\text{Break-even point} = \frac{\text{net capital investment}}{\text{savings/unit}}$$

Many companies have minimum payback requirements to allow expenditures whose break-even point is less than a given amount of time (or other measurement). The break-even point has taken the name of payback from this minimum payback usage. Hence, the more common term is payback, and management is more likely to ask for the payback of a particular energy option.

It is important to note that break-even analysis is not restricted to time as a base. Production and energy consumption are also good bases. For example, the break-even point can be calculated in terms of the amount of product manufactured such as dollar savings per pound of output.

When questions concerning future operations are appropriate, a more sophisticated method of analysis is justified, particularly if the minimum payback period exceeds several years. A complete, long-term analysis is well worth considering in determining energy savings.

3.3.2 Marginal cost analysis

Marginal (or incremental) cost analysis is more a concept than an economic model. The marginal concept has predominant use in the economic community and has popular use in making decisions. It is primarily used in calculating cogeneration sales to the utility and in the development of special rates or contracts. Marginal cost analysis is simply the determination and use of the next increment of the cost of money or cost of electric energy. It is usually prudent to consider costs of the next increment of power, production, investment, and money.

3.3.3 Life cycle costing

Life cycle costing (LCC) is the evaluation of a proposal over a reasonable time period considering all pertinent costs and the time value of money. The evaluation can take the form of present value analysis, which this subclause will use, or uniform annual cost analysis.

The LCC method takes all costs and investments at their appropriate points in time and converts them to current costs. Inflation is assumed equal for all cost factors unless it is known to differ among cost items. Items for consideration include the following:

a) Design cost
b) Initial investment
c) Overheads
d) Annual maintenance costs
e) Annual operating costs
f) Recurring costs
g) Energy costs
h) Salvage values
i) Economic life
j) Tax credits
k) Inflation
l) Cost of money

3.4 Time value of money

3.4.1 Determining the cost of money

A dollar today is worth more than a dollar in the future because today's dollar can generate profit. Most companies use the cost of borrowing and the return on investment to determine the cost of money. The energy engineer should work with the appropriate financial people to determine the cost of money.

Inflation, the rate of price increase, is a very important concept in the time value of money. For example, inflation may direct a decision to buy now or next year. If an item increases in cost by 20% a year and money can be borrowed at 10% a year, one should buy now. However, if the reverse is true, it might be wise to wait a year to purchase. In simplified economic evaluations, inflation is assumed to have equal effects on all alternatives; it should only be included in more complex analyses or if one or several items increase in cost at significantly different rates.

The time value of money is important to the engineer because the engineer shall be able to evaluate alternatives by translating a dollar of expense or investment at various times to equivalent amounts. To accomplish this task, the next subclause will develop conversion factors that translate future dollars or payments into their present values and vice versa.

3.4.2 Calculating the time value of money

The following list of terms is used in subsequent discussions:

annuity: A series of equal amounts evaluated at the end of equal time periods (usually 1 yr) for a specified number of periods.

capital investment: The amount of money invested in a project or piece of equipment (this includes labor, material, design, and debugging monies).

compound interest: Interest that is applied to both the accumulated principal and interest. For example, a 12% annual interest rate compounded at 1% each month on $1.00 is $(1.01)^{12}$ or $1.13, which results in a 12.7% effective (simple) annual interest rate.

constant dollars: The worth of a dollar amount in a reference year, including the effect of inflation. Constant dollars are used in economic indicators.

current dollars: The worth of a dollar today.

depreciation: The mathematical distribution of the capital investment over a given period of time, which may or may not concur with the estimated useful life of the item.

discount rate: The percentage rate used by a corporation that represents the time value of money for use in economic comparisons.

future worth (value): The value of a sum of money at a future time.

initial cost: Synonymous with capital investment. Capital investment is a more proper and clearly recognized term.

present worth (value): The value of an amount discounted to current dollars using the time value of money.

For the purpose of this subclause, it is assumed that the initial investments are made at the beginning of the first year, and the future expenditures are made at the end of the year. Hence, $1.00 invested now at a 10% annual interest rate will grow to $1.10 at the end of the first year (i.e., Year 1 in the formulas), and $1.00 in operating and expense dollars for that year will have a $0.91 present value ($1.001/1.1 = $0.91). The present worth is the value at Year 0 and the future worth is the value at the end of the nth year. The entire conversion process is based on two basic calculations: the present worth of a single amount and the present worth of an annuity.

The single present worth factor is used to find the present worth of a single future amount. The present worth factor (PWF) is determined by the following equation:

$$PWF = \frac{1}{(1 + i)^n}$$

where

i = Interest or discount rate expressed as a decimal
n = Number of years

The present worth is calculated by multiplying the future amount by the present worth factor or

$$PW = PWF \times FW$$

where

PW = Present worth
PWF = Present worth factor
FW = Future worth or future amount

For example, suppose that a company has the choice of refurbishing an existing induction heating unit for $10 000 (and replacing it in five years for $120 000) or buying a new unit for $100 000. The company's discount rate is 12%. The choice is whether to spend $10 000 now and $120 000 in five years or to spend $100 000 now. To compare the alternatives, the $120 000 future amount is converted to today's dollars.

Alternate A
Present: = $10 000 PW

Future: $= \dfrac{1}{(1+0.12)^5} \times 120\ 000$

= $68 100 PW
Total: = $78 000 present worth

Alternate B
Present: = $100 000 PW
Future: = 0
Total: = $100 000 present worth

In this case, there is a clear $22 000 benefit in renovating the unit and paying a larger amount for a new machine in five years. Neglecting the time value of money would have led to the wrong decision.

The present worth of an annuity factor (PAF) converts a series of future uniform payments to a single present worth amount. The uniform payments are made at the conclusion of a series of equal time periods. The mathematical description of this factor is as follows:

$$ PAF = \frac{(1+i)^n - 1}{i(1+i)^n} $$

= Present worth of an annuity factor

and

$$ PW = PAF \times AP $$

where

AP = Annuity amount (or equivalent annual payment)

Suppose the induction heater in the previous example had an energy cost of $20 000/yr but the new unit only cost $15 000/yr to operate. The annuity factor will provide a base of comparison by including the annual energy cost as a single present worth amount. While the net cost or savings can be used, the actual amounts for each alternative should be used to reduce errors and clarify results.

Alternate A
Current expense: $10 000 PW
Future capital investment: 68 100 PW

Future energy costs =

$$\frac{(1+0.12)^5 - 1}{0.12\,(1+0.12)^5} \cdot \$20\,000 \; = \; 72100 \text{ PW}$$

Total: $150 190 PW

Alternate B
Current capital investment: $100 000 PW
Future energy costs =

$$\frac{(1+0.12)^5 - 1}{0.12\,(1+0.12)^5} \cdot \$15\,000 \; = \; 54000 \text{ PW}$$

Total: $154 000 PW

While the choice is still Alternate A, intangible factors could change the decision since the difference is only 2.7%.

The future worth factor (FWF) converts a single current dollar amount to a future amount. The future worth factor for a single present value is the reciprocal of the single present worth factor or

$$FWF = (1 + i)^n$$

$$FW = FWF \times PW$$

The future worth of an annuity factor (FAF) converts an annuity to a single future amount.

$$FAF = \frac{(1+i)^n - 1}{i}$$

and

$$FW = FAF \times AP$$

The uniform annuity factor (UAF) is used to convert a single present amount to a series of equal annual payments. The uniform annuity factor is the reciprocal of the present worth of an annuity factor (PAF).

$$UAF = \frac{i\,(1+i)^n}{(1+i)^n - 1}$$

and

$$AP = UAF \times PW$$

It is frequently necessary to open a savings account for a future purchase, and the amount saved is called a sinking fund. The annuity (sinking fund) required to accumulate some future amount is determined by using the sinking fund annuity factor (SAF). The factor is simply the reciprocal of the future worth annuity factor (FAF) or

$$SAF = \frac{1}{(1+i)^n - 1}$$

The preceding factors are summarized in table 3-1 and their use is graphically displayed in figure 3-1. However, most studies will use only the present worth of a single future amount and the present worth of an annuity factor. The next subclause describes the use of these factors in an economic analysis of energy options.

Table 3-1—Time value factors

Symbol	Name	Description	Formula*
AP	Annuity payment	Equal amounts of money at the ends of a number of periods	—
FAF	Future worth annuity factor	Converts an annuity to an equivalent future amount	$FAF = \frac{(1+i)^n - 1}{i}$
FW	Future worth	The dollar amount (of an expense or investment) at a specific future time	—
FWF	Future worth factor	Converts a single present amount to an amount at a future point in time	$FWF = (1+i)^n$
PAF	Present worth annuity factor	Converts an annuity to a single present amount	$PAF = \frac{(1+i)^n - 1}{i(1+i)^n}$
PW	Present worth	The single value or worth today	—
PWF	Present worth factor	Converts a future amount to an amount today	$PWF = \frac{i}{(1+i)^n}$
SAF	Sinking fund annuity factor	Converts a future amount into an equivalent annuity	$SAF = \frac{i}{(1+i)^n - 1}$

*The two variables have the following definitions:
n = Number of years in the evaluation period
i = Interest rate or other cost of money factor used

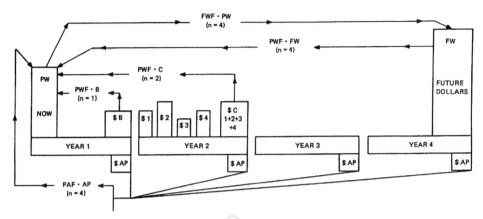

Figure 3-1—Time value chart

3.4.3 Life cycle cost example

The use of the time value of money formulas presented in 3.4 and summarized in table 3-1 is best illustrated by example. While the information presented in this example is representative of actual costs and equipment performance, it is intended only for illustration.

A 100 hp outdoor dip-proof (ODP) motor, which runs 5000 h/yr at an operating load of 75% of design, goes out of service. The plant engineer must choose one of the following:

— *Option 1.* Rewind the failed motor.
— *Option 2.* Purchase a new motor of normal operating efficiency.
— *Option 3.* Purchase a new motor of higher operating efficiency.

Option 1, motor rewind, costs $1500 plus $200 installation, and results in a motor operating efficiency of 90.9% for a first year energy operating cost of $23 400, but will last for only 10 years. Option 2, purchasing a new normal-efficiency motor, costs $2100 plus $200 for installation, results in a motor operating efficiency of 91.9% for a first year energy operating cost of $23 100, and will last for 15 years. Option 3, purchasing a new high-efficiency motor, costs $2500 plus $200 for installation, results in a motor operating efficiency of 94.8% for a first year energy operating cost of $22 400, and will last for 15 years.

For the purpose of this illustration, it is assumed that there is a 30-year investment horizon and that each option is exclusive of the other. The rate of inflation is assumed to be 4%/yr and the corporate discount rate is set at 20%. The salvage value of each option at the end of the 30 years is assumed to be equal for all three options, and so is not considered in this illustration. Finally, the tax effects of the alternatives (such as depreciation) are not considered.

The life cycle costs of Option 1 consist of three motor rewindings (in Years 0, 10, and 20) and 30 years of operating costs. The life cycle costs of Option 2 consist of two normal-efficiency motor purchases (in Years 0 and 15) and 30 years of operating costs. The life cycle costs of Option 3 consist of two high-efficiency motor purchases (in Years 0 and 15) and 30 years of operating costs.

The present worth of the capital and installation costs of the future motor rewindings and future motor purchases are estimated using the future worth factors, FWF (to account for the effects of inflation), and the present worth factors, PWF (to discount these future expenses into present day amounts).

Present worth of capital and installation costs for Option 1:
    1st rewind cost = $1 500 + $200 = $1 700
    Future worth factor, 4% inflation, 10 years = $(1 + 0.04)^{10} = 1.480$
    Present worth factor, 20% discount rate, 10 years = $1/(1 + 0.20)^{10} = 0.162$
    2nd rewind cost = $1 700 × FWF × PWF = $406
    Future worth factor, 4% inflation, 20 years = $(1 + 0.04)^{20} = 2.191$
    Present worth factor, 20% discount rate, 20 years = $1/(1 + 0.20)^{20} = 0.026$
    3rd rewind cost = $1 700 × FWF × PWF = $97
    Present worth of 1st, 2nd, and 3rd motor rewinds = $1 700 + $406 + $97 = $2 203

Present worth of capital and installation costs for Option 2:
    The first normal-efficiency motor cost = $2 100 + $200 = $2 300
    Future worth factor, 4% inflation, 15 years = $(1 + 0.04)^{15} = 1.801$
    Present worth factor, 20% discount rate, 15 years = $1/(1 + 0.20)^{15} = 0.065$
    2nd normal-efficiency motor cost = $2 300 × FWF × PWF = $269
    Present worth of 1st and 2nd motor costs = $2 300 + $269 = $2 569

Present worth of capital and installation costs for Option 3:
    1st normal-efficiency motor cost = $2 500 + $200 = $2 700
    Future worth factor, 4% inflation, 15 years = $(1 + 0.04)^{15} = 1.801$
    Present worth factor, 20% discount rate, 15 years = $1/(1 + 0.20)^{15} = 0.065$
    2nd normal-efficiency motor cost = $2 700 × FWF × PWF = $316
    Present worth of 1st and 2nd motor costs = $2 700 + $316 = $3 016

The present worth of 30 years of operating costs is estimated using the present worth annuity factor, PAF. It is assumed that electricity rates will escalate at the rate of inflation (4%). In order to use the present worth annuity factor, the discount rate must be adjusted for the effects of inflation. A "real" discount rate that is netted of inflation is defined as follows:

$$\frac{(1 + i) \times (1 + f)}{1}$$

where

    i = The cost of money used (in this case, the discount rate = 20%)
    f = The rate of inflation (in this case, 4%)

Thus

$$\frac{(1+0.20) \times (1+0.04)}{1} = 0.154$$

Present worth annuity factor for 30 years =

$$\frac{(1+0.154)^{30} - 1}{0.154 \times (1+0.154)^{30}} = 6.41$$

Present worth of operating cost, Option 1 = PAF × annual operating costs
= 6.411 × \$23 400 = \$150 017

Present worth of operation cost, Option 2 = PAF × annual operating costs
= 6.411 × \$23 100 = \$148 094

Present worth of operating cost, Option 3 = PAF × annual operating costs
= 6.411 × \$22 400 = \$143 606

The total life cycle cost of each option is the sum of the present worth of the capital and installation cost of each option and the present worth of operating costs for each option.

Total life cycle cost, Option 1 = PW capital and installation + PW operating costs
= \$2 203 + \$150 017 = \$152 220

Total life cycle cost, Option 2 = PW capital and installation + PW operating costs
= \$2 569 + \$148 094 = \$150 663

Total life cycle cost, Option 3 = PW capital and installation + PW operating costs
= \$3 016 + \$143 606 = \$146 622

The life cycle cost analysis shows that Option 3 has the lowest life cycle cost of the three options considered. Based on this analysis, the plant engineer orders the purchase of high-efficiency motors to replace the failed motor.

3.4.3.1 Relationship to other economic methods

The life cycle cost calculations illustrated are consistent with respect to the considerations described in 3.3.1 and 3.3.2.

With respect to 3.3.1, for example, the break-even point or payback period for Option 3 compared to Option 1 can be calculated as follows:

$$\text{break-even point} = \frac{\text{Option 3 first cost} - \text{Option 1 first cost}}{\text{Option 1 annual energy cost} - \text{Option 3 annual energy cost}}$$

$$= \frac{\$2700 - \$1700}{\$23\ 400/\text{yr} - \$22\ 400/\text{yr}} = 1\ \text{yr}$$

The break-even point for Option 2 compared to Option 1 is:

$$\text{break-even point} = \frac{\text{Option 2 first cost} - \text{Option 1 first cost}}{\text{Option 1 annual energy cost} - \text{Option 2 annual energy cost}}$$

$$= \frac{\$2300 - \$1700}{\$23\ 400/\text{yr} - \$23\ 100/\text{yr}} = 2\ \text{yr}$$

In this case, the break-even analysis yields results that are consistent with the life cycle cost analysis, namely, that Option 3 is preferable to Options 1 and 2. Life cycle and break-even cost analyses will not, however, always lead to consistent results. Generally speaking, the results of these two types of analysis will be consistent whenever discount rates are high and break-even times are short. For economic analyses of alternatives in which break-even times exceed 3–5 years, the use of life cycle cost analysis is recommended as the superior of the two approaches.

With respect to the marginal cost analysis described in 3.3.2, the engineer shall endeavor to use marginal values in all aspects of a life cycle cost analysis. In many cases, the use of marginal values is often self-explanatory. For example, the entire illustration presented in 3.4.3 is marginal in the sense that the engineer is faced with a decision regarding an increment to the operation of the facility, i.e., what to do about a failed motor. Each option considered is evaluated in that framework based on a marginal discount rate and the marginal costs of electricity as determined by the rate tariff (e.g., there are no customer or fixed monthly charge savings associated with any of the options considered).

3.4.3.2 Sensitivity analysis

The value of the engineer's recommendations is enhanced if the economic analysis presented in support of an energy saving measure is robust with respect to reasonable changes in assumptions. For example, the rationale for most energy saving activities is the substitution of capital invested today for savings that accrue in the future. Thus, the value of the investment is based on expectations regarding operating cost savings in the future. It is prudent to consider whether the investment will remain attractive under different assumptions.

We shall consider two particular sensitivities, based on the illustration presented in 3.4.3: What if the cost of the high-efficiency motor is higher than anticipated? What if the efficiency of the high-efficiency motor is lower than that claimed by the manufacturer? With the modern computer program/spreadsheet, this type of analysis can be performed easily.

A particularly valuable approach to sensitivity analysis is the calculation of break-even points. The break-even analysis presented in 3.4.3.1 is an example in which the break point is calculated for the time at which cumulative energy savings exactly offset increased first costs.

In the context of the two sensitivities, we will calculate the point at which the life cycle costs analysis yields identical total life cycle costs for Option 3 and Option 1.

Using the methods developed in 3.4.3, it can be shown that Option 3, the high-efficiency motor, will remain cost-effective relative to Option 1, the motor rewind, up until the point at which the total capital and installed cost of the energy efficient motor exceeds $6 262, or more than twice the original estimate. Similarly, holding the original cost estimate fixed, Option 3 remains cost-effective relative to Option 1 until the efficiency of the high-efficiency motor drops below 91.54%, or 3.26% below the manufacturer's rated efficiency.

These two calculations illustrate how the engineer can "bound" the cost-effectiveness of his estimates by examining the limiting case of the break points for selected key assumptions. In this example, the choice of Option 3 over Option 1 appears to be robust since significant deviations from the original cost and performance assumptions are required to make Option 3 less cost-effective than Option 1.

3.4.3.3 Levelized cost analysis

Levelized cost analysis is an alternative economic model that is wholly consistent with life cycle cost analysis. Instead of calculating the cumulative present worth or life cycle cost for each energy saving alternative, levelized cost analysis calculates an annual measure of capital and operating costs that remains constant over time. The present worth of levelized costs over the life of an investment is exactly equal to the present worth of the full capital and operating costs of the investment.

Levelized costs are useful because they allow the engineer to directly compare the value of increased capital costs (for a more energy efficient investment) to the energy savings resulting from that investment. In effect, the use of levelized costs allows the engineer to calculate break-even points for alternative investments that account simultaneously for both the capital and operating costs of the investments.

Levelizing capital costs requires no more than calculating the reciprocal of the present worth of annuity. That is, the present worth of an annuity calculates the present value of a stream of future payments or annuities that is constant over time. The reciprocal of the present worth annuity factor of PAF (see table 3-1) multiplied by the capital cost of the investment yields the annual value of a level stream of annuities whose present value equals that of the investment.

Returning to the illustration in 3.4.3:

PAF, discount rate 15.4%, 10 years = $[(1 + 0.154)^{10} - 1]/[0.154 \times (1 + 0.154)^{10}] = 4.943$

PAF, discount rate 15.4%, 15 years = $[(1 + 0.154)^{15} - 1]/[0.154 \times (1 + 0.154)^{15}] = 5.736$

Levelization factor, discount rate 15.4%, 10 years = 1/PAF = 0.202

Levelization factor, discount rate 15.4%, 15 years = 1/PAF = 0.174

The levelized first cost of Option 1 = Capital and installation cost × (1/PAF)
= $1 700 × 0.202 = $343

The levelized first cost of Option 2 = Capital and installation cost × (1/PAF)
= $2 300 × 0.174 = $400

The levelized first cost of Option 3 = Capital and installation cost × (1/PAF)
= $2 700 × 0.174 = $470

The annual operating costs remain as given in 3.4.3.

First, consider the differences in the levelized first costs of each investment. The levelized first cost of Option 2 if $57 more than Option 1 ($400 – $343), and that of Option 3 is $127 more than Option 1 ($470 – $343). The meaning of these differences is that, in order to be cost-effective, the efficiency of Option 2 relative to Option 1 must lead to annual operating cost savings that exceed the annualized difference in their capital and installation costs, which is $57. Similarly, in order to be cost-effective, the efficiency of Option 3 relative to Option 1 must lead to annual operating cost savings that exceed the annualized difference in their capital and installation costs, which is $127.

Now, consider the differences in annual operating costs. The annual operating costs of Option 2 are $300 less than those of Option 1 ($23 400 – $23 100), so Option 2 is cost-effective relative to Option 1 (since annual operating cost savings exceed $57). The annual operating costs of Option 3 are $1 000 less than those of Option 1 ($23 400 – $22 400), so Option 3 is cost-effective relative to Option 1 (since annual operating cost savings exceed $127). In addition, Option 3 is also more cost-effective than Option 2 since the difference in annual operating costs, $700 ($23 100 – $22 400), exceeds the difference in levelized capital and installation costs, $70 ($470 – $400).

Levelized cost analysis is based on life cycle cost analysis and can be useful in assessing first cost versus operating cost trade-offs. In all cases, life cycle cost analysis will lead to the most comprehensive treatment of all cost factors. Specific consideration for the use of levelized cost analysis includes:

a) Treatment of investments with unequal lifetimes—this will only be an issue when the rate of escalation for the investments differs for each alternative. In these cases, a single present worth must be calculated for each alternative over a fixed investment horizon before levelizing to total amount.

b) Treatment of annual operating costs that escalate at different rates—when annual operating costs escalate at a rate different than that of future replacements for the investment alternatives (in these examples, both were assumed to escalate at the rate of inflation), it will be necessary to calculate the present worth of the future annual operating costs and then levelize them using the same factor used to levelize the capital and installation costs of the alternatives.

3.5 Utility rate structures

3.5.1 Electric tariff

A tariff is filed by each electric company and approved in its filed form or as modified after rate hearings by the regulatory body. Each tariff has two sections:

a)   Rules and regulations
b)   Rate schedules

Rules and regulations are conditions under which a utility will supply power to a customer. These include billing practices, rights-of-way, metering, continuity of service, power factor, line extensions, temporary service, and many other details.

Rate schedules are the prices for electric service to different classes of customers. The four common classes are residential, commercial, industrial, and street or area lighting. There may be several rate schedules available on each customer class that are usually based on load magnitude. Special rate schedules and individual contracts are also common. A typical rate schedule for commercial and industrial customers usually contains most of the following elements: rate availability and characteristics; net rates for demand, energy, and power factor; minimum charges; payment terms; terms of the contract; off-peak service; untransformed service; and riders. Each electric rate is usually contracted for a period of 1 yr and customers are entitled to the cheapest available rate, providing they meet the service characteristics specified in that rate. The following subclauses describe the elements in a rate schedule and subsequent subclauses will cover detailed examples.

3.5.2 Rate structure elements

The usual rate structure establishes monthly charges for kilowatt demand, energy (kWh), and the power factor, which are added to comprise a base rate. Rate structures also include a fuel or energy adjustment charge, which is applied to all kilowatthours consumed and then added to the base rate to obtain the total charge. The load factor, described more fully in Chapter 4, also affects the utility bill.

The demand component is designed to allow the utility to recover the capital costs associated with the construction of generating stations, substations, and transmission and distribution lines capable of meeting the customers' demand requirements. In most cases, a customer pays for the average demand in the highest 15-min or 30-min energy usage period during each billing period. Since there are 2880 fifteen-minute periods in a month, it is easy to see the reason for controlling demand. The utility shall supply sufficient capacity to meet this one period out of the total 2880 periods each month. Hence, both utilities and customers benefit from good demand control.

Energy charges are much easier to understand since a customer pays for the number of kilowatthours used to do the work required. A customer normally pays for all kilowatthours used. The energy component of a bill primarily recovers fuel costs but it also recovers operation and maintenance costs such as expendable materials, salaries and wages, gasoline, and tools.

Most utilities charge for reactive power usage (kilovars) for at least the very large users. The reactive power supplied to motors and transformers is paid for in some manner since the power company sizes its facilities to generate and transmit these kilovars. The total requirement is determined by the vector sum of the real and reactive power, so the term *power-factor clause* is commonly used. Methods used to calculate the reactive charge vary from a very clear charge per kilovar hour or per kilovar demand to what appear to be hidden means. The reactive charge can be reduced or eliminated by installing power-factor correction equipment—normally static capacitors. A comprehensive coverage of power factor correction is covered in IEEE Std 141-1993, IEEE Recommended Practice for Electric Power Distribution for Industrial Plants (IEEE Red Book), and IEEE Std 241-1990, IEEE Recommended Practice for Electric Power Systems in Commercial Buildings (IEEE Gray Book).

The load factor is the ratio of the average kilowatt demand to the peak kilowatt demand. Utilities prefer a constant, nonvarying load or a 100% load factor where the average usage and the peak usage are the same. Many tariffs are structured to encourage better load factors.

The last common element of a rate structure is a fuel or energy adjustment clause (EAC). The purpose of this billing procedure is to enable a utility to recover its fuel costs quickly in a market where fuel costs fluctuate widely within short periods of time. The main objective of the EAC is to eliminate frequent and costly rate cases, an expense that is borne by each customer. The charge is normally applied to all kilowatthours used.

The following is an explanation of the rate forms most often used today. The rate structure atmosphere is changing so quickly that the following information can be outdated in a short period of time. Furthermore, the regulatory climate in each state is so different that these rate forms may no longer be available in some states.

a) *Declining block rate.* The rate for the first kilowatt and kilowatthour is typically the highest cost per kilowatt or kilowatthour. Hence, an initial kilowatt or kilowatthour block is billed at the highest rate. Additional consumption beyond this first block is then billed at a lower rate. There is no limit to the number of possible blocks, but it is unusual to see more than four.

   This rate form was developed because utilities found that as a customer's consumption increased, the relative cost to provide the electric service decreased. The reduction in service costs was then reflected in a lower charge per unit as usage increased.

b) *Demand energy rates.* The determination of demand was covered earlier in this subclause. When the calculation of "demand" includes the effects of load factor, the rate is called a load factor, hours use, or demand-energy rate. In this rate form, the number of kilowatts in one or more of the first energy blocks is determined by kilowatt demand and a predetermined number of hours use, as shown in the example in 3.6. The number of kilowatthours in subsequent energy blocks is determined in the same manner. Thus, the larger the ratio between the average kilowatt and peak kilowatt, the more kilowatthours are billed in the lowest energy block and the lower a power bill becomes.

$$\text{Load factor} = \frac{\text{average kW (or kVA)}}{\text{peak kW (or kVA)}}$$

$$= \frac{kW_A}{kW_P} \text{ or } \frac{kWh}{kWd \times h/mo}$$

$$\text{Hours use} = \frac{kWh}{\text{peak kW}}$$

$$= \frac{kW_A}{kW_P} \times \text{hours per month}$$

c) *Seasonal rates.* Power company yearly load patterns vary from one company to another. Some utilities experience a summer peak, while others see a winter peak. Other companies have yearly peaks that have no seasonal correlation. To discourage wasteful use of electricity in the peak seasons, some utility regulatory commissions require a higher charge during the peak seasons. In some rate schedules, the highest kilowatt demand during the peak season, or any month, determines the minimum kilowatt billing demand for the next 11 months or the next off-season months. This method of seasonal billing, sometimes called a ratchet clause, should encourage a large customer to use demand control or load management techniques during the peak season. The concept is based on the fact that the cost to provide service during the peak season is greater than at other times of the year.

d) *Time-of-day rates.* These rates are becoming more popular. The charge for a kilowatt or a kilowatthour is less for time periods other than the peak for the utility. Some utilities have "on peak," "off peak," "shoulder" rates, and weekend rates; and the variety of different time-of-day rates will probably continue to expand. There are a variety of ways that these rates are administered. Sometimes the pricing is directly associated with the time period and other times a credit is given for usages outside the peak period.

e) *Interruptible rates.* While some individuals use *interruptible* and *curtailable* interchangeably, this recommended practice will use interruptible to mean a rate based on the premise that the utility turns off the electric supply to a facility under predetermined circumstances or agreements.

As a rule, interruptible rates are considerably lower than general service rates and, hence, have definite economic advantages. Sometimes the number of interruptible hours per year, or the number of hours per interruption, or both, are limited by the rate schedule. The customer shall weigh the benefit of the greatly reduced electrical costs against the losses associated with a complete shutdown. Sometimes this rate involves a special contract with the features detailed in item f).

f) *Curtailable rates.* In a curtailable rate structure, the customer makes predetermined, voluntary load reductions upon request by the utility. This rate structure usually involves some formal agreement between the user and the utility. The agreement usually involves such important criteria as

1) The time period between the power company request for a load reduction and the reduction

2) The magnitude of the reduction

3) The maximum number of curtailments per year

4) The maximum length of each curtailment

5) The total number of curtailable hours in a year

This type of rate has advantages for the utility in that it can shed load quickly when critical power shortages occur. It should be noted that the customer may not have to completely shut down to obtain a rate reduction. Generally, an interruptible rate will be lower than a curtailable rate. However, the rate may include an extremely heavy penalty charge for failure to curtail on request.

3.5.3 Proposed electric rate structures

Conservationists, environmentalists, politicians, social scientists, and others have been and will continue to look for ways to reduce the growth in electricity usage and the cost to residential customers. These special interest groups have influenced public utility commissions. Many factions believe that existing rate structures discriminate against certain customer classes and are a hindrance to their cause. Specifically, they believe the declining rate blocks and the lower unit cost for electricity enjoyed by commercial and industrial customers cause waste and unfairly discriminate against small users of electricity. Their natural solution is to change the rate structures. The purpose of this subclause is to familiarize the energy engineer with the more frequently proposed new rate structures. Because new rate forms are often not cost based, companies, consultants, and engineers should participate in the rate-making process by intervening in rate cases. Utility commissions and the political system should have the technical input of the engineering professions in the decision-making process.

The four common rates are as follows:

a) *Flat rates.* In this rate, all users of electricity would pay the same amount per kilowatthour for all kilowatthours used. In some cases, demand charges would also be eliminated.

b) *Inverted rates.* This rate form is the reverse of declining block rates. The unit cost would rise with higher usage. The first kilowatthours consumed would cost less per unit than the last kilowatthours used.

c) *Marginal rates [marginal cost pricing (MCP)].* In this system, the utility would charge each customer based on the actual added cost imposed on the utility by that customer's usage. The power company would first calculate how much each customer adds to its operating costs, then it would anticipate how its system would be expanded to meet growing demand and assess all customers in proportion to their contribution to that expanding demand. A variation of MCP is what rate experts refer to as long-range incremental costing (LRIC). Pricing is based on the expected cost to produce electricity at some point in the future.

d) *Lifeline rates.* The basic concept of lifeline rates is to lower bills of low users of electricity who are assumed to have low income. This is accomplished by inverting the rate structure for residential customers only. The revenue shortfalls are made up by higher rates for other customers.

The Public Utility Regulatory Policies Act of 1978 (PURPA) created new procedures for rate making by establishing Federal standards for rate redesign and by setting up new classes of intervenors. PURPA suggests that rate reform should encourage one or more of the following (in addition to encouraging cogeneration): conservation of energy, efficient use of utility facilities and resources, and equitable rates for electric customers. The effect of PURPA should be to encourage utility rate intervention by individuals. This action will probably cause additional proposals for new rate structures, and rate cases will be more drawn out.

3.6 Calculating the cost of electricity

A few examples will simplify the seemingly complex nature and wide variety of electric rates. Throughout the 1990s one can expect wide use of block rates, demand usage rates, and fuel charges. Virtually all large utilities include a provision for reactive (var) charges in selected rates by using power factors or some form of demand or block rate, or both. The flat rate is not covered due to its simplicity. The utility or public utility commission will provide rate details for a specific situation.

3.6.1 Block rate with var charge example

Tables 3-2 and 3-3 show rate schedules as they might be received from commissions. The rate schedule is essentially a contract with a utility and should be read and understood. The riders and general rules and regulations are also part of the contract. For this example, assume that the plant's July electrical consumption is 2520 kW demand, 1 207 200 kWh, and 896 kvar, which produces a power factor of 94.2. Furthermore, the plant is billed on Schedule A (table 3-2), and the fuel charge is 1.5 cts/kWh (or 15 mil). The calculation of the electric bill is shown in table 3-5 and is described in the ensuing paragraphs.

There are two blocks (tables 3-2 and 3-3) for the kilowatt charge and a flat rate for the kilovar (reactive demand) charge. Notice the higher charge for summer usage which indicates that this is a summer peak utility. The first 50 kW is billed at $4.83/kW and the remaining 2 470 kW is billed at $3.80 for a total charge of $9 627.50. It is important to note that the demand is determined for the current month only. The minimum charge on this rate is $11.00 plus fuel. The entire 896 kvar usage is billed at $0.20/kvar for a total reactive charge of $179.20.

All three blocks for kilowatthour charge are used for this load. The large portion of the kilowatthours in the last block is an indication that a different rate for higher usage may be available. The first 40 000 kWh are billed at 2.654 cts/kWh, the next 60 000 kWh are billed at 2.094 cts, and the remainder are billed at 1.524 cts. The total kilowatthour charge is then $19 191.73.

The flat rate fuel charge of 1.5 cts/kWh is applied to the entire 1 207 000 kWh usage. The fuel charge of $18 108 is then added to the demand and energy charges for a total bill of $47 106.43. The average cost of electricity is 3.9 cts/kWh.

Table 3-2—Schedule A

Applicable to any commercial or industrial consumer having a demand equal to or in excess of 30 kW during the current month or any of the preceding 11 months.

Monthly rates:	Summer	Winter
(1) Kilowatt demand charge ($ per kW)		
For the first 50 kW	4.83	4.01
For all excess over 50 kW	3.80	2.98
(2) Reactive demand charge (cts per kvar)		
For each kvar of billing demand	20.0	20.0
(3) Kilowatthour charge (cts per kWh)		
For the first 40 000 kWh	2.654	2.354
For the next 60 000 kWh	2.094	1.794
For all excess	1.524	1.274

(4) Seasonal rates: The winter rates specified above shall be applicable in seven consecutive monthly billing periods beginning with the November bills each year. The summer rates shall apply in all other billing periods.

(5) Fuel cost adjustment: The above kilowatthour charges shall be adjusted in accordance with the fossil fuel cost adjustment, Rider No. 6.

(6) Other applicable riders: The rates specified above shall be modified in accordance with the provisions of the following applicable Riders:

Primary metering discount: Rider No. 2
Supply voltage discount: Rider No. 3
Direct current service;   Rider No. 5

Minimum charge: $11.00 per month or fraction of a month plus fuel cost adjustment.

Maximum charge: If a consumer's use in any month is at such low-load factor that the sum of the kilowatt demand, reactive demand, and kilowatthour charges produces a rate in excess of 11.0 cts per kWh, the bill shall be reduced to that rate per kilowatthour of use in that month plus the fuel cost adjustment charge but not less than the minimum charge.

Special rules:
(1) Combined billing: Where two or more separate installations of different classes of service on the same premises are supplied separately with service connections within 10 ft of each other, the meter registrations shall be combined for billing purposes, unless the consumer shall make a written request for separate billing.
(2) Schedule transfers: If for a period of 12 consecutive months, the demand of one installation or the undiversified total demand of several installations eligible for combined billing in each such month is less than 30 kW, subsequent service and billing shall be under the terms of the general commercial schedule for the duration that such scheduling is applicable.
(3) Reactive billing demand:
(a) If the kilowatt demand on any class of service is less than 65 kW for three-phase installations or 75 kW for single-phase installations, the reactive billing demand shall be zero.
(b) If the kilowatt demand is 65 kW or higher for three-phase installations or 75 kW or higher for single-phase installations, the reactive billing demand shall be determined by multiplying the monthly kilowatt demand by the ratio of the monthly lagging reactive kilovoltampere hours to the monthly kilowatthours.
(4) Service interruption: Upon written notice and proof within ten days of any service interruption continuing longer than 24 h, the company will make a pro rata reduction in the kilowatt demand rate. Otherwise, the company will not be responsible for service interruptions.

Table 3-3—Schedule B

Applicable to any consumer having a demand of less than 10 000 kW and using more than 500 000 kWh per month during the current month or any of the preceding 11 months. No resale or redistribution of electricity to other users will be permitted under this schedule.

Monthly rates:	Summer	Winter
(1) Kilowatt demand charge ($ per kW)		
For the first 50 kW	4.83	4.01
For all excess over 50 kW	3.80	2.98
(2) Reactive demand charge (cts per kvar)		
For each kvar of billing demand	20.0	20.0
(3) Kilowatthour charge (cts per kWh)		
For the first 40 000 kWh		
For the next 60 000 kWh	2.654	2.354
For the next 200 kWh per kWd but not less than 400 000 kWh	2.094	1.794
For the next 200 kWh per kWd	1.524	1.274
For all excess	1.144	0.944
	1.004	0.794

(4) Seasonal rates: The winter rates specified above shall be applicable in seven consecutive monthly billing periods beginning with the November bills each year. The summer rates shall apply in all other billing periods.

(5) Fuel cost adjustment: The above kilowatthour charges shall be adjusted in accordance with the fossil fuel cost adjustment, Rider No. 6.

(6) Other applicable riders: The rates specified above shall be modified in accordance with the provisions of the following applicable Riders:

Primary metering discount:	Rider No. 2
Supply voltage discount:	Rider No. 3
Consumer's substation discount:	Rider No. 4

Special rules:
(1) Submetering or redistribution prohibited: This schedule is applicable only where all of the electricity supplied is used solely by the consumer for his own individual use.
(2) Schedule transfers:
(a) If for a period of 12 consecutive months, the kilowatthour use in each such month is less than 500 000 kWh, subsequent service and billing shall be under the terms of Schedule A when this schedule is applicable.
(b) If in any month the maximum 30 min kW demand exceeds 10 000 kW, the consumer shall contract for service under Schedule B beginning with the next succeeding month.
(3) Reactive billing demand: The reactive billing demand shall be determined by multiplying the monthly kilowatt demand by the ratio of the monthly lagging reactive kilovoltampere hours to the monthly kilowatthours.
(4) Service interruption: Upon written notice and proof within ten days of any service interruption continuing longer than 24 h, the company will make a pro rata reduction in the kilowatt demand rate. Otherwise, the company will not be responsible for service interruptions.

3.6.2 Demand usage rates example

In the preceding subclause, the large number of kilowatthours in the last block indicated the possibility of a better rate. The usage is more applicable to the rate schedule noted in tables 3-3 and 3-4; therefore we will use the same usage figure on Schedule B. This rate schedule combines the block rate and demand usage rate. In addition, the customer can benefit from owning the equipment on his property (see table 3-4, Rider no 4). The cost benefit of owning equipment to supply the 2 500 kW (2 640 kVA) plus load is $9 000/yr. The kilowatthours calculated for each block are multiplied by the appropriate rate to determine the total kilowatthour charge. For this schedule, the charge is $16 760.69, which is $2 431.04 less than the amount using the previous schedule. The fuel charge is the same. The average cost per electricity is then 3.64 cts/kWh. These calculation are shown in table 3-6 and are explained in the following paragraphs.

Table 3-4—Riders

Rider No. 1—Fuel adjustment for special contracts: The cost of fuel as used in Rider No. 1 to Tariff PUCO No. 11 shall be the delivered cost of fuel as recorded in Account Nos. 501 and 547. Such fuel cost will be reported to the commission on a routine basis. Any proposed change in the type of fuel to be purchased, source of supply, or means of transportation that is estimated to increase or decrease the cost of fuel per million Btu by $0.01 or more shall be submitted to the commission for approval. Unless the commission shall take positive action within 15 working days to disapprove a proposal of an applicant, such proposal shall be deemed to have been approved.
Rider No. 2—Primary metering discount: If the electricity is metered on the primary side of the transformer, a discount of 2% of the primary meter registration in each of the company's electric schedules in which this rider is applicable will be allowed for electricity so metered.

Rider No. 3—Supply voltage discount: A discount on the monthly kilowatt demand charges in each of the company's electric schedules in which this rider is applicable will be allowed when the supply is entirely from 132 kV overhead circuits or 33 kV overhead circuits (for the purpose of this rider, 33 kV overhead shall include 13.8 kV overhead transmission circuits fed directly from a power plant bus):
 	
                            Discount per kW of
     Class of supply   Demand billed per month
     132 kV overhead $0.30
      33 kV overhead $0.10

Rider No. 4—Consumer's substation discount: If the consumer elects to furnish and maintain or lease, or otherwise contract for all transforming, switching, and other equipment required on the consumer's premises, a discount of $0.30/kW or demand billed will be allowed on the monthly kilowatt demand charges in each of the company's electric schedules in which this rider is applicable.

Table 3-5—Block rate example

Usage: 2520 kW; 896 kvar, and 1 207 200 kWh				Rate: Schedule A Table 3-3
(1) Kilowatt demand charge			Subtotal	
50 kW × $4.83	=	$ 241.50		
2470 kW × $3.80	=	$9 386.00		
			$9 627.50	
(2) Reactive demand charge				
896 × $0.20	=		$179.20	
(3) Kilowatthour charge				
40 000 × $0.02654	=	$1 061.60		
60 000 × $0.02094	=	$1 256.40		
1 107 200 × $0.01524	=	$1 6873.73		
Total charge	=		$19 191.73	
(4) Fuel charge				
1 207 200 × $0.015	=		$18 108.00	
(5) Total electric charge				
kW	=	$9 627.50		
kvar	=	179.20		
kWh	=	1 9191.73		
Fuel	=	18 108.00		
Total	=	$47 106.43		

The demand charges will still total $9 806.70. The first two kilowatthour blocks will be the same at $1 061.60 and $1 256.40. The remaining 1 107 200 kWh will be allocated to the remaining blocks It should be noted that a minimum of 400 000 kWh ($6 096) is billed in the third block. The total number of kilowatthours in each demand usage block for the example is easily determined by multiplying the 200 kWh/kWd by the demand: 200 kWh/kWd × 2520 = 504 000.

To determine the hours in each block, the following is used:

Total kilowatthours	1 207 200
(40 000 in Block No. 1)	− 40 000
Balance for Block No. 2	1 167 200
(60 000 in Block No. 2)	− 60 000
Balance for Block No. 3	1 107 200
(504 000 in Block No. 3)	− 504 000
Balance for Block No. 4	603 200
(504 000 in Block No. 4)	− 504 000
Balance for Block No. 5	99 200 kWh
(99 200 in Block No. 5)	

Table 3-6—Demand rate example

Usage: 2520 kW, 1207 200 kWh, and 896 kvar	Rate: Schedule B Table 3-3

(1) Kilowatt demand charge
  $50 \times \$4.83 = \quad 241.50$
  $2470 \times \$3.80 = \underline{9\,386.00}$
  $\qquad\qquad\quad 9\,627.50$

Credit for Rider No. 4 = $756.00 ($0.30 × 2520)

(2) Reactive demand charge
  $896 \times \$0.20 = \$179.20$

(3) Kilowatthour charge
  Block No. 1   40 000   × $0.02654 = $1 061.60
  Block No. 2   60 000   × $0.02094 =   1 256.40
  Block No. 3   504 000   × $0.01524 =   7 680.96
  Block No. 4   504 000   × $0.01144 =   5 765.76
  Block No. 5   <u>99 200</u>   × $0.01004 =   <u>995.97</u>
            1 207 200 $16 760.69

(4) Fuel charge @ $1.5 cts/kWh
  $1\,207\,200 \times \$0.015 = \$18\,108.00$

(5) Total electric charge
  kW $9 627.50
  kvar   179.20
  kWh   16 760.69
  Fuel   <u>18 108.00</u>
  Total $44 675.39
  Possible credit   <u>– 756.00</u>
  Total w/credit $43 919.39

3.6.3 Important observations on the electric bill

Energy conservation demands a close look at energy costs. It is obvious from the preceding example that energy is the biggest portion of this bill. This fact is not always discernible as some rates include some minimum fuel cost in the kilowatthour portion of the rate. In the example, the fuel cost represents 40% of the electric cost. The kWh cost represents another 38% of the bill. The demand cost represents only 22% of the bill. However, in other cases, demand may represent the major part of the bill.

Since in this case each kilowatthour is associated with 78% (40 + 38) of the electric cost, a 10% reduction of kilowatthours brings four times the benefit over an equal reduction in demand. A 10% reduction in peak demand will reduce the power bill by only 2%, which can be offset by a 2.5% increase in kilowatthours. In this case, the plant engineer should proceed with caution in controlling demand and, preferably, look for ways to reduce kilowatthours even during off-peak hours.

Many articles and technical papers stress power-factor correction. In this case, the power factor is exceptionally good, but it is beneficial to see the savings achievable by power-factor correction. With an 80% power factor, the reactive demand is 1890 kvar (0.75 × 2520) and costs $378. Power factor correction to the 94% (896 kVar) level is then worth $198.80 per month (378 − 179.20) or $2 385.60/yr.

Finally, a reduction in usage will not give a proportional reduction in the electric bill. Suppose this manufacturer achieves a 10% reduction in kilowatthours, kilovars, and kilowatts, the savings is only 5% as shown in table 3-7 due to the removal of the least costly increments. All energy cost analysis should use the "tail rate" in evaluating energy savings because the reduction will only reduce units in the last (tail rate) blocks.

Table 3-7—Dollar savings from energy reduction

Energy
1 207 200 × 0.1 = 120 720 kWh saving
    99 200 kWh Block No. 4 savings = $995.97
    21 520 kWh Block No. 3 savings = <u>246.19</u>
                             $1 242.16 energy savings

.kvar "flat" = 0.1 × 179.20 = $17.92
.kW 0.1 × 2520 = 252 kW savings
252 Block No. 2 savings = $957.60

Total savings = $2 217.68 or 4.96% of pre-energy savings amount

3.7 Loss evaluation

3.7.1 Introduction

All electrical equipment has some loss; nothing is 100% efficient. These losses can vary with output levels and age, or they can remain constant. For example, conductor losses vary as the square of the load current, while the magnetic losses of a transformer are relatively constant with load (but vary approximately as the voltage squared).

In virtually all loss evaluations, a load profile shall be established by either analytical or empirical methods. Furthermore, the efficiency of the device under investigation shall be determined for each set of anticipated operating conditions. The efficiency at full load is meaningless for comparing two devices that will be operated at half load unless the losses are solely a function of load.

All losses can be classified into two types. No-load loss is the quantity of losses when the device is idling or in a standby mode. Load losses are the additional losses at each load increment. The efficiency of a device is usually given at the full-load condition, which is but one point in the efficiency spectrum for many devices. Even the single-point efficiency can have different values, depending on the standard under which the device was tested. See [B1][1].

3.7.2 No-load (or single value) loss calculation

The no-load losses are constant for transformers, motors, and adjustable speed drives. The losses do vary as a function of voltage, frequency, and temperature, but these variables are expected to remain fairly constant over the evaluation period. Energy costs on lighting systems or similar processes with constant losses can be evaluated by using the no-load loss technique and substituting the "on" values for no-load values.

The cost of no-load losses has two components: the demand cost, D_N, and the energy cost, Q_N. The demand cost is merely the diversified kilowatts or kilovoltamperes times the tail demand rate, the cost of the last block used in the demand charge. The energy cost is the hours of "on" time multiplied by the energy charge, which is the tail rate including the fuel charge.

The diversity is the per unit amount of an individual load that contributes to the billed demand. The diversity varies from zero when the load does not contribute to the plant's billing demand to one when the entire load is added to the plant's billing demand.

In mathematical form, the no-load cost of losses is as follows:

$$D_N = (\text{diversity}) \times \frac{\$}{kW} \times 12 \text{ months} \tag{1}$$

= Demand cost per year per kW of no-load loss

$$Q_N = (\text{no-load hours}) \times \frac{\$}{kW} \tag{2}$$

= Energy cost per kW of no-load loss

or equation 2 can be rewritten:

$$Q_N = \frac{\text{no load hours}}{\text{Day}} \times \frac{\text{days}}{\text{week}} \times \frac{\text{weeks}}{\text{year}} \times \frac{\$}{kWh} \tag{3}$$

or

$$Q_N = \frac{\text{no-load hours}}{\text{week}} \times \frac{\text{weeks}}{\text{year}} \times \frac{\$}{kWh} \tag{4}$$

---

[1]The numbers in brackets correspond to those of the bibliography in 3.8.

3.7.3 Load loss calculation

Load losses are more complex only because they vary over the evaluation period; so they shall be developed in increments. No-load losses can be combined with load losses (e.g., in cases where it is not possible to separate the losses or when it is desirable to look at total losses as a single entity). It is usually easier to treat load and no-load losses separately for transformers, because load losses can be mathematically expressed in terms of load.

Load losses have a demand and an energy component. The demand component includes the device's effect on the plant's electrical peak. The demand-loss cost for 1 yr is then

$$D_L = \sum_{i=1}^{12} (\text{diversity})_i \times (\$/\text{kW})_i \times (P_i) \tag{5a}$$

where

i = Month
P_i = Unitized level of load losses in the ith period

If the cost, relative power level, and diversity are constant, the equation is

$$D_L = 12 \times D \times DC \tag{5b}$$

= Demand cost per year per kW of load loss

where

D = Diversity
DC = Demand cost

Since the energy cost is a function of the load cycle, the load cycle shall be determined. Meters can be installed on existing equipment to obtain actual values. If actual data is not available, a load schedule or profile can be obtained by metering similar processes or by theoretical analysis. It is usually wise to group loads in terms of hours. This results in small loss of accuracy and great ease of calculation. This calculation is much more cumbersome if time-of-day rates are applicable, because the load cycle needs to be associated with the proper time of day.

The energy cost is then the sum of energy used times the cost of energy. The sum of usage is simply the sum of the load levels times their associated "on" times. The number of hours per year is then multiplied by the tail energy rate (including the fuel charge). In some cases, the calculation may require summing several different types of load cycles during a year. In the simplest case of constant daily load, the following equation applies:

$$Q_L = 365 \times \frac{\$}{\text{kWh}} \times \sum_{i=1}^{n} t_i P_i \tag{6}$$

= Annual \$/kW load loss

67

where

n = Number of periods
t_i = Duration of the ith period, hours
P_i = Unitized level of load losses in the ith period

In the following subclauses, the value of P_i is unitized in terms of load level and Q is put in terms of dollars per kilowatt per year. By finding the value of 1 kW of losses, both the manufacturer and the energy engineer will know the cost of losses for a specific design/ machine and can make a logical choice for a price versus loss decision. The loss evaluation generally is made for a period greater than 1 yr but not exceeding the anticipated useful life of the equipment.

3.7.4 Motor loss evaluation with example

The load and no-load losses in a motor combine in a manner shown in figure 3-2. When no specific curve is available, one can approximate motor losses by using no-load and load losses. The no-load losses are composed of the hysteresis, eddy current, and windage and friction losses at full-load speed and temperature. The no-load losses costs can therefore be calculated by using equations 1 and 2. The load losses (the remainder) vary as the square of the motor load, and consist of stator winding and stray losses.

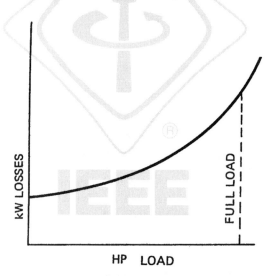

Figure 3-2—Motor losses

The motor-load losses costs can be calculated by recognizing that the major component is the I^2R losses in the winding and the armature. Sophisticated programs would consider additional adjustments due to heating effects, etc. The value of load loss at any particular load is the square of the ratio of that load to nameplate. More specifically, the value of yearly motor load losses is as follows:

$$D_L = \sum_{i=1}^{12} (\text{Diversity}) \left(\frac{HP_i}{HPR}\right) \times DC_i \qquad (7a)$$

= \$/kW of annual load loss

$$Q_L = \left(\sum_{i=1}^{n} t_i\right)\left(\frac{HP_i}{HPR}\right)^2 \times EC_i \qquad (7b)$$

= \$/kW of load loss per period

where

D_L	= Demand cost per kW of load loss
Q_L	= Energy cost per kW of load loss
HP	= Peak motor load, hp
HP_i	= Motor load for the ith interval, hp
HPR	= Rated motor horsepower
n	= Number of intervals being evaluated
t_i	= Duration of the ith interval, hours
DC_i	= \$/kW demand cost for the ith interval (usually monthly)
EC_i	= \$/kWh energy cost for the ith interval

An example will illustrate the loss evaluation technique. Assume a motor is used in a five-day, 10 h per day process that runs 50 weeks/yr. The motor runs at 0.25 load for 2 h, 0.50 load for 4 h, full load for 2 h, and idles the remaining 2 h. The tail rate energy cost is \$0.10/kWh and the tail demand rate is \$15.00/kW. The peak motor load is coincidental with the plant's electrical peak.

a) The annual no-load energy cost is as follows:
Q_N = (10 h) × (5 days/week) × (50 weeks/yr) × \$0.1/kWh
    = \$250/kW of no-load losses
D_N = 1 kW × \$15/kW × 12 mo/yr
    = \$180/kW

b) The annual energy load loss is as follows:
Q_L = [(2 h) (0.25 load)^2 + 4 h (0.50 load)^2 + 2 h (full load)^2] × (5 days/week)
× (50 weeks/yr) × (\$0.1/kWh) = (0.125 + 1 + 2) × 5 × 50 × 0.1
    = 3.125 × 5 × 50 × 0.1
Q_L = \$78/kW of load loss

NOTE—No load-loss entry is required for the 14 h when the motor is off or for the 2 h when it idles.

c)   The annual demand load-loss cost is:
$D_L = 1 \times 15 \times 12 = \$180/kW$

A no-load and load motor loss reduction of 1 kW each is worth \$688 (250 + 180 + 78 + 180). With a five-year life and a 20% cost of money, each kilowatt reduction in both load and no-load losses is worth \$2058 more in purchase price for the aforementioned loading and electric rate.

3.7.5 Transformer loss and example

The transformer loss evaluation is almost identical to the motor evaluation. Important differences include the fact that losses are a squared function of kilovoltampere load, the no-load losses occur continuously, and the load usually increases each year. Additional sophistication can be added to show the effects of temperature due to load level and voltage level. The transformer can also be loaded well above nameplate in certain situations and it has longer life than a motor. Larger transformers have pumps, fans, and other auxiliary equipment whose energy costs must be included in the loss evaluation.

The equations are similar to the motor equations and the no-load demand equation is identical. The equations for transformer loss evaluation are as follows:

$D_N = $ (see equation 1)

$$Q_N = 24 \times 365 \times EC = 8760 \, EC \tag{8}$$

$$D_L = \left(\sum_{i=1}^{n} (\text{diversity}) \right) \left(\frac{kVA_i}{kVA_n} \right)^2 \times DC_i \tag{9}$$

$$Q_L = EC \times \sum_{i=1}^{n} t_i \left(\frac{kVA_i}{kVA_n} \right)^2 \tag{10}$$

where

DC_i	= Demand cost in \$/kW for the ith interval
EC_i	= Energy cost in \$/kWh for the ith interval
D_N	= No-load demand loss cost per kilowatt per year
Q_N	= No-load energy loss cost per kilowatt per year
D_L	= Load demand cost per kilowatt per year
Q_L	= Load energy cost per kilowatt per year
kVA_n	= Transformer nameplate rating, kVA
kVA_i	= kVA load for the ith interval
i	= An interval of constant load

The effect of load growth can easily be added to the above equations by the following equation:

$$\text{load-loss cost in Year } n = (D_L + Q_L) \times [(1+g)^{n-1}]^2 \qquad (11)$$

where

g = The per unit rate of load growth

This multiplication factor does not apply to the no-load cost of losses.

EXAMPLE: Consider the purchase of a 5000 kVA, 34 500–4160 V, three-phase transformer. The peak load is 3 000 and should grow at a rate of 5% per year. The tail-rate demand is $10 and the tail-rate energy is 0.10/kWh, and they are expected to increase 12%/yr. Transformer peak and billing peak are coincidental. The factory operates in two shifts for 5 days each week all year. On evenings and Sundays the load is approximately 40% of peak. During the 16 h of production, 4 h see 60% of peak, 4 h see 80% of peak, and the remaining 8 h are at 3000 kVA (note that kVA is used and not kW). Furthermore, the load curve is identical for all working days (which is usually not the case).

This transformer has a 30-year life at the specified loading, but if the load grows at 5% each year, this transformer will be replaced much sooner than 30 years if it is changed out at nameplate (5 000 kVA). The years until the load reaches nameplate value can be calculated quite easily as follows.

$$\frac{5\ 000}{3\ 000} = (1 + 0.05)^n$$

Since

$$(1 + 0.05)^n = \frac{5\ 000}{3\ 000}$$

$$n \times \ln(1.05) = \ln\left(\frac{5}{3}\right)$$

$$n = \frac{\ln\left(\dfrac{5}{3}\right)}{\ln(1.05)} = \frac{0.51083}{0.04879}$$

$$= 10.46 \text{ or approximately } 10.5 \text{ yr}$$

Therefore, the evaluation period should be 10.5 yr if it is desirable to change the unit out at nameplate loading.

a) No-load losses for Year 1 (assuming equal load all year and constant rates):
$D_N = 1/\text{mo} \times 12 \text{ mo/yr} \times \$10/\text{kW} = (\$120/\text{kW})/\text{yr}$
$Q_N = 24/\text{day} \times 365 \text{ days/yr} \times 0.1 = (\$876/\text{kW})/\text{yr}$
Total no-load loss for Year 1 is $120 + 876 = \$996/\text{kW}$

b) The load curve is as follows:

1) Evenings and weekends are $0.4 \times 3000 = 1200$ kVA for 8 h per day $(24 - 16 = 8)$ for 6 days and 24 h on Sunday for a daily equivalent of $48 + 24 = 72$ h per week.

2) The other levels are $0.6 \times 3000 = 1800$, $0.8 \times 3000 = 2400$, and 3000. At $6 \times 4 = 24$ and $6 \times 8 = 48$ h per week, respectively.

c) Load losses for Year 1:

$$D_L = 1 \times 12 \times \left(\frac{3000}{5000}\right)^2 \times \$10$$

$$= (\$43.20/\text{kW})/\text{yr}$$

using weekly loads

$$Q_L = \left[72\left(\frac{1200}{5000}\right)^2 + 24\left(\frac{1800}{5000}\right)^2 + 24\left(\frac{2400}{5000}\right)^2 + 48 \times \left(\frac{3000}{5000}\right)^2\right] \times 52 \times \$0.10$$

$$= (4.1472 + 3.1104 + 5.5296 + 17.28) \times 52 \times 0.1$$
$$= 30.0672 \times 52 \times 0.1$$
$$= (\$156.35/\text{kW})/\text{yr}$$

The load losses for Year 1 cost $199.55.

d) The losses in Year 9 are as follows:
No-load losses change only by inflation or $\$996 \times (1.12)^9 = \$2\,762$
The load losses change by both inflation and load or $\$(199.55)\, [(1.05)^8]^2 \times (1.12)^9$
$= \$199.55\, (1.48)^2 \times (2.77) = \$1211$

It should be noted that the no-load losses in the example were over four times the load losses for Year 1. This points out that energy may be wasted in sizing transformers too large for the anticipated load (a detailed calculation would be required to verify this conclusion).

3.7.6 Other equipment

The method used in 3.7.3 and 3.7.4 can be used to evaluate any item or process cost. One needs only to determine the losses from no-load to full-load and the load cycle. Loss evaluation can apply to conductor sizing, rectification equipment, variable speed drives, lighting systems, controls and sources, and even different types of processes.

3.8 Bibliography

Additional information may be found in the following sources:

[B1] Bonnett, A. H., "Understanding Efficiency in Squirrel-Cage Induction Motors," *IEEE Transactions on Industry Applications,* vol. 1A-16, no. 4, July/Aug. 1980, pp. 476–483.

[B2] Brown, Robert J., and Yanuck, Rudolph R. *Life Cycle Costing,* Commonwealth of Pennsylvania, 1979.

[B3] *Dranetz Field Handbook for Power Quality Analysis,* Edison, NJ: Dranetz Technologies Inc., 1991.

[B4] Grant, E. L., and Ireson, W. G. *Principles of Engineering Economy,* 4th ed. New York: The Ronald Press Company, 1960.

[B5] IEEE Std C62.41-1991, IEEE Recommended Practice on Surge Voltages in Low-Voltage AC Power Systems (ANSI).

[B6] Kovacs, J. P., "Economic Considerations of Power Transformer Selection and Operation," *IEEE Transactions on Industry Applications,* vol. 1A-16, no. 5, Sept./Oct. 1980, pp. 595–599.

[B7] McEachern, Alexander. *Handbook of Power Signatures,* Foster City, CA: Basic Measuring Instruments, 1989.

Chapter 4
Load management

4.1 Definition of load management

Load management is the control of usage of electrical or other forms of energy by reducing or optimizing the amount of such usage and the rate of such usage (*demand*).

In the case of electrical power systems, served by a utility or private generation system or both, electrical energy usage in kilowatthours (kWh), reactive energy usage in kilovarhours (kvarh), real power demand in kilowatts (kW), apparent power demand in kilovoltamperes (kVA), reactive power demand in kilovars (kvar), power factor (usually a penalty), and load factor are the items to be controlled or managed. As a general approach, those items that are billable or can be cost-controlled are those that are incorporated into the energy management system.

The item most often controlled is electrical demand, as it is the most susceptible to savings on a short-term basis. Electrical demand, often representing, in part, the cost of additional generation to the utility, is usually a relatively high-cost item; demand peaks can introduce abnormally high costs for the entire billing period (usually a month), or in the case of a ratchet clause, for many such periods.

Many people narrowly define load management as demand control, but good engineering dictates control of electrical usage 24 hours each day and 365 days each year. This concept demands load curtailment even during the lowest usage times of the plant's load cycle.

Energy control represents the reduction in the total use of kWh or sometimes kVA. Energy control is most often accomplished by demand control methodology; however, automated load shedding, lighting level reduction, time control of energy-using appliances or equipment, and interactive control of parameters determining the rate of expenditure (a thermostat serves as a simple example) are the most common approaches to energy control. In contrast to demand control, the only time that the use of excess energy is recognized is when the utility bill is received and analyzed—perhaps weeks or months after the event.

The application of load management concepts requires an understanding of utility rates, auditing, and metering, as well as a basic knowledge of the process and load being controlled. The engineer must first audit and meter the system and then determine which of the electrical loads can be reduced. With this knowledge, the load management program can begin.

4.2 Demand control techniques

Trimming power peaks without sacrificing the quality or quantity of production and basic services forms the essence of demand control.

The fundamental principle of demand control is fairly simple. It is necessary to determine at what time of the day and on which days the peaks occur and then determine which loads are in use at that time. Next, the magnitude of the loads must be determined and decisions must be made as to which operations can be curtailed or deferred to reduce the demand peak and the power bill. The subsequent subclauses describe the various types of effective means of controlling plant demand.

4.2.1 Methods of manual demand control

Maximum benefits can be attained from a plant power survey focused on those areas of the plant where it will pay to have continuous records made through the installation of permanent instrumentation. Total plant records will show how each piece of equipment contributes to the total load picture and will reveal whether equipment is operating within specifications. Areas of energy waste should be identified.

If certain heating, ventilating, and air conditioning (HVAC), or other nonessential systems in the plant can be switched off when the building is unoccupied or for a few minutes during the peak demand period, a timer may be a very effective demand controller. Outdoor and indoor display lighting systems are also candidates for time-clock control. Electric process heat systems can also be timed or staged. All time-clock-based systems have limitations. Unless an astronomic or seven-day type is used, the controller must be reset frequently. A power interruption can require subsequent time reset unless the timer has a mechanical spring backup or a battery backup power source in the case of electronic timers. Newer programmable timing controllers can provide hundreds of patterns and time changes during the year.

With proper metering, demand can be controlled manually by simply watching the meters. Ideally, local readings should be transmitted to a common point where a staff operator (or, preferably, some device) can observe the rate of consumption. The staff operator initiates load removal by switching off the noncritical load to keep the demand under a predetermined level. A reliable communication system or a remote control should be installed for the operator to initiate the switch-off and switch-on action.

The advantages of manual control are that it shows how much can be accomplished by cutting demand, investment is small, and it enables management to think out the problems they will eventually have to face if they go on to some form of automatic load shedding.

The essence of all but the simplest manual control is provision of metering and alarms that are immediately available to the operators. Daily or weekly readings of utility meters, which are often remotely located, are seldom adequate. Rather, readings, both predictive and totalized, should be made available at the operator's location. Logging of the information, either manually or automatically, will encourage operators to stay alert. If such control is important, alarms should be provided to indicate the need for action; casual observation of meters is not adequate for indicating the need for action.

4.2.2 Automatic controllers

When a complex, fine-tuned operation is desired, a more sophisticated, automatic demand controller should be installed.

Automatic controllers can be categorized by operating principle: instantaneous demand, ideal rate, converging rate, predicted demand, and continuous integral. Some controllers are off-shoots of these five basic versions; others are hybrids embracing more than one operating principle. Installation costs will vary depending on controller location and the number and location of controller loads.

Most demand controllers require pulse signal inputs derived from the utility's demand meter. One pulse set indicates the usage rate and the other pulse set indicates the demand interval. The controller then observes each interval for rate of usage. The following detailed descriptions explain how each type of controller uses this information to control demand.

4.2.2.1 Instantaneous demand

With an instantaneous demand controller [figure 4-1(a)], action is taken when instantaneous demand exceeds the established prescribed setpoint value. A setpoint value is determined for the demand interval. Straight line accumulation or constant usage is presumed. When one-fourth of the demand interval has transpired, accumulated demand or kilowatthour usage should be no more than 25%. Loads are switched in and out of service in accordance with this criteria. This mode of operation might result in short cycling of loads. In any demand control system, short cycling can be effectively damped out by simply installing cycle timers in the control circuits of problem equipment or by having logic in the control that performs the same timing function.

4.2.2.2 Ideal rate

With an ideal rate controller [figure 4-1(b)], ultimate demand limit is prescribed and a slope is established to define when usage indicates that this limit is likely to be exceeded. The ideal rate controller does not begin from zero at the start of the demand interval but from an established offset point that takes into account nondiscretionary loads. Slope of the *ideal rate of use curve* is that defined by this offset point and a chosen maximum demand. The offset provides a buffer against unnecessary action early in the demand interval, thereby reducing cyclic equipment operation.

4.2.2.3 Converging rate

The converging rate controller [figure 4-1(c)] works like the ideal rate controller, but it operates on an accumulated usage curve whose upper limit is defined by the specified maximum demand. It also employs an offset to minimize nuisance operation early in the demand interval. But unlike the ideal rate controller, which established parallel rate-of-use lines for loads, the converging rate controller load lines are not parallel. They converge at the maximum demand point to permit vernier-like control toward the end of the demand interval when accumulated registered demand might be critically near the setpoint.

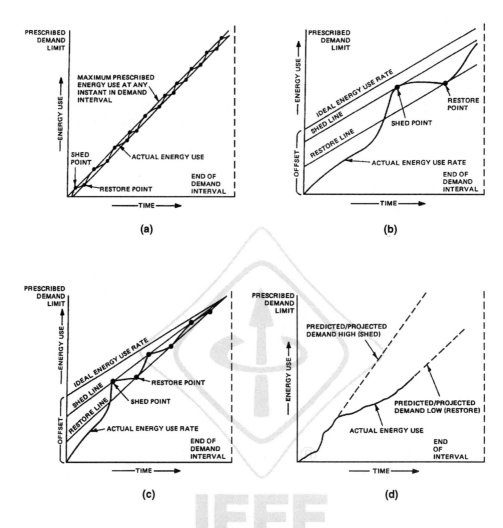

Figure 4-1—Automatic demand control principles
**(a) Instantaneous demand principle (b) Ideal rate principle (c) Converging rate
principle (d) Predicted demand principle (e) Continuous integral principle**

4.2.2.4 Predicted demand

With a predicted demand controller [figure 4-1(d)], average usage is monitored periodically
through the demand interval and compared with the instantaneous usage at that particular
moment. This information is used to continually develop a curve of predicted usage for the
remainder of the interval. If the predicted curve indicates that the target setpoint might be
exceeded, action is taken.

4.2.2.5 Continuous integral

The continuous integral controller [figure 4-1(e)] monitors power usage continuously, rather than only when a time pulse signal is transmitted by the utility company's demand meter. When action is called for, the controller activates a satellite (remote) cycle timer, which sheds loads for a predetermined, fixed period. If further action is called for, other timers are activated until usage is brought in line with the desired objective. Because satellite timers, once activated, will shed loads through a complete cycle and overlap the demand intervals, short-cycle operation is reduced.

4.2.3 Microprocessor-based system

Using a keyboard input to a computer, the plant engineer can specify the maximum demand that can be tolerated, based on previous experience. The microprocessor continually monitors the plant's electric consumption and, by one of the aforementioned methods, determines if the demand limit will be exceeded. If demand is well below the limit, control action is not taken. If the computer predicts that demand may exceed the limit, the preselected loads can be automatically turned off to reduce demand or an alarm can alert the building engineer who then makes a manual adjustment decision. It provides several levels of load-shedding priority. Some loads are designated *low priority*; these will be shed in round-robin rotation as needed. Others can be placed in a separate *high priority* category, which indicates they will not be shed until the supply of low-priority loads is exhausted. Limiting and load shedding can be increased during peak periods, with demand limits relaxed during low-cost, off-peak times. Two of the most valuable features of this system are its logging and printing capability and its ability for in-house reprogramming.

With time-of-day control, equipment managed by the microprocessor-based system operates only when needed. This system can cycle the various building loads and can vary the load's duty cycles according to the time of day, staggering equipment off times to reduce electrical demands. Flexible load cycling thus retains full equipment capacity for fast warm-up or cool-down, while reducing demand and energy expense under routine operating conditions.

With the tremendously increased memory capabilities of the microprocessor-based computers, usually referred to as the personal computer (PC), the mainframe and minicomputers have for the most part been superseded for the typical new energy management system.

The microprocessor has enabled the development of an entire series of control systems with embedded computer technology; that is, small pre-programmed computers are built into control devices. Many of the remote devices contain built-in or embedded computers, particularly for providing a communication channel to the central unit. The modern energy management system may be a combination of this and the PC technology.

The microprocessor-based technology has become so powerful that it is common to combine most of the building control systems including energy management, security, fire detection and alarm, control of mechanical and electrical system, data logging, and lighting control into the one computer. In larger systems, the distributed computer concept, where a series of small computers report to the central control unit, provides additional reliability and segregation of functions.

4.3 Utility monitoring and control system

A central in-plant utility monitoring and control system should consider the total plant site, including fuel usage in the production of process steam, hot water, chilled water and electricity, energy usage in the production process in maintaining the plant environment, and the cost of purchased electricity. This utility management system consists of three hierarchical tiers: the operating, supervisory, and management planning levels (figure 4-2). The data within this hierarchy requires more manipulation and refining as it progresses from a lower to a higher level.

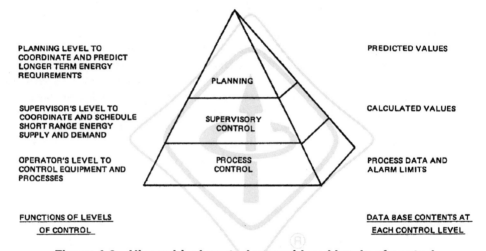

Figure 4-2—Hierarchical control pyramid and levels of control

The uses for utility management systems are as diverse as the process they serve. The more obvious ones are described in 4.3.1 through 4.3.6.

4.3.1 Energy distribution

By monitoring power production and purchases, engineers can recommend changes in the distribution or purchase of energy.

4.3.2 Monitor energy consumption

Constant monitoring and evaluation of energy usage by department or area can prevent extraordinary consumption.

4.3.3 Methods of conservation

Computer analysis of plant conditions (cycling of exhaust fans, setting of air conditioners, etc.) can substantially minimize demand and, sometimes, consumption.

4.3.4 Load shedding

Selective shedding of low-priority loads can minimize demand peaks. Load shedding is also valuable when utilities place absolute ceilings or penalties for excess use on the amount of energy supplied.

4.3.5 Cogeneration

A computer-based energy controller is beneficial for a plant using process steam to generate power for internal use. If by-product gas is to be burned as part of the fuel, the controller determines the economic mix of purchased oil with waste gas needed to produce the steam necessary for efficient power generation. The controller calculates that level of generation to be maintained under each operating condition and considers how the contract with the utility affects the mix.

4.3.6 Maintenance prediction

Early detection of rising temperatures, abnormal currents, or other operating irregularities during normal monitoring by the computer can signal a need for maintenance before equipment is damaged. The savings in maintenance costs can be significant. With advance warning, equipment shutdowns can be scheduled to avoid costly disruptions of the production process.

4.4 HVAC and energy management

4.4.1 HVAC monitoring and control

HVAC monitoring and control includes the automatic monitoring of HVAC equipment and also provides operating personnel with information on the status of these systems and selected components. Temperature, dewpoint, humidity, pressure, flow rate, and other key operating parameters are continuously monitored and displayed upon command or when any abnormal or alarm condition occurs.

An HVAC facility system also provides for the remote control of necessary functions for the operations of HVAC equipment. From the operator-machine or man-machine interface (OMI/MMI), fans can be switched on or off and their speed can be adjusted, dampers and their control valves can be positioned, pump speed can be controlled, equipment can be started and stopped, control points can be adjusted, and all other functions necessary to properly operate and monitor the mechanical equipment of the facility can be controlled.

HVAC facility systems generally are programmed for several operating modes. These programs are developed by the energy management and facility operating groups. Their data should be incorporated into the software to ensure HVAC operations meet code, occupancy, and energy conservation needs. These HVAC programs will account for seasonal needs as well as day, night, and holiday occupancy in each of the buildings, or building areas, that comprise the facility. In the case of large areas in which occupancy varies greatly over short time periods, the programs may have to include real-time or hourly control of HVAC components and possibly a major portion of the lighting.

The types of equipment typically supervised by an HVAC control system are as follows:

— Air-handling equipment
— Steam absorption chillers
— Direct-fired absorption chillers
— Boilers
— Electric-motor driven water chillers (centrifugal, reciprocating or helical screw)
— Steam-turbine driven water chillers
— Diesel-engine driven water chillers
— Air compressors
— Air-cooled condensers
— Dampers
— Evaporators
— Fans
— Heat pumps
— Heat exchangers
— Liquid tanks
— Pumps
— Refrigerators
— Sump equipment
— Valves
— Control switches (electric/pneumatic, pneumatic/electric)
— Reheat devices
— Cooling towers
— Ice-making equipment
— Ice storage systems
— Exhaust fans
— Water softening equipment

The HVAC conditions and quantities to be monitored or controlled may include the following:

— Optimized start
— Supply air temperature and water temperature reset
— Temperature dead-band operation
— Enthalpy changeover
— Demand limiting
— Damper position

- Flow rates
- Fuel supply and consumption
- Gas volume
- Humidity and dewpoint
- Real and reactive electric power demand and consumption
- Line current and voltage(s)
- Liquid level
- Equipment running time
- Equipment wear (revolutions or cycles)
- Leaks and oil spills
- Fan speed
- Degree days of heating and cooling
- Power failures and irregularities (main, auxiliary, control)
- Pressure
- Programmed start/stop operations
- Status of miscellaneous equipment and systems
- Temperature
- Toxic gases and fluids
- Hazardous gases, dusts, and fluids
- Combustible gases (e.g., methane in sumps and manholes, hydrogen in battery rooms)
- Valve position
- Wind direction
- Wind velocity
- Solar energy available (kJ/m^2)
- Daylight available (lx)
- Solar collector tilt angle
- Holiday scheduling
- Run-time reduction
- Night temperature set back
- Central monitoring
- Optimized fresh air usage to meet indoor air quality standards ·
- Trend logging
- Pump speed
- Steam flow
- Chilled water flow and temperature
- Energy of heating or cooling consumed (Btu or MJ)
- Indoor air environmental quality

4.4.2 Direct digital control

Traditionally, mechanical systems for buildings have been designed with automatic temperature control (ATC) for HVAC systems. Considerable experience has been gained both with ATC systems as well as microcomputer applications for process controls. Thus, micro-computer technology offers engineers a powerful tool for the control of HVAC systems.

4.4.2.1 Automatic temperature control

In closed-loop control, a sensor provides information about a variable (e.g., temperature) to a *controller* that actuates a *controller device*, such as a valve, to obtain a desired *setpoint*. The output of the controller should operate the controller device to maintain the setpoint (for example, by modulating a chilled water flow through a coil) even if air or water flow rates or temperatures change. This should happen on a continuous basis and should be fast enough to maintain the setpoint, in which case the controller is said to be operating "in real time."

The creation of a comfortable environment by heating, cooling, humidification, and other techniques is a real-time process that requires closed-loop control. HVAC systems require many control loops. A typical air-handling unit (AHU) needs at least three control loops (one each for fresh air dampers, heating coil, and cooling coil) plus accessory control devices to make them all work in harmony. The way in which these control loops operate has a major effect on the amount of energy used to condition the air.

4.4.2.2 Computer control techniques

When the controller in a closed-loop system is a digital computer, then it is called "direct digital control."

This seems the obvious way to apply a computer to a control loop. However, most computers in control applications today are *not* applied in this way. Until recently, most computers were principally used as *supervisory systems* to supervise the operation of an independent control system.

A supervisory computer monitoring the ATC system and capable of resetting the controller setpoint has some very basic limitations, as follows:

— The most sophisticated supervisory computer cannot improve the operation of the control loop because the controller is really in command. Any deficiencies or inaccuracies in the controller will always remain in the system.

— Interfacing the computer to a controller that is frequently a mechanical or electromechanical device is expensive and inaccurate.

— The computer's sensor and the actual controller's sensor may not agree, leading to a good deal of confusion or a lack of confidence in one system or the other.

4.4.2.3 Direct digital control (DDC) computer programs (software)

A computer's power is in its programming software. When applied to automatic temperature control, properly designed programming software offers dramatic benefits, as follows:

— Control system design is not "frozen" when a facility is built. Alternative control techniques can be tried at any time at little, if any, additional cost.

— With software configured control, all control panels can be identical, which facilitates installation, checkout, and maintenance. One standard DDC computer can control virtually any HVAC equipment.

— The control system can be improved with programming enhancements in the future. No additional equipment or installation normally will be required.

— Comfort and operating cost tradeoffs are easily made by the flexibility to modify the operating parameters in the control system. Optimum energy savings can be realized without sacrificing occupant comfort.

Flexible software programs should allow changing not only setpoints but control strategies as well. Control actions, gains, loop configurations, interlocks, limits, reset schedules, and other parameters are all in software and should be able to be modified by the user at any time without interrupting normal system operations.

With DDC, an operator, via the program, may access all important setpoints and operating strategies. Accuracy is assured by the computer. Control loops can be reconfigured by revising the loop software, with no rewiring of control devices. Reset schedules can be changed just as easily. For example, heating setpoints and strategies can be set in the summer with complete assurance that the DDC system will perform as expected when winter arrives.

4.4.2.4 DDC loops

Typically, DDC closed loops consist of sensors and actuators, in addition to digital computers, as the controllers. Certain design features should be used to obtain optimum performance from DDC loops. Sensors for DDC loops are very important, since the computer relies on their accuracy to provide the precise control that an HVAC system operator needs. A 1 °F change in some temperatures, such as chilled water, can affect energy consumption by a couple of percentage points, so that a control system with even 1 °F of error is not fully controllable in terms of energy use. So as not to waste the precision of the DDC, quality sensors should be used that do not require field calibration and do not have to be adjusted at all to interface with the DDC computer. Control setpoints are thereby achieved with optimum accuracy under all conditions at all times.

With the computer performing DDC, the traditional problems of temperature fluctuations and inefficient operation can be eliminated. Proportional-integral-derivative (PID) control techniques provide for the fast, responsible operation of controlled devices by reacting to temperature changes in three ways:

— The difference between setpoint and actual temperature (proportional)
— The duration that the difference has persisted (integral)
— The rate that the actual temperature is changing (derivative)

PID saves energy and increases accuracy simultaneously by eliminating hunting and offset and by decreasing overshoot and settling time.

All digital computers work with binary (either on or off) information. Since it is necessary to modulate controlled devices (e.g., motors that operate dampers or valves), a complicated interface device (transducer) is often employed. A better method to use, which has been perfected in much more demanding process applications, is pulse-width modulation (PWM). The computer's binary outputs are directly connected to a modulating device. PWM uses bidirectional (open/close) pulses of varying time duration to position controlled devices exactly as required to satisfy demand. Wide pulses are used for major corrections, such as a change in setpoint or start-up conditions. The pulse width becomes progressively shorter as less correction is required to obtain the desired control setpoint.

4.4.2.5 DDC energy management

Many strategies have been developed to effectively manage and save energy in HVAC system operation. DDC systems can be intelligently integrated with temperature control functions in the same computers, in such a way that energy reductions are achieved without compromising the basic temperature control functions. This will also eliminate the need to supplement a conventional ATC system with an add-on energy management system (EMS), which will save equipment, installation, and maintenance costs.

4.4.2.6 DDC distributed networks

Implementing DDC in an entire facility with numerous HVAC equipment can be accomplished with any number of computer and process control systems. Starting with a basic control loop, a system can expand to control an entire facility.

A DDC computer should be capable of handling a number of control loops (four to eight is typical). Accessory on/off control and monitoring functions should also be controlled by the same computer. Each computer should be capable of independent operation and be able to perform all essential control functions without being connected to any other computer. This suggests that each separate major HVAC equipment (such as an air handler, boiler, or chiller) has its own DDC computer, in the same way that each would have independent conventional control panels. These are then connected together with a local area network (LAN) for communications. This results in a truly distributed processing network in which each computer can perform all control functions independently.

Twisted-pair, low-voltage control wiring (foil shielded) is an economical choice for the interconnections, although coaxial cable or fiber-optic cable systems can be used if they are installed in the facility to provide a variety of communication services.

Somewhere in this LAN, a "window" is required to allow for the staff operator to interface with the DDC computers. This is accomplished with a different type of computer, connected to the network at any location, which provides access to the DDC computers. All control setpoints and strategies can be programmed from this access computer, and all sensor readings can be monitored.

4.4.2.7 Network protocols

The specific set of coded instructions that enable microprocessor devices connected to the LAN to communicate are network protocols. The most common network protocols are the peer-to-peer type (e.g., Ethernet or ARCNET) and the IEEE 802.4 token-passing open system. Although these networks are advertised as "open systems," most manufacturers have specific message structures that are proprietary. Thus, integration of several manufacturers' protocols over the same network necessitates sharing of proprietary information.

The American Society of Heating, Refrigerating and Air-Conditioning Engineers (ASHRAE) has developed an open protocol that will enable the exchange of data between devices made by different manufacturers. Called BACnet (for Building Automation and Controls Networks), the protocol has been published as ANSI/ASHRAE 135-1995 [B1].[1]

4.4.2.8 System integrity

A DDC system can be designed for high reliability and for much shorter mean time to repair (MTTR) than a conventional ATC system. The major design requirements are as follows:

— *Independent control computers.* In a distributed processing network, these computers ensure that the failure of one computer will not adversely affect the operation of other computer systems.

— *Remote data-link diagnosis.* Allows the computer manufacturer's factory experts to telephone into the DDC system and troubleshoot control problems.

— *Universal computer replacement.* Requires that all control computers be identical, regardless of the HVAC equipment being controlled.

Since the access computer in a distributed network is not capable of any real control, it may not need any special backup system. Remote data link diagnosis can quickly pinpoint an access computer problem, and repair does not have to be immediate to maintain environmental comfort. All HVAC systems are under the control of the independent DDC computers, which should continue to function normally.

System integrity considerations should also include what happens when a computer fails. A safe condition has to exist when this happens. Therefore, whenever a DDC computer is used, all standard safety devices (i.e., for overload, smoke control, freeze protection, etc.) should remain in the system with the computer. These are usually very simple devices that have been proven in many years of HVAC system design, and are not rendered obsolete when a computer is used for direct digital control of the system.

4.5 Economic justification for load management systems

A load management system can provide substantial savings, either through reduced energy costs, or through increased production without corresponding increases in energy requirements. For complex systems, return on investment (ROI) for decreased costs and increased revenue may not alone be sufficient to justify the initial investment within the company's specified time. Many companies can justify energy management systems on an ROI basis

---

[1]The numbers in brackets correspond to those of the bibliography in 4.6.

when the concept of lost production due to shortages or cutoffs is quantified and factored into the ROI analysis.

Today's management may make investments to optimize the use of energy. Some industries may be required by the government to reduce energy demands. Other industries will have to prove that their use of energy already meets or exceeds the industry's standards for efficiency in terms of energy consumed per unit or product.

Additional benefits can be realized from a computer-based energy management system by allowing the computer to perform other duties. The data acquisition and reporting capability of the computer may be used to monitor and record the operation of pollution control equipment. Its ability to schedule preventative maintenance may be used to protect all major plant equipment, not just that connected with energy production or use. Load management may prove economical where done in conjunction with these other functions.

4.6 Bibliography

Additional information may be found in the following sources:

[B1] ANSI/ASHRAE 135-1995, BACnet: A Data Communication Protocol for Building Automation and Control Networks.[2]

[B2] Batten, G. L., Jr., *Programmable Controller—Hardware, Software and Applications.* TAB Professional and Reference Books, 1988.

[B3] Chen, Kao, and Palko, Ed, "An Update on Rate Reform and Power Demand Control," *IEEE Transactions on Industry Applications*, vol. 1A-15, no. 2, Mar./Apr. 1979.

[B4] Dacquisito, J. F., "Beating Those Power Demand Charges," *Plant Engineering*, Nov. 1971.

[B5] Hansen, A. G., "Microcomputer Building Control Systems Managing Electrical Demand on Energy," *Building Operating Management*, July 1977.

[B6] Hugus, F. R., "Shipbuilding and Repair Facility Controls Demand To Reduce the Cost of Electricity," *Electrical Construction and Maintenance*, July 1973.

[B7] IEEE Std 241-1990, IEEE Recommended Practice for Electric Power Systems in Commercial Buildings (IEEE Gray Book) (ANSI).[3]

[B8] IEEE Tutorial Course 91EH0337-6 PWR, Fundamentals of Supervisory Systems, 1991.

---

[2]ASHRAE publications are available from the Customer Service Dept., American Society of Heating, Refrigerating and Air-Conditioning Engineers, 1791 Tullie Circle, NE, Atlanta, GA 30329, tel (404) 636-8400.

[3]IEEE publications are available from the Institute of Electrical and Electronics Engineers, 445 Hoes Lane, P.O. Box 1331, Piscataway, NJ 08855-1331.

[B9] IEEE Tutorial Course 88EH0280-8-PWR, Distribution Automation, 1988.

[B10] Jarsulic, N. P., and Yorksie, D. S., "Energy Management Control Systems," *Energy Management Seminar Proceedings*, Industry Applications Society, 77CH1276-51A, Oct. 1976 and Oct. 1977.

[B11] Quinn, G. C., and Knisley, J. R., "Controlling Electrical Demand," *Electrical Construction and Maintenance*, June 1976.

[B12] Maynard, T. E., "Electric Utility Rate Analysis," *Plant Engineering*, Nov. 1975.

[B13] Meckler, Milton, "Energy Management by Objective," *Buildings*, Cedar Rapids, Iowa, Nov. 1977.

[B14] Niemann, R.A., "Controlled Electrical Demand," *Power Engineering*, Jan. 1965.

[B15] Ochs, H. T., Jr., "Utility Rate Structures," *Power Engineering*, Jan. 1968.

[B16] Palko, Ed., "Saving Money through Electric Power Demand Control," *Plant Engineering*, Mar. 1975.

[B17] Palko, Ed., "Preparing for the All-Electric Industry Economy," *Plant Engineering*, June 1976.

[B18] Peach, Norman, "Do You Understand Demand Charges?," *Power*, Sept. 1970.

[B19] Peach, Norman, "Electrical Demand Can Be Controlled," *Power*, Nov. 1970.

[B20] Rekstad, G. M., "Why You'll Be Paying More For Electricity," *Factory Management*, Feb. 1977.

[B21] Relick, W. J., "Using Graphic Instruments To Hold Down Electric Power Bill," *Plant Engineering*, May 1974.

[B22] Talukdar, S. and Gellings, C.W. *Load Management*. New York, NY: IEEE Press, 1987.

[B23] Wright, A., "Keeping That Electric Power Bill Under Control," *Plant Engineering*, June 1974.

Chapter 5
Energy management for motors, systems, and electrical equipment

5.1 Overview

5.1.1 Scope

This chapter gives recommendations and procedures for the conservation of electric energy in the application, design, operation, and maintenance of electrical apparatus and equipment. It is applicable to commercial, institutional, and industrial facilities.

5.1.2 Rationale

The following table shows the percent of the industrial sector fuel requirements consumed as electricity in the first column. The kilowatthours of electricity consumed is shown in the third column and the percentage is in the second column. Note that 26% of the fuel is consumed as electricity and that 69% of the electricity is consumed for motor drives.

Fuel as electricity		Electricity consumption in the industry	
Energy sector	Percent	kWh 10^9	Percent
Electric drives	20	580.9	69
Electrolytic	3	100.8	12
Direct heat	2	90.4	11
Other	1	71.6	8
Total	26	843.7	100

The potential for energy savings in industry's use of electricity are estimated to be the following (Smith 1976 [B250][1]):

— Immediate savings by operational and maintenance changes: 5–10%
— Near-term: some investments and process equipment changes: 5–10%
— Long-term: major investments and process equipment changes: 5–10%

Annex 5A shows the electricity consumption in industry by standard industry classification and by component.

In the commercial sector, the electrical energy use pattern looks as follows (EPRI EU 3024 [B94]):

---

[1]The numbers in brackets correspond to those of the bibliography in 5.7.

— Lighting: 42%
— Cooling: 30%
— Heating: 11%
— Ventilation: 9%
— Other: 8%

5.1.3 Objective

This subclause indentifies specific processes in which good design, operation, and mainte-nance may result in a lower energy utilization consistent with achieving goals of productivity and employee health and morale.

5.1.4 Application

The application of this chapter is as follows:

a) 5.1 is an overview of the chapter
b) 5.2 addresses systems and equipment
c) 5.3 applies to electric motors
d) 5.4 addresses transformers and reactors
e) 5.5 addresses power factor correction
f) 5.6 provides a bibliography
g) 5A–5I are annexes to the chapter

5.2 Systems and equipment

5.2.1 Electrical distribution system efficiency

Electrical designers must have knowledge of the loads to be served by the electrical distribu-tion system presently and in the future to be able to select the safest and most economical system. Certain basic factors must be considered as recommended in IEEE Std 141-1993[2] and IEEE Std 241-1990.

The proper types of power distribution systems must also be selected as described in the two IEEE Color Books previously mentioned.

Five arrangements have found general usage. They are as follows:

— The radial system
— The secondary-selective system
— The primary-selective system
— The looped primary system
— The secondary-network system

---

[2]Information on references can be found in 5.6.

Each of these systems has power losses in the transformers, cables, equipment bus bars, and protective overcurrent devices. Efficiency characteristics are those related to overall efficiency of the system from power supply points to utilization devices under both heavy load and light load operating conditions. This means accounting for the power losses and the transformer voltage regulation.

The results of the Electrical Design Library publication, "Power Distribution Systems" [B156], indicate that the most efficient systems are the following:

— Banked secondary radial
— Primary-selective network
— Primary-selective spot network

5.2.1.1 Industrial plants

Primary and secondary voltage selection are important due to the reduction in power losses available by use of higher system voltage—this is so because power losses and attendant heating decrease as the square of the nominal system voltage.

Power losses in the electrical distribution system within factories, plants, and buildings occur through the operation of equipment inefficiently and through a distribution system design with losses in conductors and transformation equipment. Additional losses occur in the distribution system (in adjacent magnetic materials) as the electrical distribution system provides energy to the equipment loads within the facility.

These losses in the electrical distribution system appear as listed:

a) Conductor losses (I^2R)
b) Magnetic material losses
c) Rotating equipment friction and windage losses
d) Stray load losses

The evidence of these losses may be seen in the plant distribution system by measuring the voltage at the service entrance equipment and at the load. The voltage drop difference between the measurements is directly traceable to the efficiency of the electrical distribution system.

The National Electrical Code® (NEC®) (NFPA 70-1996 [B173]) gives requirements for voltage drops in Sections 210-19 (a) and 215-2. ANSI C84.1-1989 [B26] also contains voltage drop requirements.

Load flow and voltage drop calculation should be performed routinely in the design process so that the proper material, equipment, and tap setting may be specified for the distribution transformers within the facility.

Load flow and voltage drop calculations may be performed on personal computers using readily available software and following the procedures in IEEE Std 399-1990.

Where the voltage drop calculation must be performed manually, follow the procedures in the NEC® Handbook, IEEE Std 141-1993, or IEEE Std 241-1990.

5.2.1.2 Reducing losses in conductors

The I^2R losses in electrical conductors can be reduced by selecting an increased wire size in cabling and by using a heavier cross section in busbars. The economic incentive can be determined by analyzing the duty factor, the load factor, the electricity price, and any changes in conduit size due to increasing the conductor cross-sectional area or wire gauge. Economic analyses should include the savings in air-handling and air-cooling due to the reduced heat losses in the conductors. The "Copper Busbar Design Guide" computer program [B81] allows users to find the most economical size busbar in a system.

5.2.2 Personnel transportation systems

    a)   Elevators (lifts)
    b)   Escalators (moving stairs); power walks/power ramps

5.2.2.1 Elevators

Elevator installations are of several types with unique features as shown in table 5-1.

Table 5-1—Elevator characteristics

Type	Speed (ft/min)	Typical travel (ft)

1) Hydraulic
  a) Inground oil hydraulic
  b) Holeless oil hydraulic
  c) Roped oil hydraulic
	
	50–200
	50–200
	200–400
		
		70
		30
		80

2) Traction
  a) Low speed
  b) Medium speed
  c) High speed
	
	75–150
	200–450
	500–1800
		
		—
		—
		—

5.2.2.1.1 Hydraulic elevators

The power demand depends upon the speed, weight, and acceleration of the elevator car and load. The corresponding hydraulic power demand, measured in horsepower (hp), is directly proportional to the pressure and flow of oil. The pressure is determined by the gross load and the jack size.

Energy conservation opportunities for hydraulic elevators are as follows:

— Performing passenger traffic studies carefully to select the appropriate speed and car capacity

— Installing microprocessor elevator controls as discussed in 5.2.2.1.2

— Installing high-efficiency motors for the pumping unit

— The most practical rated speed for a hydraulic elevator is between 90 ft/min and 125 ft/min

— Roped hydraulic application utilizes a 2:4 roping configuration for mechanical advantage thereby reducing required torque

5.2.2.1.2 Traction elevators

Traction elevators are of a geared design or a gearless design, depending upon the rise (height), the speed, and the handling capacity.

a) Geared machines consist of a traction drive sheave directly connected to a worm and gear speed reducer driven by a high-speed ac or dc motor. As generally applied, they are found in buildings up to 30 floors and operating at speeds to 450 ft/min for passenger car loads up to 3500 lb. Geared systems generally use one of the following drives:

1) Motor-generator set with variable-voltage dc operation and geared hoisting machine (200+ ft/min)

2) Single-speed ac geared hoisting machine (speed to 150 ft/min)

3) Special elevator drive system utilizing an electronic ac drive and eddy current braking

4) Conversion of ac to dc using *silicon-controlled rectifiers* (SCRs) for the supply to the dc variable-voltage geared hoisting motor

5) *Variable-voltage variable-frequency* (VVVF) four-quandrant control *pulse-width-modulated* (PWM) drive systems

b) Gearless hoisting machine systems generally are used for the high-speed applications in high-rise buildings up to 128 floors for passenger capacities up to 4000 lb. DC gearless hoisting machines may be supplied from the same systems as for geared systems.

Energy conservation opportunities for traction elevators are as follows:

— Carefully performing an elevator system and traffic study assures passenger comfort and convenience, with assurance of maintaining a marketable building and energy conservation.

— Installing more efficient hoisting motors that consume less energy.

— Installing SCRs to supply the dc hoisting machines instead of using motor-generator sets. The SCRs result in a power savings of 10–35% compared to the use of motor-generator sets since the electronic control eliminates high starting currents and long idling power losses. SCR drives operate with full three-phase regeneration of power back into the system during operation. SCR drive efficiency is 95% (Westinghouse Elevator Co. Bulletin B-6150 [B235]).

— Installing microprocessor elevator controls.

— Installing geared systems using helical gear, which exhibit substantial efficiencies over traditional worm gear. The drawback is tendency to wear.

— Installing VVVF with full regenerative power, which is the most efficient of the three drive systems.

Older elevators will run better and more efficiently and provide better service simply by replacing the outdated switch and relay control system with microprocessor controls.

Microprocessor collective control systems are capable of interpreting every aspect of elevator operation: velocity, position, direction, passenger travel, car weight, waiting time, door operation, car assignment, energy use, and diagnostics—all in "real time." With this feedback the controller can issue changes in less than a second. The microprocessor can analyze traffic patterns and compute the best method of moving cars. The microprocessor should be capable of equalizing the loading over time. This will help to lengthen the drive and machine life cycles and their associated efficiencies. The microprocessor control should be sensitive to "full-load up" and "empty-load down" situations where the torque required to move the elevator into motion is greatest. Anticipating peak traffic times is critical.

5.2.2.2 Escalators/power walks/power ramps/passenger conveyors (data courtesy of Schindler Elevator Corp.)

Energy savings are possible on existing systems by adjusting speed and operating hours to passenger traffic and by use of high efficiency and/or multispeed motors. Speed adjustment will allow the unit to reduce energy consumption by operating at a very low speed when there is no traffic.

Energy conservation opportunities for escalators and power walks are as follows:

— Use of high-efficiency motors

— Adjusting operating hours to traffic demand

— Using variable speed ac drives that would allow the unit to slow down when the escalator has no load. The escalator is then gently accelerated when passengers are present, using automatic sensors as input to the controller. This can provide a superb quality ride and energy savings up to 30%.

— Modifying the operating mode for new escalators that allow the standard single-speed motor to operate in a wye configuration for light loads and to switch to the normal delta configuration as load increases. This produces energy savings but also gives a noticeable speed change when the switch is made. Use of additional sensors is recommended to make the transition before passengers enter the unit. The solution is less costly than a variable-speed drive but the performance is not as good.

5.2.3 Material-handling systems (SEC of Victoria [B200])

Material-handling systems are selected using the following rules to obtain the greatest productivity:

a)   Unnecessary movement should be eliminated.

b)   Processing should be carried out while the material is in motion whenever possible.

c)   Materials should be moved in a straight line whenever possible.

d)   A conveyor should be used where there is a continuous and regular flow of goods.

e)   Speed of movement should be consistent with both productivity and safety.

f)   Maximum productivity is obtained when the rate of flow and weight of equipment is high.

g)   The correct type of equipment should be used; e.g., a fork lift and a reach truck are similar but one may have advantages over the other in a specific application.

h)   Stores should be arranged (products to be stored) so that all space is used and so that stacking and retrieving is a direct operation.

5.2.3.1 Fluids

Some type of pump or fan will generally be used to move fluids. Considerations include the following:

— Density
— Viscosity
— Corrosiveness
— Abrasiveness
— Temperature
— Contamination

The energy necessary to move the fluid will be provided using one or more of these principles:

— Centrifugal force: blower fan
— Volumetric displacement: piston, gear, diaphragms, propeller, air lift, and rotary
— Transfer of momentum: ejector pump
— Mechanical impulse: axial turbine and peristaltic pump
— Electromagnetic pump in which there are no moving parts

Some materials are very viscous at low temperatures but can be made free-flowing when warmed by either of the following:

— Heating the storage tank
— Wrapping the pipework with heat tracing

The energy conservation opportunities begin with the proper selection of the pumping principles to match the fluids characteristics.

5.2.3.2 Solid materials

Solid materials include the three types shown in the following table:

Type of material	Size	Example
Particulate	Fine Medium Coarse	Flour Wheat Coal
Discrete	Small Medium Large	Electronic components Hats Car engines
Finished	Small Medium Large	Sacks of sugar Refrigerator Power transformer

Particulate material transportation methods include the following:

a) Air carriage. Fine light materials can be blown along ducts and collected with a filter.
b) Motor driven conveyor. Common types are
   1) Continuous rubber, metal, or magnetic belts
   2) A loop of buckets
   3) A series of rollers
   4) A screw inside a tube
   5) A continuous overhead chain to suspend the product
c) Vibrating beds or *linear induction motors* (LIMs)
   (A LIM is a form of non-rotating electric motor that produces thrust as distinct from rotary motion. The wound unit is usually referred to as the "primary." The wound primary units, shaped like a flat oblong package, can be connected to produce thrust in either direction. Normally the moving element [referred to as the "secondary"] is usually built into, or becomes part of, the product being conveyed. It takes the form of a strip of aluminum, iron, brass, copper, or a combination of these materials. Electromagnetic forces induced from the primary into the secondary produce energy in the form of mechanical thrust. This thrust is proportional to the square of the voltage providing the primary is covered by the secondary. LIM conveying systems include a Walt Disney Theme Park people moving system and the Houston, Texas, Intercontinental Airport Terminal-to-Terminal Train. LIMs are capable of moving package or bulk materials from ounces to thousands of pounds.)

Discrete material transportation methods include the following:

a) Conveyor belt

b) Overhead conveyor with underslung carriers

c)   Pallets or platforms transported by elevating trucks (fork lift trucks) or with the aid of a tug

d)   Air cushion lifts may be used in place of wheels on platforms to reduce tractive resistance.

e)   *Automated storage and retrieval systems* (AS/RSs) and *automatic guided vehicles* (AGVs) can be a major factor in a fully integrated material handling system.

f)   Electric hoists, cranes, and electro-magnets are useful for point-to-point delivery of heavier objects.

Transportation for finished products may be by any of the conveying systems previously described.

Technologies that should be investigated include the following:

a)   Use systems integration to reduce the time and energy in material handling.

b)   Use of *adjustable speed drives* (ASDs) may be extremely efficient where a wide range of motor speed is required and when several drive motors are used for the same movement. Variable speed, regardless of load and accurate control throughout the entire speed and torque ranges, is possible.

c)   Use LIM systems where practical, since they have no rotating parts yet provide smooth and rapid material transport. Losses due to the inefficiency of gears, clutches, bearings, or shafts are avoided.

d)   Powered drums for driving belt conveyors can be used for the package handling, bulk handling, and for sanitary applications in food processing. Powered drums are available from fractional hp to 400 hp. The powered drum can replace the traditional drive for a belt conveyor where power has been transmitted by a belt or a chain.

e)   Battery-powered electric vehicles. Consider the following:

   1)   The size and type of truck for the job.

   2)   Efficiency of conversion of dc to motive power.

   3)   Efficiency of the battery chargers, including completely automatic charging features.

   4)   Matching the motor kW rating (hp rating) with the task.

   5)   Scheduling and matching truck usage to the task.

   6)   Peak demand of charger: Evaluate magnetic amplifier circuit design as compared to ferroresonant design.

5.2.4 Ultraviolet heating of materials (SEC of Victoria [B214])

Ultraviolet (UV) radiation is a form of energy transfer from one body to another, generally, without any solid material substance in the intervening space. This energy transfer is carried out by electromagnetic waves. UV is in the electromagnetic spectrum from 0.012–0.4 µm.

These are three bands of UV radiation:

a)   Near UV (UV-A band)
b)   Middle UV (UV-B band)
c)   Far UV (UV-C band)

Combinations of UV wavelengths, particularly those from UV-A and UV-B bands, have particular industrial importance in some materials producing quick curing of various surface coatings and inks. UV can be a major conserver of energy. Contrasted with conventional drying processes for curing inks on products as much as 80% energy savings are achievable.

5.2.5 Infrared heating of materials (SEC of Victoria [B206])

Infrared (IR) transfers heat to the surface of a product by radiation. Heat transmission by radiation is a process occurring in ovens and furnaces for obtaining product temperatures ranging from 30 °C (e.g., for drying of timber coatings) to 1300 °C (e.g., for metal billet heating).

IR radiation is in the electromagnetic spectrum from 0.7–400 µm.

There are four bands of IR radiations:

a)   Short IR (0.7–2.0 µm)
b)   Medium IR (2.0–2.8 µm)
c)   Long IR (2.8–4.0 µm)
d)   Far IR (4.0–400 µm)

Such elements produce radiant and convected heat in approximate proportions of two to one. These conditions are often ideal for batch operations but are rarely suitable for on-line production because of the increased convection losses.

Heating economics should be compared on the basis of the usable heat content rather than the hypothetical potential heat value of the source.

Where radiant heat is indicated, a source that generates a maximum of IR and a minimum of convected heat should be used. Electric IR systems employing tungsten filament lamps and gold-plated reflectors operate at radiant efficiencies approaching 80%, whereas fuel-fired IR systems are inherently incapable of operating with efficiencies exceeding 30%.

Products with a suitable configuration, i.e., reasonable rates of surface area to mass, should be evaluated for electric IR heating.

The characteristics of IR heat sources include the following:

Infrared heat sources	Power density (kW/m^2)	Relative energy (percent)	
		Radiation	Convection and conduction

1) Tungsten filament lamps
  a) Glass bulb
    1) Type G30
    2) Type R40
  b) T3 Quartz lamp
	5
	20
	270
		80
		80
		86
			20
			20
			14

2) Nichrome spiral windings
  a) Quartz tube
  b) Metal sheath
	50
	80
		55
		50
			45
			50

3) Low-temperature panel heater
  a) Burned nichrome
  b) Metallic salt
	—
	—
		40–30
		40–30
			40–70
			40–70

5.2.6 Resistance heating of materials (SEC of Victoria [B211])

Resistance heating is used in the production process of manufacturing industries to heat solids and fluids and in the commercial building industry to heat hot water.

The following should be considered to improve energy efficiency for resistance heating of materials:

a)   Designing the right scheme for the production cycle

b)   Installing high quality precise instruments for best repeatability, extremely low drift out-of-calibration, and an accuracy traceable to government temperature standards

c)   Installing the appropriate thermal insulation to minimize heat losses from the process

d)   Applying electrical energy only when the product is present to be heated; for example, using sensors, such as photocells, to monitor the production line conveyor and initiate the heating controller

5.2.6.1 Space heating

Space heating equipment for industrial and commercial applications may be either of these:

—   Central electric heating
—   Zonal electric heating

Central electric heating systems generally consist of either electric boilers or electric heating coils installed in air-handling units. In either case, heat is provided to a central system for distribution to the occupied spaces.

The central electric space heating equipment can be controlled to turn on/off or to set temperatures back for the entire space served for demand control.

In larger buildings requiring central heating systems of more than several hundred kilowatts, it may be possible to use high-voltage electric or electrode boilers and eliminate transformers with their associated energy losses. (Refer also to 5.2.11.)

In addition to energy conservation, thermal control is required for the following purposes:

— Personnel protection against hot exposed surfaces and radiant heat
— Control of process temperature
— Protection of adjacent equipment through reduction in radiant heat
— Fire protection
— Reduction of noise from the process
— Reduction of condensation

5.2.6.2 Heat tracing

Heat tracing systems are designed for freeze protection and process temperature maintenance on drainlines, waterlines, safety showers, sprinkler systems, processing systems, manifolds, valves, laboratory tubing, and anywhere temperature must be maintained under varying ambient conditions.

Electric heat tracing systems include

— Resistance heating cables
— Resistance heating tapes
— Impedance heatings
— Skin-effect current tracing

Energy losses are affected by

a) *Pipe insulation.* Pipe insulation is the single most important item in conserving energy in a heating system—the insulation with the lowest *k*-value of thermal conductivity should be used consistent with cost and other physical characteristics. (Refer to the NICA Insulation Manual [B175].)

1) *Water and insulation.* Insulation that has become water soaked will have 16 to 20 times greater heat loss. Insulation systems subject to moisture should be waterproofed and sealed against moisture entry. The insulation must be dry before the sealing jacket is added. Leaks in valves and traps should be repaired. If moisture entrapment cannot be avoided, a closed cell insulation should be used in this area. Small drain holes should provide an easy exit for moisture.

2) *Oversizing insulation.* Traditionally the insulation selected is one size larger than the pipe to accommodate the heat tracing. The pipe insulation should close around the pipe and the heater so there is no gap in the insulation. Use of regular size insulation reduces the convective heat losses; however, the type of insula-

tion—soft or rigid—must be considered since rigid insulations can be broken if tightly strapped.

3) *Review existing installations.* A pipe insulation system may be a candidate for upgrading.

b) *Location of the heater on the pipe.* Parallel-type heaters are sometimes spiralled around the pipe to achieve the design wattage per foot of pipe. With single-heater-type systems run straight along the pipe, location can affect efficiency. A cable installed on the bottom of the pipe is more efficient than a cable installed on the top of the pipe by up to 3%. However, the method of installing the cable below a line tangent to the top of the pipe has advantages in installation technique and in reducing the possibility of electrolysis and moisture absorption of some heater types.

c) *Wind chill* can affect heat loss on pipe heating systems as follows:

1) *Outside the insulation.* Wind can remove the still-air film surrounding the pipe insulation. The air film varies with pipe diameter and surface of lagging material. Heat loss varies from 5.5% on small diameter pipes to 9% on 12 in diameter pipes, as compared to losses in no-wind conditions.

2) *Within and beneath the insulation.* If insulation is improperly sealed, air will flow within and beneath the insulation causing severe losses; this condition must be repaired to restore effective operation. In vertical installations there will also be a "chimney effect" that can be avoided by baffling under the insulation.

d) *Thermal coupling of electric tracing systems.* An adhesive backed foil "tape" as a heat transfer aid can improve efficiency by about 3% on the oversized insulation installation and by 7–18% in regular-sized insulation installations. Generally, the better conductive coupling through the use of metal braids and foils, the more efficient the heater.

5.2.6.3 Snow melting

An ASHRAE Class III (industrial) snow melting system may use power at a demand up to 131 W/ft^2 according to ASHRAE 1991 HVAC Applications Handbook [B15]. Snow melting systems are generally of these types:

a) Electric heating cables directly embedded in the pavement

b) Polyolefin pipe systems embedded in the pavement through which glycol solution, heated by electric boilers, is pumped

c) Electric IR systems installed on overhead poles

Energy conservation measures begin with the initial pavement design where techniques can be used to reduce back losses in the pavement.

Energy reducing operating measures should provide fully automatic control for the embedded systems. The control initiates heating based upon snowfall, air temperature, and surface conditions.

Electric IR systems may include automatic controls with temperature/moisture sensors so that the system can be turned on in advance of a storm so as to reduce buildup of snow.

5.2.6.4 Industrial thermal insulation

Insulation is one of the most cost-effective methods of reducing energy consumption in thermal processes. A considerable amount of energy is wasted where the thermal insulation is inadequate or nonexistent.

A quick method to determine the heat loss rate from the surface of a process, pipe, duct, or equipment surface, even when the type of insulation or thickness is unknown, is given by the following equation:

$$Q = (T_s - T_a) \cdot f \qquad (W/m^2)$$

where

T_s is the surface temperature (°C)

T_a is the ambient temperature (°C)

f is the surface coefficient ($Wm^{-2} K^{-1}$)
  where typical values of K are
  — Bright metal surface: 5.7
  — Dull metal surface: 8.0
  — Bare or rough surface: 10.0

Insulation material selection, application, installation, and procedures for determining the economic thickness of insulation are contained in the NICA Insulation Manual [B175].

Floating insulation should be considered for installation in pickling, chlorinating, anodizing, plating, and hot-water-rinsing processes since the heat dissipation from the exposed liquid is so great that thermostatic control quite often serves no useful purpose. Floating insulation, e.g., in the form of balls, can greatly improve the thermal efficiencies of such processes.

Floating insulation reduces heat losses by about 70% (at 90 °C) for a single layer. Evaporation losses are reduced about 90%.

A single layer of polypropylene balls covers about 80% of the surface over which it extends. A double layer provides 100% coverage.

As a guide, heat losses from open tank surfaces are as follows:

Solution temperature (°C)	Losses (kW/m^2)
50	2.2
60	3.5
70	5.5
80	8.0
90	12.0

5.2.7 Radio frequency heating of dielectric materials (SEC of Victoria [B210])

The induction and dielectric techniques can be used to generate heat within the material itself. These techniques result in efficient heating by enabling

— A high concentration of heat energy
— Selectivity in location of heat application
— Accurate control of heat duration

Dielectric heating includes both of the following:

— Radio frequency dielectric (RFD)
— Microwave

Dielectric heating is used for heating non-metals such as timber, textiles, plastics, adhesives, etc. Heating is produced by subjecting the material to a high-frequency electromagnetic field. The dielectric properties of a material depend largely on its molecular structure. Dipolar molecules, although electrically neutral, have separate positive and negative charges. These charges try to follow the alternating electric field and, in so doing, generate heat.

The rate of heat generation is a function of the loss factor of the material (i.e., its ability to absorb electromagnetic energy) as well as the voltage and frequency applied across the electrodes.

Heat generated is proportional to loss factor · voltage^2 · frequency.

As a rule, the lowest practical frequency should be used that produces the desired result. However, if the voltage necessary to produce the required heat exceeds the dielectric strength of the material, then a higher frequency is employed to avoid possible destruction of material or electrode flashover.

Industrial dielectric heating can replace other processes and result in overall energy savings. Production can start almost immediately after switching on. Little power is used during a standby period. Power consumption is also self-limiting in drying applications. There is also no time lost in conducting heat from the surface to the center of the material.

Selective heat location within a product is possible as some materials absorb radio frequency energy more easily than others. Hence selective heating or moisture extractions is possible.

Optimum efficiency in industrial dielectric heating requires consideration of the following:

— Electrode construction
— Calibration and tuning of the radio frequency (RF) generator
— Calibration of the airgap between electrodes and workpiece
— Loss reduction in the electrodes and connecting strips

5.2.8 Tungsten halogen heating (SEC of Victoria [B213])

Tungsten halogen heating is an *ultra-high* efficient form of IR heating with a radiation generating ratio of 0.86. Tungsten halogen heaters provide effective use of energy by heating the people and equipment, not the surrounding air.

The lamps can provide a radiant heat intensity of 40 W/linear cm plus visible light at a rate of 7.5 lm/W. Lamp life is 5000 h when operated at full power output (even longer when operated at a reduced output) and in a normal industrial environment the heater lamp should last approximately 5 years. Luminaires have luminous intensity distributions of 30, 60, 90, and 100 degrees, symmetrical and asymmetrical, and power ratings up to 5 kW.

Energy management techniques for tungsten halogen heating are as follows:

a) Use clear quartz lamps unless visible glare is objectionable or unless extreme fixture vibration would cause breakage. Red filter sleeves are available to reduce the glare but maintain the high intensity heat.

b) Where clear quartz lamps are used, the visible light may contribute to the ambient and task illuminance requirements. Lighting sensors can integrate this visible light into the power setting of an electronic fluorescent lighting system.

c) Reflectors must be kept sufficiently clean to avoid absorption of heat that will result in lower radiant efficiency and also the ultimate destruction of the reflector.

d) Use only clear quartz or ruby red lamps for snow melting.

e) Turn on the system prior to storms to raise the temperature of the pavement above the freezing point.

f) Use output-power-type controls. The output power varies as the 1.5 power of the lamp voltage. The following should be considered when applying controls:

    1) Since IR heaters do not heat the air primarily, thermostats should not be used to control the radiant heat level.

    2) Full voltage-half voltage controllers can apply phase-to-phase and phase-to-neutral voltage, when used on single-phase systems.

3)   Star-delta controllers will apply phase-to-neutral voltage and phase-to-phase voltage, when used on three-phase systems.

4)   Controllers that modulate the voltage to the lamps by interrupting the ac sine wave or chopping each half cycle into on and off phases provide smooth flicker-free continuous heat. The operation is similar to a light-dimming system.

5)   For automatic temperature regulation, an IR thermistor sensor can be used to sense the radiant heat from the luminaires and control the voltage modulation power controller.

5.2.9 Induction heating and melting techniques (SEC of Victoria [B203])

The types of power sources typically available are the following:

—   Line frequency   50 Hz; 60 Hz
—   Magnetic frequency multipliers   150 Hz; 180 Hz; 540 Hz
—   Rotating machines   up to 10 kHz
—   Power electronic inverters   up to 50 kHz
—   High frequency oscillators   200 kHz to 2 MHz

The power to the work-coil is the sum of the heating power required in the work-piece, the I^2R heat, plus the power used to heat the work-coil. The electrical efficiency of the process is affected by the resistance of the work-coil and this should be made of material of the lowest resistivity possible.

The equivalent resistance of the work-piece changes with power frequency, permeability of material, resistivity of material, and the number of turns in the work-coil. Being a highly inductive circuit, the power factor will be poor and some means of correction will always be needed.

The energy to heat a work-piece by induction follows the normal heat calculation: weight times rise in temperature times specific heat.

Surface heating of a product uses high power, short heating time, and high frequency.

Through-heating of a product uses low power, long heating time, and lower frequency.

For maximum energy efficiency, the following must be considered:

—   The characteristics of the work-piece
—   The work-coil's design
—   The efficiency of the frequency generator design
—   Automation of the control process
—   Coolant flows for the work-coil with high-efficiency pumping

Induction melting is performed in furnaces at frequencies from 50–10 000 Hz. Channel furnaces have large capacity, low power, and the highest electrical efficiency. Coreless furnaces have a small furnace capacity. The maximum electrical efficiency, once the metal is in the molten state, is determined by the crucible shape:

— Outside diameter, D, to work-coil lengths, L, needs to be between 0.8 and 1.2.

— Outside diameter, D, to current penetration depth, d, of the molten metal needs to be 4 or above, as shown in the following table:

D/d ratio	Percent energy absorbed
4	74
2	21
1	3

5.2.10 Induction metal-joining techniques (SEC of Victoria [B204])

Induction metal-joining techniques include soldering, brazing, and welding, and the bonding of ceramics.

5.2.10.1 Welding

The major application of induction heating is the continuous welding of steel tubing. Radio frequency (450 kHz) is used for most of these applications on which the work-coil surrounds the tubing and reduces current flow in the two abutting edges.

5.2.10.2 Soldering

Soldering describes a joining operation using nonferrous filler metals with melting points up to 500 °C. Solders are alloys of tin and lead having the property of melting entirely within a limited temperature range. They are usually referred to as soft solders. Power requirements for soft soldering at 190 °C/kg of steel are 10 kW for 10 s, for example, or 2 kW for 50 s where the time refers to the duration for the applied power to raise the steel to be joined to 190 °C.

5.2.10.3 Brazing

Brazing is used when a tougher, stronger joint is required than those usually joined by soldering. The most common brazing materials are pure copper, brass, bronze, silver, and silver alloys. The short heating times achieved with soldering are not possible with brazing, due to the higher melting temperature of the brazing alloys and the congruent greater losses due to radiation, convection, and conduction. Power requirements for silver brazing at 700 °C/kg of steel are 45 kW for 10 s, for example, or 10 kW for 50 s where the time refers to the duration for the applied power to raise the steel to be joined to 700 °C.

Efficiency depends upon the design of the work-coil, the engineering of the process (selection of soft solder or brazing alloy, frequency, time, and power), the reduction of heat losses, accuracy of instrumentation, and efficiency of the RF generator.

5.2.11 Electric steam generation

Electric boilers are used for steam generation for the following:

a)   Autoclaving
b)   Platen heating
c)   Jacket cooking
d)   Pressing clothes
e)   Degreasing
f)   Humidification
g)   Live injection

The types of electric boilers are listed in annex 5B.

5.2.11.1 Efficiency considerations (SEC of Victoria [B199])

With industrial plants of any size it is generally desirable to install a number of medium-sized electric boilers close to their steam load rather than provide a central boiler house to provide all steam requirements. The main advantage of this is to reduce pipe losses, which can be considerable. At off-peak times, when only part of the steam load is required, these losses can be greater than the steam load itself. (Where reliability is an important factor, such as with hospitals and production lines, a central boiler house is often preferred. Here reliability can often be achieved by paralleling two or more small units.)

In many industries there are periods of high and low steam demand. In large installations, boiler utilization may be improved by installing multiple units that are brought into production according to steam demand. However, to reduce the maximum demand, avoid load fluctuations over short periods, or take maximum advantage of off-peak tariffs, some form of steam storage that can be recharged during periods of low steam demand is often necessary.

In addition to parallel and tandem operation, the following should be considered:

a)   Utilize electric steam boilers to capacity when operating.
b)   Switch off when not operating.
c)   Use off-peak rates.
d)   Minimize the length of distribution piping.
e)   Shut off valves when appliances are not in use.
f)   Ensure all pipework and flanges are properly insulated.
g)   Eliminate any steam system leaks.
h)   Minimize pressure and temperatures for the process.
i)   Install timers to control work cycles.
j)   Use low-pressure steam for direct injection.
k)   Optimize mechanical water extraction before drying.

l)   Use recirculation when heating.

m)   Recover heat from flash steam.

n)   Return condensate to feedwater.

o)   Maintain steam traps and vents in good condition.

p)   Provide feedwater temperature as high as possible to the boiler.

5.2.12 Cool storage systems (data courtesy of Calmac Manufacturing Corp.)

Cool storage is a load management technique that shifts the electrical requirements for air-conditioning or process-cooling from daytime to nighttime hours. This displacement of load reduces the peak electricity demand and can result in significant electrical bill savings. Energy can also be saved in cool storage, as explained in 5.2.12.5.

Cool storage extracts heat from a storage tank or tanks at night and uses the storage device to absorb heat from the load the following day. Since the refrigeration work is done off-peak, less work is required during the day so the peak demand on the electrical system is reduced.

5.2.12.1 Full storage systems

The capacity of the storage system is large enough to enable it to supply all of the daytime cooling needs on the "design day" or hottest-weather day. The refrigeration equipment can be switched off during the peak period leaving only the chilled water pumps and air-handling unit fans running in the building air-conditioning system. Full storage is best suited for applications where a) the utility's peak period is relatively short, and b) the demand rate is high, the differential between peak and off-peak rates being substantial.

5.2.12.2 Partial storage systems

5.2.12.2.1 Load-leveling partial storage systems

Load-leveling partial storage systems supply only part of the building's peak-period cooling needs from storage. Storage capacity is sized such that parallel operation of storage and refrigeration equipment meet design-day cooling needs. This mode minimizes the size of storage and refrigeration equipment compared to other storage techniques, thereby minimizing first costs. Operating savings are smaller since only a partial reduction in cooling demand during peak hours is achieved. However, partial storage is turned into full storage during spring and fall; this benefit, along with its lower cost, attracts most buyers. Partial storage is competitive with nonstorage and can be less when cold air is used. Load-leveling partial storage is best suited for applications in which a) the utility's peak period is long, and b) the differential between peak and off-peak rates is small and demand charges are relatively low.

5.2.12.2.2 Demand-limited partial storage systems

Demand-limited partial storage systems reduce the building's maximum electrical demand to a predetermined level, normally the level of peak demand due to noncooling loads. Storage capacity is sized such that parallel operation of storage and refrigeration equipment does not cause the building's total electrical demand to exceed the maximum level. This strategy requires real-time controllers to monitor the buildings noncooling loads and control the ratio of storage-and-chiller-supplied cooling. Demand-limited partial storage is best suited for applications in which a) the building's noncooling load undergoes a large daily swing and b) the building's occupancy period is relatively short, less than 10 h/day. This strategy is particularly effective where demand charges are high and real-time building energy management controllers are installed.

5.2.12.3 Chilled water storage systems

Chilled water storage systems make use of the heat capacity of water in its liquid forms. Chilled water systems typically are designed to operate so that the stored water undergoes a temperature swing of 15–18 °F corresponding to the difference between the maximum acceptable return temperature from the building's chilled water distribution system and the minimum practical chilled water temperature.

Chilled water storage tanks require about 10–13 ft^3 per ton-hour (3.52 kWh) of refrigeration. In new construction, chilled water storage may be more cost-effective than ice storage in buildings larger than 500 000 ft^2 (46 450 m^2).

5.2.12.4 Ice storage systems

Ice storage systems can be used for any process that requires chilled water, such as

a)   Air-conditioning
b)   Process cooling, e.g., cheese, milk, chocolate
c)   Cooling of chemical processes
d)   Injection mold cooling

The main advantage of ice storage over chilled water storage is the smaller volume required. Theoretically, were it possible to cycle the entire storage volume between solid and liquid phases, an ice storage system would be about a factor of ten smaller than a chilled water system operating through a temperature swing of 15 °F. In practice, the volume savings are limited to about a factor of five to eight.

Two types of ice storage systems are available: *static* (ice-building) and *dynamic* (ice-shucking or ice-harvesting). In the static systems, ice is formed on the cooling coils in the storage tank itself. There are two types of static ice systems: a) ice made on the evaporator and b) ice made remote from the evaporator with a coolant. Remote ice allows the compressor to operate by day, while 25% glycol/water allows a secondary coolant to melt off ice as needed. These packaged units can be connected to a building air-conditioning system directly. Dynamic ice storage systems make ice in chunk or crushed form and deliver it for storage in large bins. Another dynamic ice storage system uses refrigerant plates or tubes that are vertically suspended above the storage tank. Water from the tank is run over the plates and ice sheets form on the surface. Periodically the ice is harvested by injecting hot refrigerant inside the plates for a few seconds causing the ice sheet to break off and fall into the tank below.

Field-erected ice storage systems are made up to 173 000 ton-hours, while factory assembled is up to 1500 ton-hours (5280 kWh). Modular systems are popular up to 50 000 ton-hours (176 000 kWh). Tank size is typically 2–4 ft^3 (0.057–0.113 m^3) per ton-hour (3.52 kWh) of storage. Ice storage systems use all types of refrigeration-compressors, including centrifugal, rotary, or reciprocating, depending upon capacity requirements.

In buildings smaller than 500 000 ft^2 (46 450 m^2), ice storage is normally more cost-effective than chilled water storage.

5.2.12.5 Energy considerations

The energy performance of a thermal energy storage system can result in significant reduction in overall energy consumption. A report in the May 1993 *ASHRAE Journal* [B21] documents the retrofit of a Texas Instruments Co. electronics manufacturing plant in Dallas to a full-shift thermal energy storage system. The 10-year-old 1 100 000 ft^2 (102 109 m^2) factory retrofit achieved a reduction in peak electrical demand of 30.2%; and a reduction in annual cooling electricity usage of 28.3%. Ice storage systems go to a lower evaporator temperature when freezing to ice at night if water-cooled (wet bulb), but air-cooled (dry bulb) *does not* and is often more efficient at night making ice than standard air-conditioning by day. Air-cooling constitutes 80% of *non-water* cooling. A report in the December 1991 *ASHRAE Journal* [B25] explains this clearly, and also shows that cold duct air, at 42–45 °F (7 °C) or less, making use of ice storage, saves 45–50% of fan motor power with smaller ducts, and results in lower relative humidity, more condensate, cleaner water, improved indoor air quality, and savings of cost, space, and energy. (See [B239].)

When ice storage systems are used there may be a reduction in kWh consumption of 20% due to air-cooled night condenser temperature and cold air. Retrofitting rooftop units with central storage systems, sized at 60% of the air-conditioning load, also can give 30% demand saving and 15–28% energy savings.

5.2.13 Solar-assisted heating systems

5.2.13.1 Solar-assisted electric water heaters

Solar-assisted electric water heaters can be used for the following purposes:

— To heat water during the day that is consumed or returned to storage. When there is not enough daily solar energy, the electric heating element in the storage tank switches on when the facility is on the off-peak tariff.
— To pre-heat water during the day that is circulated for use and then instantaneously boosted in temperature at the point-of-use by electric instantaneous heating units.

Solar-assisted electric water heaters mostly use flat-plate solar collectors. Water flows from the storage tank through the collectors where it is heated by the sun, then back into the tank where it is stored for later use. The solar collectors work best if they continuously track the sun; otherwise, they should be mounted at an appropriate fixed-tilt angle.

Of utmost importance in the selection of solar-heating-system components are the efficiency of the collector at the heating range required, the thermal insulation of the storage tank, and the thermal insulation of the piping.

The amount of energy, or total solar radiation, available to the collector is the sum of three components:

— Direct (from the sun)
— Diffuse (from the sky)
— Reflected (from other surfaces)

The heat loss parameter measures heat lost to the environment in degrees Fahrenheit per square foot per hour per British thermal unit ($°F/ft^2/h/Btu$). The heat loss parameter for service water heating can range from 0.05 at 70% efficiency to 2.35 at 25% efficiency, depending upon the climate at the site and desired hot water delivery temperature.

For example, assume that the temperature of water is to be raised 100 °F on a day when the radiation is 300 Btu/h/ft^2, the heat loss parameter is 0.3, and the solar collector efficiency is about 35%. In this example, a 40 ft^2 collector will be able to produce

$$300 \text{ Btu/h/ft}^2 \cdot 40 \text{ ft}^2 \cdot 35\% \text{ efficiency} = 4200 \text{ Btu/h}$$

$$\frac{4200 \text{ Btu/h}}{8.34 \text{ lb/gal} \cdot 100 \text{ °F rise}} = 5 \text{ gal/h}$$

During the entire day, a collector angled at 30°, at 40° N latitude, can produce (on January 21st)

$$1600 \text{ Btu/ft}^2/\text{day} \cdot 40 \text{ ft}^2 \cdot 35\% \text{ efficiency} = 23\,100 \text{ Btu}$$

The quantity of water that can be raised by 100 °F is

$$\frac{23\,100 \text{ Btu}}{8.34 \text{ lb/gal} \cdot 100 \text{ °F}} = 27.7 \text{ gal}$$

Annually, the solar collector is capable of replacing electricity in the amount of

$$\frac{2025 \text{ Btu/ft}^2/\text{day} \cdot 365 \text{ days/year} \cdot 40 \text{ ft}^2}{3412 \text{ Btu/kWh}} \cdot 35\% \text{ efficiency} = 3033 \text{ kWh/year}$$

5.2.13.2 Solar-assisted heat pump systems

Solar-assisted heat pump systems utilize solar panels with commercial heat pumps in an integrated system of heating conditioned space:

a)   Solar-heated water is pumped direct to indoor draw-through fan coil units without being stored.

b) Solar-heated water, previously stored, is pumped to the indoor fan coil.

c) Hot gas from the compressor circulates to the indoor fan coil, heating the indoor air and condensing to hot liquid. Liquid passes through an expansion valve, which converts liquid to cold gas. Heat from the solar collector hot water on the stored solar-heated water is absorbed by the cold gas.

5.2.14 Compressed-air systems (SEC of Victoria [B196])

Production of compressed air in industry consumes approximately 10% of all electrical energy. Compressed-air production, distribution, and end-uses can be inefficient when not carefully controlled. A typical compressed-air system wastes approximately 15% of the electrical energy consumed. Further, 80% of the remaining energy is discarded as heat that could be easily reclaimed for space or process heating. Heat-recovery equipment can capture heat from the air compressor or the aftercooler.

Compressor types are shown in figure 5-1. The compressor in most factories will usually be one of the following types:

a) Positive displacement
   1) Reciprocating
   2) Rotary vane
   3) Rotary screw
b) Continuous-flow
   1) Centrifugal

The choice of compressor type depends upon the compressed-air-usage pattern and site details.

Fresh air is drawn into the compressor by way of an air filter and is usually compressed to a pressure of 101.5–116 lbf/in^2 (700–800 kPa or 7–8 bar) for distribution around the factory. The compression process generates a lot of heat and the compressor may be cooled by air, water, or oil.

Table 5-2 compares compressor types.

The full-load and part-load performance of compressor types varies considerably as shown in table 5-3.

Air consumption of pneumatic tools and appliances is shown in table 5-4.

5.2.14.1 New compressed-air systems energy savings

When designing a new compressed-air system, optimum efficiency can be achieved by

— Selecting the appropriate compressor for the application;
— Operating the compressor efficiently for high part-load/full-load performance;
— Selecting pneumatic tools and appliances that consume the least ft^3/min for the task.

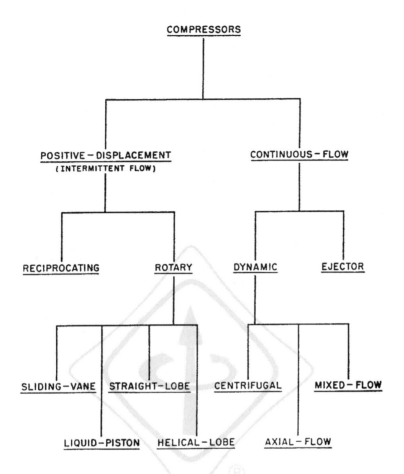

Source: Gibbs 1969 [B120].

NOTE—The terms "rotary vane" is synonymous with the sliding-vane rotary type, and the term "rotary screw" is synonymous with the helical-lobe rotary type.

Figure 5-1—Types of air compressors

Efficiencies of an existing system can be improved by a five-step audit procedure. These steps are as follows with typical results shown in table 5-5:

Step 1)   Measure the energy consumption of the compressed-air system.
Step 2)   Determine the air leakage.
Step 3)   Identify pressure loss, air intake loss, pipe resistance loss, and other losses.
Step 4)   Identify opportunities for waste heat recovery.
Step 5)   Estimate the total energy and cost savings to the compressed-air system.

Table 5-2—Performance of positive displacement type compressors

Criteria	Reciprocating	Rotary vane	Rotary screw
Efficiency at full-load	High. Shows a significant advantage for multistage machines.	Medium to high	High
Efficiency at part-load	High	Poor below 60% full load	Poor below 60% full load
Capacity	Low to high	Low to medium	Medium to high
Pressure	Medium to very high	Low to medium	Medium to high
Maintenance	Many wear parts	Few wear parts	Very few wear parts

Table 5-3—Power requirements of positive displacement type compressors

Criteria	Reciprocating	Rotary vane	Rotary screw
Full load, power requirement at 101.5 lbf/in^2 (700 kPa)	0.189–0.212 kWh/ft^3/min 0.40–0.45 kW/L/s	0.165–0.189 kWh/ft^3/min 0.35–0.40 kW/L/s	Single-stage: 0.179–0.203 kWh/ft^3/min 0.38–0.43 kW/L/s Two-stage: 0.142–0.165 kWh/ft^3/min 0.30–0.35 kW/L/s
No load, power as a percent of full load	30–40%	25–60%	10–25%

NOTES
1—To convert from ft^3/min to L/s, multiply by 0.47195.
2—The other components are listed in 5C.2 of annex 5C.

Compressed-air energy consumption is measured by one of the following two methods shown:

Method 1) A kilowatthour meter is used to measure the average power over a test period, e.g., one month.

Method 2) If the compressor is operating at a steady load, the power can be measured directly with a kilowatthour meter, or measurement of the compressor-circuit three-phase currents can be used to determine compressor energy consumption:

Power (watts) = $(3)^{1/2}$ (volts) (amperes) (power factor) where the values are "system voltage" and "average phase current." Power factor is usually 0.85 at full load, reducing to 0.7 at half load and to about 0.2 at no load.

Monthly energy use (kWh) = power (kW) · number of operating hours per month.

Table 5-4—Typical air consumption of pneumatic tools and appliances

Tool or appliance type	Size	ft³/min of free air	L/s of free air
Drills	7 mm 10 mm 13 mm 25 mm 50 mm 75 mm	10–16 15–20 25–30 60–80 80–120 100–130	4.7–7.5 7.1–9.4 11.8–14.1 28.3–37.7 37.7–56.6 47.2–61.4
Grinders	50 mm diameter Up to 150 mm diameter	20–25 50–60	9.4–11.8 23.6–28.3
Sanders and polishers		10–45	4.7–21.1
Torque wrench for nuts up to	7 mm 13 mm 25 mm 38 mm	10–15 25–35 40–55 50–70	4.7–7.1 11.8–16.5 18.9–26.0 23.6–33.0
Screwdrivers		7–25	3.3–11.8
Nut runners		10–30	4.7–14.1
Spray guns	Small Medium Large	1–5 5–12 12–25	0.47–2.4 2.4–5.7 5.7–11.8
Blow guns		5	2.4
Air motors	Allow 34–40 ft³/min (16–19 L/s) per unit power at maximum power for motors below 1.5 hp (1.1 kW) and 25–30 ft³/min (12–14 L/s) per unit power for larger motors.		

NOTE—To convert from ft³/min to L/s, multiply by 0.47195.

The full-load and no-load test points can be determined and the power input at any operating point as shown in figure 5-2.

The energy-to-supply air leaks can be more accurately measured at night or at lunch time when all the production plant using compressed air may be shut down. Small air demand items, such as controls, may be left running. With multiple compressor installations on a common air system, the air leakage test of the whole system should be performed with only one compressor running. Two methods of air leakage tests are as follows:

Method 1)

    a)   Let the compressor build up the system pressure until it reaches the shut-off point and the compressor stops. If the compressor never reaches the shutoff point, the air leakage must be very high.

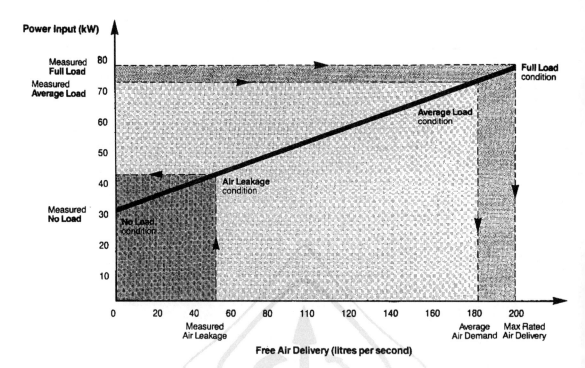

Source: SEC of Victoria [B196]

NOTES

1—An accurate assessment of compressor power usage requires no-load and full-load tests to be conducted. For the **no-load test**, shut the valve between the compressor and the air receiver. Measure the power consumed by the compressor. For the **full-load test**, expel some air from the air receiver. Measure power consumed by the compressor as it builds up air pressure.

2—This graph is an approximation based on electrical measurements obtained as described in Note 1.

Figure 5-2—Air compressor performance graph

b) For a period of 30 min, record the time the compressor runs and the time that it is off.

c) The quantity of air leaking from the system can be estimated as follows:

Quantity of air to supply leaks

$$= \frac{\text{rated free air delivery (ft}^3/\text{min}) \cdot \text{time on (min)}}{\text{time off} + \text{time on (min)}}$$

$$= \frac{\text{ft}^3/\text{min} \cdot T_{\text{on}}}{T_{\text{off}} + T_{\text{on}}}$$

Table 5-5—Energy distribution in a typical compressed-air system

Energy distribution	Before energy audit	After plant improvements (without heat recovery)	After plant improvements (heat recovery replaces other heating cost)
Useful work	12%	12%	12%
Pipework friction and air leaks	5%	2%	2%
Motor losses	5%	2%	2%
Waste heat	78%	67%	Zero
Energy saved	Zero	15%	15%
Energy recovered as heat	Zero	Zero	67%

Method 2)

   a) Run the compressor until the pressure in the air receiver builds up to its maximum operating pressure.

   b) Switch off the compressor and measure the time taken for the receiver pressure to drop by, say, 15 lbf/in^2, (the pressure should drop enough to allow the compressor to restart for the second half of the test).

   c) Switch on the compressor and record the time taken to build up the 15 lbf/in^2 pressure loss.

   d) The quantity of air leaking from the system can be estimated as follows:

Quantity of air to supply leaks

$$= \frac{\text{rated free air delivery (ft}^3\text{/min)} \cdot \text{time for pressure to rise (min)}}{\text{time for pressure to drop (min)} + \text{time for pressure to rise (min)}}$$

or

$$\frac{\text{ft}^3\text{/min} \cdot T_{\text{rise}}}{T_{\text{drop}} + T_{\text{rise}}}$$

5.2.14.2 Compressed-air energy cost savings recommendations

This subclause contains the major energy cost savings recommendations and lists other energy conservation techniques that can result in further savings.

   a) Design and selection of equipment:

      1) Do not purchase or retain oversized compressors.

2) Consider running costs of compressors when purchasing new equipment, especially when average air demand will be much less than the rated air delivery.

A compressor with a 50% greater purchase price than its competitor could end up saving more than its total purchase price over a five-year period when compared with a poor part load performance.

3) Ensure adequate receiver capacity for peak demand periods.

The receiver acts to smooth out the air demand of the factory. The more peaky the demand, the larger the receiver should be. Typically, the receiver should be able to store at least 0.28252 ft^3 (8 L) of volume for every 2.11887 ft^3/min (L/s) free air delivery at full load of 101.5 lbf/in^2 (at 700 kPa). If demand is extremely variable, this volume should be increased three times. Although this may seem large, the cost of a receiver is much less than a compressor it may replace. A receiver can save capital costs by reducing the capacity of the compressor that may have to be installed to meet the peak load. It can also reduce running costs by allowing the compressor to unload or remain switched off at times of low air demand.

4) Select all flexible air hoses and distribution pipework to achieve low air velocity. Do not have them excessively long.

Making sure that the velocity of air within the distribution pipework is low ensures that there are no excessive pressure drops within the system. As a guide, it is recommended that the velocity of the compressed air in the distribution pipework does not exceed 20–32 ft/s (6–10 m/s).

5) Arrange all air distribution pipework to slope down to suitably located draining points, preferably fitted with automatic traps.

A well-drained air distribution system reduces to a minimum the water that is carried over to tools and processes. Excessive water can cause corrosion of tools and damage to products.

6) Ensure distribution pressure loss does not exceed about 7.25 lbf/in^2 (50 kPa) at times of peak use.

By suitably selecting pipework and hoses, selecting appropriate fittings and valves, and keeping filters clean, etc., the pressure loss within the distribution system can be minimized. Minimal pressure drops will allow lower pressure to be established at the compressor, with savings in energy requirements.

7) Provide air dryers that produce the required pressure dew point for the application, which should not be lower than necessary. In the compressed-air system, determine the ambient temperature around the pipework, then determine the required dew point temperature to prevent condensation.

Air dryers should produce the pressure dewpoint required for the application and not lower than necessary. Demands for exceptionally dry air may be met by installing air dryers on the branch line for the special application.

8)   Make sure there is an efficient aftercooler separator to condense and remove as much oil and water as possible. The aftercooler may be built into a packaged system and therefore may not be visible.

The aftercooler condenses and removes moisture from the air, and thus prevents equipment or product damage. The aftercooler also provides an excellent source of recoverable heat.

b)   Operation and maintenance

1)   Repair all air leaks promptly; they waste a lot more expensive air than may seem apparent.

A typical factory may be wasting up to 20% of its compressed-air costs on supplying air leaks.

2)   Install a pressure regulator at each point of air usage and adjust it down to the minimum pressure required for efficient and reliable operation.

3)   After reducing the pressure required at each point of use, adjust the operating pressure of the compressor to the minimum possible above the highest point of use pressure.

A typical system pressure reduction of 14.5 lbf/in^2 (100 kPa) produces about 8% reduction in power requirement at the compressor. A reduction in distribution air pressure also produces a major reduction in leaks.

4)   Switch compressors off when not in use.

Consider automatic time switching of the compressor plant. Compressors operating outside of production hours are probably supplying leaks in the system only. If there is a continuous requirement for air, consider using the smallest compressor available. It may be economical to install a small dedicated compressor for nonproductive hours use only.

5)   Load and unload compressors progressively for multiple compressor installations. Switch off as the load decreases. Arrange sequencing to keep compressors with high no-load requirements fully loaded.

Sequencing control units are available from most compressor manufacturers. These units provide automatic control and achieve the most economic operations. They can be set to allow compressors with poor part load performance to run as full loaded as possible.

6)   Eliminate all misuse of compressed air.

Using compressed air for cooling or continual blowing is wasteful. A fan would be more economical. Blow guns should only be operated at low pressure, 29 lbf/in^2 or less (200 kPa or less); otherwise, they will not only waste energy but could also be dangerous. If blow off must be done with compressed air, then fit a properly designed nozzle.

7)   Utilize wasted heat from compressor for heating water or air. The heated water or air can then be used for space heating or process use.

This heat can be extracted, depending upon the compressor type, from the after-cooler, the intercooler (if fitted to two stage machines), and the oil cooler. Most compressor manufacturers can supply an add-on heat recovery package.

8) Ensure that inlet air to the compressor is drawn from a cool source (but do not extend air intake to excessive length in order to achieve this).

Cooler inlet air is more dense and thus increases the delivered volume of compressed air. Typically, a 5.4 °F (3 °C) reduction in air intake temperature produces a 1% reduction in energy consumption.

9) Clean compressor intake filters regularly.

10) Investigate whether air cylinders with air-actuated return stroke can be operated with a lower return air pressure.

The reduced air pressure leads to a lower air demand and energy saving.

11) Isolate or eliminate redundant distribution pipework.

Redundant pipework sections will almost certainly have air leaks and, therefore, waste air unnecessarily.

12) Set up a regular maintenance program and monitor the trend.

Regular maintenance helps to ensure efficient operation. Records in the form of graphs help to show any deteriorating trends in compressed-air-equipment performance.

5.2.14.3 Water-cooled compressors

A typical heat recovery system transferring heat from the compressor lubricating oil to water obtains water at temperatures up to 176 °F (80 °C). When the hot water demand is reduced, the compressor oil is directed to the oil cooler rather than the plate heat exchanger.

The heated water can be used for the following:

a) Space heating
b) Preheating boiler feedwater
c) Hot water for showers and washrooms
d) Industrial cleaning processes

5.2.14.4 Air-cooled compressors

A typical heat recovery system for transferring heat from the compressor, motor, and oil cooler to heat air obtains air at temperatures up to 122 °F (50 °C).

The heated air can be used for the following:

a) Space heating
b) Drying following painting, varnishing, and washing processes
c) Air curtains

d)   Preheating air for industrial boilers using fossil fuel

To calculate the amount of heat available, multiply the horsepower of the compressor by 42 and get the Btu/min, then multiply by 60 to get the Btu/h. For example, a 50 hp compressor generates $42 \cdot 50 \cdot 60 = 126\,000$ Btu/h. The temperature rise of the air is then determined from the quantity 100 ft^3/min/hp of air passing through the ventilator:

$$126\,000 \text{ Btu/h} \div \left(\frac{100 \text{ ft}^3/\text{min/hp} \cdot 50 \text{ hp} \cdot 60 \text{ min/h}}{14 \text{ ft}^3/\text{lb of air}} \cdot 0.23 \ (°\text{F} \cdot \text{Btu/lb}) \right) = 25.5 \ °\text{F}$$

5.2.15 Refrigeration equipment

5.2.15.1 General

Refrigeration is a process of transferring energy from a low-temperature source to a higher temperature sink, by circulating a refrigerant through an expansion or metering device from a high-pressure side containing a condenser to a low-pressure side containing an evaporator.

5.2.15.2 Efficiency definitions

Efficiency definitions that describe the performance of systems and equipment for refrigeration and air-conditioning are as follows:

a)   *Coefficient of performance (COP).* Describes the ratio of the refrigeration delivered versus the rate of energy input, in consistent units, of a complete operating system under designated conditions:

$$\text{COP} = \frac{\text{kW thermal refrigeration delivered}}{\text{kW electric energy input}}$$

b)   *Energy efficiency ratio (EER).* Describes the ratio of the refrigeration effect (output) in units of Btu/h, to the electrical input in units of watts:

$$\text{EER} = \frac{\text{Btu/h (output)}}{\text{W (input)}}$$

c)   *Seasonal energy efficiency ratio (SEER).* Describes the ratio of the total refrigeration effect over an entire cooling season to the energy consumption in the same period for maintaining the cooling:

$$\text{SEER} = \frac{\text{Btu/h (output for entire season)}}{\text{Wh (input for entire season)}}$$

d)   *Ton of refrigeration:*

1 ton-hour of refrigeration $= 12\,000$ Btu/h $= 3.517$ kWh $= 12.66$ MJ

e)   *kW/ton:* $\dfrac{\text{kilowatts of electrical input to the chiller}}{\text{tons of refrigeration output}}$

5.2.15.3 Types of refrigeration systems

The five types of refrigeration systems are as follows:

a)   Mechanical refrigeration
b)   Absorption refrigeration
c)   Thermoelectric refrigeration
d)   Steam-jet refrigeration
e)   Air-cycle refrigeration

5.2.15.4 Electrically driven mechanical refrigeration

Electrically driven mechanical refrigeration is the term that describes the vapor compression refrigeration cycle of a refrigerant chilling system. In this cycle there are four stages:

a)   Compression
b)   Rejection of heat
c)   Expansion
d)   Absorption of heat

5.2.15.5 Components

The major components for the vapor compression cycle areas follow:

—   Compressor
—   Refrigerant
—   Condenser
—   Expansion valve or metering device
—   Evaporator

Other components for the refrigeration cycle may include

—   Economizer
—   Purge unit

5.2.15.5.1 Compressors

Compressors in the mechanical refrigeration cycle are of two types:

a)   Dynamic compressors—for example, the centrifugal compressor—increases the refrigerant vapor pressure by transferring the axial momentum of the impeller to the vapor followed by a conversion of this vapor to a temperature rise.

b)   Positive displacement compressors, including reciprocating, rotary screw, and trochoidal, increase the refrigerant vapor pressure by reducing the volume of the compressor chamber, thereby transferring work energy.

5.2.15.5.2 Refrigerants

Refrigerants are the heat-transfer fluid. They are selected based upon safety, flammability, toxicity, ozone depletion potential, global warming effect, efficiency, and other factors.

The efficiency of the refrigerants most commonly used range from a COP of 5.02 for R-11 used in centrifugal chiller compressors to 4.67 for R-22 used in reciprocating and screw compressors.

5.2.15.6 Refrigeration system descriptions

(See annex 5D.)

5.2.15.7 System arrangements

Systems can be single or multiple.

5.2.15.7.1 Multiple arrangements

Chillers may be arranged in parallel or in series.

a)   *Parallel arrangements.* For chilled and condenser water circuits connected in parallel, assuming two units of equal size, each will reduce in capacity to about 50% of total system capacity, at which point the lag unit could be shut down by an automatic sequence control.

b)   *Series arrangements.* Since each chiller is cooling 100% of the circulated water quantity through 50% of the required temperature range, the input power per unit of refrigeration decreases since each machine is operating at a lower temperature differential. Refer to the chiller manufacturer's performance data.

5.2.15.7.2 Variable-water volume chillers system

When the parallel arrangement of chillers is installed, a bypass or cross-connection between the chilled water supply and return piping can be used to separate the chilled water system into primary and secondary piping loops that can operate as a variable-volume water system. The primary water flow is relatively constant while the secondary water flow is varied as the flow demand is reduced through two-way valves controlling water through the cooling units.

5.2.15.7.3 Separate chiller pumping

When dual temperature (hot and chilled) water systems are serviced by a single pump the chiller pressure loss (head loss) is continually applied against that pump, even though the chiller may only be used 4–6 months annually in some applications. By installing a chiller bypass connection and a separate smaller chiller pump, the excess pump head is eliminated during the heating months.

5.2.15.8 Energy saving features and equipment for centrifugal chiller and refrigeration systems

5.2.15.8.1 Centrifugal chiller capacity control

Capacity control for part-load performance and shutdown are important features for energy efficiency. Centrifugal compressors may be operated at less than full load by swirling the refrigerant through setting *prerotation vanes* (PRVs) before the impeller. This technique is used where the driver operates at constant speed. Adjustable speed control of the driver, as with an *adjustable speed drive* (ASD) may also be used separately or in combination with PRV control. Table 5-6 shows the power required by each technique.

Table 5-6—Centrifugal compressor part-load performance

System volumetric flow (%)	Power input (%)	
	Speed control	Vane control
111	120	—
100	100	100
80	76	81
60	59	64
42	55	50
20	51	46
0	47	43

Source: 1992 ASHRAE Handbook [B16].

In practice, a microprocessor computes the optimum motor speed as well as the optimum PRV position. Essentially, the motor speed is reduced with the PRVs wide open as the chiller load falls off. At the point where further reduction in load prevents the compressor from developing the necessary pressure for the system, further reduction is achieved through the use of the PRVs.

Multiple chiller units, with one chiller equipped with PRV control and the second chiller equipped with PRV/ASD control, would usually designate the PRV/ASD-equipped unit as the "lead" or "upstream" unit.

Other centrifugal compressor capacity control techniques are as follows:

a) Suction throttling
b) Adjustable diffuser vanes

c)   Movable diffuser walls
d)   Impeller throttling sleeves
e)   Combinations of these with PRV/AFD control

5.2.15.8.2 Multiple-unit sequencing control

This controller automatically cycles chillers to permit the most efficient operation of each chiller during system part load.

5.2.15.8.3 Demand limit control

This controller measures the facility power curve (kW demand) and reduces demand peaks by modulating the chillers through control of the PRV, or PRV/ASD, without switching off cycling motors. This control reduces the integrated demand over the demand interval as metered by the utility.

5.2.15.8.4 Remote leaving chilled water temperature reset control

This controller resets the chilled water temperature according to the actual air-conditioning or process requirement.

5.2.15.8.5 Remote current limit reset control

This control scheme allows the electrical demand to be varied in discrete steps.

5.2.15.8.6 Return chilled water-temperature control

This control provides for a constant return chilled water temperature, allowing the leaving chilled water temperature to rise at part load.

5.2.15.8.7 Centrifugal refrigeration system economizer

Adding an economizer to a centrifugal refrigeration system increases the refrigeration-cycle efficiency. Flash-cooling liquid refrigerant from the condenser in an economizer, before passing the refrigerant to the cooler (evaporator), increases the cooling capacity of the total system and decreases the amount of power consumed.

5.2.15.8.8 Feed-forward control

By monitoring the temperature and humidity in the controlled spaces, special controllers, referred to as *feed-forward controllers*, allow the chiller to maintain a setpoint without "chasing the load" or waiting for the thermal inertia in the chilled water system.

5.2.15.8.9 Dual-centrifugal-compressor chillers

This consists of one chiller package with two centrifugal compressors. A single unique common refrigerant circuit permits either compressor to develop up to 60% of the system full-load design capacity. If the full-load design of a dual-compressor chiller were selected at 0.75 kW per ton of refrigeration, then at 60% system load a single-compressor operation will equal only 0.55 kW per ton of refrigeration.

5.2.15.9 Reciprocating compressors

5.2.15.9.1 Parallel operation

Parallel operation of reciprocating compressors has some advantages at part load.

5.2.15.9.2 Capacity control

Capacity control of reciprocating compressors may be achieved by

a)   Controlling suction temperature

b)   Controlling discharge pressure

c)   Returning discharge gas to suction

d)   Adding re-expansion volume

e)   Changing the stroke

f)   Opening on cylinder discharge port to suction while closing the port to discharge manifold

g)   Changing compressor speed

h)   Closing off cylinder inlet

i)   Varying suction valve opening

5.2.15.10 Large rotating-vane compressors

Capacity control from 100% down to 20% with proportional power reduction can be accomplished by valves in both end covers.

5.2.15.11 Helical rotary compressors

Single-screw and twin-screw compressors may be capacity-controlled by variable compressor displacement (slide valve control). When cooling demand decreases, the slide valve, a moving segment of the compressor rotor housing, causes refrigerant gas to bypass back to suction, without compression. Screw compressor performance at reduced speed is usually significantly different than at the rating point and requires careful analysis.

5.2.15.11.1 Economizers

Screw compressors are available with a secondary suction port, positioned between the primary compressor suction and discharge port, which improves compressor-useful refrigeration and efficiency when used with an economizer.

5.2.15.11.2 Variable-volume ratio

Twin-screw compressors are available with volume ratio control that adjusts to the most efficient ratio for whatever system pressures exist.

5.2.15.12 Incentives

Customer incentive programs and rebate programs for installing or retrofitting more efficient air conditioners, heat pumps, and chillers are offered by many utilities. The staff of the Edison Electric Institute (EEI) can provide information on utilities' programs and their qualifications. It is important to consider these programs' requirements before designing, specifying, stocking, or repairing equipment in order to qualify for payments.

5.2.15.13 Codes

The BOCA® National Energy Conservation Code of 1993 [B45] requires HVAC performance specifications. Designers are required to select equipment and to specify systems that comply with the following:

— Energy efficiency ratios (EERs) for equipment
— Coefficients of performance (COPs) for equipment
— Steady-state efficiencies of equipment
— Air-transport factors for systems
— Water-transport factors for systems
— Controls requirements for systems
— Insulation requirements for equipment and building systems

5.2.16 Heat-recovery systems

Heat-recovery systems are classified according to the following applications:

— Process-to-process
— Process-to-comfort
— Comfort-to-comfort

Waste-heat recovery depends upon these conditions being satisfactory:

— There must be sufficient waste heat of suitable quality to recover.
— There is a positive use for the heat once it has been recovered.

Properly designed waste-heat-recovery devices, incorporated into new plant design or retrofitted, can yield substantial savings on heating and cooling equipment and on operating costs. Sometimes the incorporation of heat recovery is the only means of stretching existing heating and cooling equipment capacities to meet the demands of a new building addition or process.

5.2.16.1 Efficiency of heat recovery

Heat-recovery equipment is described in terms of thermal efficiency or recovery factors. An example, when considering air-to-air systems, is as follows:

Under normal flow conditions, that is with equal mass flow rates on both supply and exhaust, and "clean" operating conditions, the thermal efficiency is defined as

$$\eta = \frac{T_{sl} - T_{se}}{T_{ee} - T_{se}}$$

where

T_{sl} is the supply air leaving temperature
T_{se} is the supply air entering temperature
T_{ee} is the exhaust air entering temperature

5.2.16.2 Indirect evaporative cooling

Pre-cooling of supply air in the summer months in comfort-to-comfort application enables the waste-heat-recovery devices to be used throughout most of the year.

In dry hot weather, the sensible heat-recovery devices may use the technique referred to as *indirect evaporative cooling*. Indirect evaporative cooling reduces the supply air temperature without increasing the moisture content. This is accomplished by spraying a mist of water vapor into the hot exhaust airstream in front of the heat exchanger. The mist deposits on the exhaust side surface and evaporates, thereby lowering the temperature of the plates near the exhaust entrance to approximately the exhaust wet bulb temperature. The outside air entering the heat exchanger is exposed to the cooled plates and can approach approximately 25% of the difference between the outside dry bulb and exhaust side wet bulb temperatures.

5.2.16.3 Equipment

The heat recovery process functions by these techniques:

—   *Parallel flow.* Both fluids flow in approximately the same direction.
—   *Counterflow.* The two fluids flow in opposite directions.
—   *Crossflow.* The two fluids flow in perpendicular directions.

Products incorporating these techniques include the following:

a)   Thermal wheel or rotary regenerative heat exchanger
b)   Run-around-coil
c)   Heat pipe or dual-phase medium heat exchanger
d)   Recuperator
e)   Heat exchangers: flat-plate-type; spiral-plate-type; tube-type

    1)   Air-to-air plate heat exchanger
    2)   Liquid-to-liquid plate heat exchanger
f)   Heat pump

Devices a) through e) are used in applications where the air being used as the heat source is at a sufficiently high temperature to provide energy for the heat sink, usually cold fresh air, without the need to be raised in temperature. Device f), the heat pump, is used to upgrade waste heat to a sufficiently high temperature at which it can be used effectively. It is an alternative to the other devices where the waste heat source is at a temperature too low to be used by direct heat exchange. Several of these heat-recovery devices are described in the following subclauses: 5.2.16.3.1–5.2.16.3.4.

5.2.16.3.1 Thermal wheel

A rotary heat exchanger is usually a packed cylinder that rotates about 20 r/min within an airtight casing that bridges the duct between the exhaust and supply air.

A thermal wheel of the non-hygroscopic type transfers only heat between the two air streams while the hygroscopic type can transfer both heat and moisture between the two air streams. A purge section cleanses the wheel during each revolution to avoid cross-contamination.

The heat-recovery efficiency varies from 65% to 85% depending upon type and design.

5.2.16.3.2 Run-around coil (liquid-coupled indirect heat exchanger)

Finned air-to-water heat exchangers are installed in the ducts between which heat is to be exchanged. A water or water/glycol circuit is used to transfer heat from the warm exhaust air to the cooler incoming supply air. The coil loop recovery units provide heat transfer efficiencies of about 50%. Some additional energy is required for the pump and for the increased fan power requirements resulting in a system efficiency of about 40%.

Advantages of the run-around coil are as follows:

a)   Multiple supply and exhaust systems can be combined into a single loop.
b)   Exhaust air systems and supply air systems may be remote from each other.
c)   There is no possibility of cross-contamination between supply and exhaust air streams.
d)   It can be retrofitted into existing or new systems.

5.2.16.3.3 Heat pipe

The heat pipe, a passive heat exchanger used for gas-to-gas heat recovery, is available in two types:

—   *Horizontal.* A wick within the tube transfers heat by wick action.

—   *Vertical.* Heat from a warm lower duct is transferred to a cold upper duct by phase-change of the refrigerant.

131

Advantages of the heat-pipe recovery units are as follows:

— Sensible heat recovery
— Efficiency of about 50–60%
— No cross-contamination

5.2.16.3.4 Liquid-to-liquid plate heat exchangers

Whenever heat can be usefully recovered from a liquid process stream and used without being upgraded, then the plate heat exchanger should be considered. In the plate heat exchanger, pressurized hot and cold fluids are directed in counterflow thin streams through alternative metal portion usually stainless steel, consisting of gasketed metal plates. Plate heat exchangers are available as

— Single pass
— Multi-pass
— Multi-duty

Advantages of the plate heat exchanger are as follows:

— Heat transfer coefficiency of 80–90%

— Suitability for small spaces

— Totally enclosed unit

— High differential pressures allowed so either side can handle low or high pressures

— Capability of transferring energy between two liquid flows with a temperature difference as small as 2 °F (1 °C)

Applications for plate heat exchangers include the following:

— *Free-cycle cooling, chiller bypass.* Free-cycle cooling allows the chiller to be bypassed whenever the available city or ground water dips below 50–55 °F (10–13 °C). In free-cycle cooling, river, pond water, or cooling-tower water is pumped through a plate heat exchanger on the cold side, cooling clean uncontaminated water in a closed loop on the hot side. Available water is cold enough to allow free-cycle cooling for up to eight months out of the year in many locations in the U.S.

— *Cooling tower isolation.* In order to allow open cooling-tower water to circulate through the chilled water system, it must be chemically treated. The use of a plate heat exchanger and a fresh-water closed-circuit cooling system allows the chemical water treatment to be reduced. Cooling-tower water can be used for free-cycle cooling whenever the temperature for the water drops below 46–48 °F (8–9 °C).

— *Waste-heat recovery*

— *Geothermal heating*

— *District heating or cooling*

— *Solar heating*

— *Water source heat pump*

5.2.16.3.5 Electric industrial heat pumps

An electric heat pump is a refrigeration machine that extracts heat from a heat source, such as air or water, and transfers it to another medium. A nonreversible heat pump (heat only) is used for service water and process heating.

Heat pumps for industrial-process heating recover waste heat, elevate its temperature, and return it for process use. They differ from heat pumps used for space heating and cooling in several ways. They tend to be larger than their commercial counterparts, typically over 200 kW. They run base loaded and accumulate high annual hours of use, normally amounting to more then 5000 kWh for each kilowatt connected.

Several types of heat pumps are described in the following paragraphs. The overwhelming majority of industrial process heat pumps are *closed-cycle* or *open-cycle* systems.

a) *Closed-cycle heat pumps.* Most closed-cycle heat pumps use a water-to-water or cascade-refrigerant-to-water configuration. Closed-cycle heat pumps have coefficients-of-performance of 4 to 6 and provide recovery of otherwise wasted heat.

b) *Open-cycle heat pumps.* Open-cycle heat pumps, also known as mechanical vapor recompression (MVR) systems, use the material being processed directly as the working fluid instead of transferring heat through the medium of a separate refrigerant. The process vapor (e.g., overhead vapor in a distillation column) is delivered to the compressor that raises its temperature and pressure. The heated vapor is delivered back to the process. Source water is flash-evaporated, and the resulting vapor is compressed to required steam pressures.

   Open-cycle systems are being applied in industries where large quantities of water vapor are produced in evaporation and drying, such as in drying timber and other wood products and in food industries (milk products, sugar, distilled spirits, grain drying, and starch production).

c) *Heat-recovery heat pumps.* Heat-recovery heat pumps present another opportunity to save energy. Combined with conventional chillers, boilers, and thermal storage, these heat pumps are able to meet the heating, cooling, and water heating needs of most large commercial buildings.

d) *Double-bundle heat-recovery chillers with storage.* In this system, a storage tank is added to store heat during occupied hours by raising the temperature of water in the tank during unoccupied hours. Water from the tank is gradually fed to the evaporator providing load for the compressor and condenser that heat the building during off hours.

e) *Double-bundle, heat-reclaim heat-recovery chillers.* There are two models of double-bundle heat-reclaim chillers:

1) Standard double-bundle heat-recovery chiller operates similar to a conventional chiller except there are two circuits in the condenser to reject heat. One of these is used for space or service water heating. The other circuit permits conventional heat rejection, as through a cooling tower. Thus with a leaving hot water temperature of 105 °F, the condensing water temperature will be about 115 °F. All the heat rejection will be at this temperature—about 115 °F—even when only a small amount of heat is required by the system. A typical double-bundle heat-reclaim system's ability to operate at low loads in heat-recovery mode is minimized. Since there is only one compressor for year-round use, the compressor impeller must produce the maximum head required. Since the heat-recovery mode requires higher condensing pressures and temperatures than the cooling-only mode, the impellers of standard centrifugal chillers adapted for heat reclaim become energy wasters during the cooling-only season, and power consumption can increase by 5–15%.

2) A heat-reclaim chiller with booster performance operates similarly to a conventional chiller except for its separate heat-rejection condenser and a separate heat-recovery condenser. This allows separate refrigerant circuits and two different condensing temperature levels. The heat-rejection condenser can operate at 75 °F, while the heat-recovery condenser operates at 115 °F to supply 105 °F leaving hot water, for example. Only the reclaimed energy is raised to the higher temperature. This equipment is capable of supplying hot water temperatures up to 120 °F. Main compressor impeller is selected for optimum performance during the cooling mode, as for a conventional chiller; however, in this equipment there is also a booster compressor to produce the higher temperature needed in the heat-recovery condenser. Overall system performance is more energy-efficient compared to standard double-bundle heat-recovery chillers.

f) *Multistage heat-transfer (cascade) systems.* Cascade systems typically combine an electric chiller and a water-to-water heat pump. The chiller's condenser becomes the heat pump's heat source. This permits the heat pump's condenser water to be delivered to the heating system or heating coils at temperatures of 130–140 °F (or warmer) whenever there is a cooling load.

The heat pump and chiller operate independently. The chiller is controlled by the cooling load and operates at the lowest cooling-tower water temperatures and best efficiency that ambient conditions allow. The heat pump is controlled by the heating load and consumes only the power necessary for heating.

g) *Multiple-chiller cascade systems.* In larger systems, two chillers can be arranged in series and the condensers are also piped in series. In such systems the heat pump capacity is often one-third the chillers' refrigeration rating.

5.2.17 Laser processing

Lasers used in industrial production are of two types:

a) Solid-state devices that deliver energy in pulses
b) Gaseous lasers that are used for continuous power

Laser-light power-source intensity can range from a few watts to more than 25 kW. The efficiency of solid-state lasers varies from 0.5–1.5% while the efficiency of gaseous lasers varies from 5–10%.

Laser-processing applications are cutting, drilling, welding, surface treatment, and scribing. Laser processing may provide faster operation and for some applications may be the only technique available since the laser can start and stop anywhere and is extremely precise.

Development on improving the laser efficiency and power is continuing.

5.2.18 Electron beam heating

Electron beam (EB) heating employs a directed and focused beam of electrons to heat materials for welding and heat treatment. Welding units range from 7.5–60 kW capability and operate at about 85% efficiency. EB welding speed is about five to fifteen times faster than *metal inert gas* (MIG) and *tungsten inert gas* (TIG) systems. EB heat treating allows control of the power density and energy input into the material for concentration in the heat-affected zone, thereby saving energy over other systems.

5.2.19 Electroslag processing

Electroslag processing is used to make high-purity metal shapes for high-alloy and specialty steels. When used to produce ingots for further fabrication the process is referred to as *electroslag remelting* (ESR), and when used for finished products the process is referred to as *electroslag casting* (ESC).

The metal to be processed is in the form of an electrode. It is lowered into a synthetic refractory slag contained within a water-cooled mold. DC is then passed from this electrode through the slag pool to the base plate. As the heat melts the electrode molten droplets fall into the slag pool where impurities are chemically removed. The pool of molten metal is then solidified under controlled conditions to produce a highly refined grain structure.

Power for ESR furnaces may be 50% more than for *vacuum arc remelting* (VAR); however, ESR permits higher production rates resulting in lower per-unit energy consumption. ESC is a simple-step process that uses less energy than competitive processes for finished products. Efficiency improvements in the conversion of ac to dc and in the rate of furnace heat losses are possible areas for reducing energy consumption.

5.2.20 Electrical discharge machining and electrochemical machining

Electrical discharge machining (EDM) and *electrochemical machining* (ECM) are metal-removal processes primarily used for the specialized, precision, low-volume production of complex shapes.

EDM uses an electric arc to remove metal from the workpiece. The electric arc is a medium-voltage pulsating dc. The workpiece and tool are immersed in a dielectric fluid, such as transformer-grade mineral oil, and the tool is moved by a servomotor over the workpiece. EDM units range up to 10 kVA and removal rates, up to 25 in³/h. EDM uses low-voltage non-pulsating dc. An electrolyte solution is pumped under high-pressure between the tool and the workpiece. ECM units up to 150 kVA are used for removal rates to 150 in³/h.

Efficiency improvements in the conversion of ac to dc and in the control of the process are possible areas for reducing energy consumption.

5.2.21 Plasma processing

Plasma processing uses plasma torches from 15 kW to over 6 MW that operate up to 10 000 °F. Plasma processing is used in the direct reduction of iron ore for the production of sponge iron and for other applications such as the destruction of toxic wastes, recovery of metals, and replacement of fossil fuel combustion in ore kilns.

Two types of plasma torches are used: *transferred arc* and *nontransferred arc*. In the transferred arc process, the material being heated is part of the electrical circuit. In the nontransferred arc process, both the positive and negative attachment points are within the torch itself. Only the plasma flame emerges from the nontransferred arc torch.

Since plasma torches can replace processes performed using conventional combustion without the environmental emission problems and dependency of fossil fuels, production efficiency and energy conservation result when this process is selected. The plasma torch operates using almost a purely resistive load and maintains a high power factor over the cycle.

5.3 Electric motors

5.3.1 General

This subclause provides basic data, suggestions, and techniques for the designer to learn and implement, which should result in more useful mechanical work from less electrical energy.

Electrical energy is converted to mechanical energy by electrical motors. The chart depicts this conversion:

Electrical energy	→	Electromechanical conversion prime movers (synchronous, induction, dc, linear induction motors, solenoids)	→	Power coupling (clutch, if used)	→	Power transmission chain (gears, belts, shafts, drives, if used)	→	Mechanical device (shaping, forming, transporting)

Mechanical energy = product of electrical energy × efficiencies

U.S. motor sales by type and horsepower are shown in table 5-7. Observe the large quantity of dc motors.

Table 5-7—Total motor sales, 1991 (thousands of units)

Motor type	Motors sold		
	Fractional hp	1–5 hp	Over 5 hp
Fractional horsepower	145 825	n/a	n/a
DC	n/a	1672	830
Single-phase ac	n/a	1309	655
Synchronous ac	n/a	0	5
Polyphase ac	n/a	1240	703
Total	145 825	4221	2193

Source: EPRI RP-2918-15 [B97].

The breakdown of the commercial and industrial motor drive consumption and predicted growth by application is shown in tables 5-8 and 5-9.

Table 5-8—Commercial base-case projections (10^9 kWh): Baseline projection before utility demand side management (DSM)

Sector	Electricity sales (10^9 kWh)		
	1990	2000	2010
Space heating	76.35	118.33	152.05
Cooling	182.65	224.03	257.13
Vent	84.42	101.93	116.86
Water heating	27.09	36.51	44.22
Cooking	24.87	29.54	33.45
Refrigeration	65.84	74.22	84.44
Outside lighting	34.05	40.16	47.15
Inside lighting	271.47	296.05	325.37
Office equipment	26.46	42.54	51.96
Miscellaneous	40.92	54.11	64.13
Total	834.13	1017.41	1176.76

Table 5-9—Industrial base-case projections (10^9 kWh):
Baseline projection before utility demand side management (DSM)

Sector	Electricity sales (10^9 kWh)		
	1990	2000	2010
Motors:	672.69	794.86	946.78
Pumps, fans, comps.	315.82	371.20	437.20
Materials handling	172.86	204.99	245.61
Materials processing	184.01	218.67	263.96
Melting	23.94	30.71	41.37
Process heating	54.71	89.23	148.74
Drying and curing	7.31	9.97	14.78
Electrolytics	91.26	104.44	121.44
Lighting	63.48	73.31	84.92
Miscellaneous	59.30	71.56	86.46
Total	972.68	1174.09	1444.50

Electric motors in the U.S. are manufactured to NEMA MG 1-1993 [B164] specifications and are listed for safety by UL 1004-1994 [B224] or the UL standard for electric motors by application. Efficiency of electric motors is determined by testing in accordance with the following IEEE standards:

— IEEE Std 112-1991 [B122]—This standard includes test procedures for integral and fractional horsepower motors.

— IEEE Std 113-1985 [B123]

— IEEE Std 114-1982 [B124]

— IEEE Std 115-1995 [B125]

5.3.2 DC motors

DC motors are used in some industrial applications when adjustable speed operation is required and in building elevator systems where traction drives are specified.

A dc motor requires a source of dc power to excite the stationary field winding and to supply power to the rotating armature winding through the brushes and commutator.

DC motors provide quick and efficient stopping through dynamic or regenerative braking. The speed of a dc motor can be smoothly controlled down to zero revolutions per minute and then immediately accelerated in the opposite direction.

5.3.2.1 DC motor types

The five dc motor types are as follows:

a)   Shunt-wound
b)   Compound-wound
c)   Series-wound
d)   Permanent-magnet (available to 5 hp)
e)   Brushless dc motors (0.5–10 hp)

5.3.2.2 Power supplies for dc motors

NEMA has classified power supplies for dc motors by an alphabetic order of increasing magnitude of ripple current. A dc motor may be operated on a power supply having a letter designation occurring prior in the alphabet to the letter stamped on the nameplate. Operating a dc motor on a power supply having a letter designation occurring later in the alphabet than the letter stamped on the nameplate requires derating the motor due to the increased losses. The power supply codes, specified in 12.65 of NEMA MG 1-1993 [B164], are shown in table 5-10.

Table 5-10—NEMA power supply codes

Description	Code
1. DC generator, battery or 12-pulse/cycle, 6-phase, full control	A
2. Six pulse/cycle, three-phase, full control, 230 V or 460 V input to rectifier	C
3. Three pulse/cycle, three-phase, semibridge, 1/2 control, 230 V or 460 V, 60 Hz input to rectifier	D
4. Three pulse/cycle, three-phase, half-wave, 460 V, 60 Hz input to rectifier	E
5. Two pulse/cycle, single-phase, full-wave, 230 V, 60 Hz input to rectifier	K

Source: NEMA MG 1-1993 [B164].

DC motor applications for energy efficiency are as follows:

a)   Select the motor size to closely match the load. An oversized motor runs inefficiently; the brush current density is low and it does not commutate and film well. Commutation problems may result from contamination or low humidity.

b)   Select an efficient power supply, preferably NEMA Class C.

5.3.3 AC motors

AC motors are manufactured as single-phase and polyphase.

Single-phase motors are generally used in applications where the power supply is single-phase and the power requirement is less than 1 hp. Where a three-phase power supply is available, motors larger than 1 hp are normally three-phase, depending upon the economics of the installation cost for the motor and controller.

To properly select and apply ac motors an understanding of the efficiency definitions is essential:

a) *Efficiency* $= \dfrac{\text{output}}{\text{input}} = \dfrac{\text{output}}{\text{output} + \text{losses}}$

   The motor output is the listed hp or kW stamped on the rating nameplate.

b) *Nominal efficiency* is a number assigned to a range of efficiencies into which a particular motor's efficiency falls. As a result, the efficiency rating stamped on the nameplate may overstate or understate the motor's actual efficiency. Nominal efficiency represents the "average efficiency" for a large population of motors of the same design. Individual motors can vary widely from the average. The efficiency of individual motors can vary widely from the average as a result of normal variations in material, manufacturing, and test procedures. Therefore, average expected and nominal values should not be accepted as guaranteed values.

c) *Minimum efficiency* is a statistical efficiency based upon the normal frequency distribution curve of motor efficiency test results.

   The minimum efficiency of a motor can be calculated from the nominal efficiency as follows:

$$\text{EFF}_{min} = \cfrac{1}{\cfrac{1.2}{\text{EFF}_{nom}} - 0.20} = \text{EFF}_{nom} \cdot \cfrac{1}{1.2 - 0.20 \cdot \text{EFF}_{nom}}$$

d) *Guaranteed efficiency* is the value set by the motor manufacturer for the given motor rating.

e) *Calculated efficiency* is the engineering design value.

f) *Full-load or 4/4 load efficiency* is not specific for evaluation purposes.

g) *Apparent efficiency* is the product of the motor power factor and efficiency. Since the power factor could be high and the efficiency low, or the reverse, this term is not specific for evaluation purposes.

NEMA specifies efficiency levels for polyphase squirrel-cage induction motors to be classified as *energy-efficient*. The nominal full-load efficiency as determined in accordance with MG 1-1993 [B164], 12.58.1 (Method B of IEEE Std 112-1991), and identified on the nameplate and shall equal or exceed the values shown in tables 5-11 and 5-12 for the motor to be classified as *energy-efficient*, Design E.

Table 5-11—Full-load efficiencies of energy-efficient motors

hp	2-pole		4-pole		6-pole		8-pole	
	Nominal efficiency	Minimum efficiency	Nominal efficiency	Minimum efficiency	Nominal efficiency	Minimum efficiency	Nominal efficiency	Minimum efficiency
Open motors								
1.0	—	—	82.5	80.0	80.0	77.0	74.0	70.0
1.5	82.5	80.0	84.0	81.5	84.0	81.5	75.5	72.0
2.0	84.0	81.5	84.0	81.5	85.5	82.5	85.5	82.5
3.0	84.0	81.5	86.5	84.0	86.5	84.0	86.5	84.0
5.0	85.5	82.5	87.5	85.5	87.5	85.5	87.5	85.5
7.5	87.5	85.5	88.5	86.5	88.5	86.5	88.5	86.5
10.0	88.5	86.5	89.5	87.5	90.2	88.5	89.5	87.5
15.0	89.5	87.5	91.0	89.5	90.2	88.5	89.5	87.5
20.0	90.2	88.5	91.0	89.5	91.0	89.5	90.2	88.5
25.0	91.0	89.5	91.7	90.2	91.7	90.2	90.2	88.5
30.0	91.0	89.5	92.4	91.0	92.4	91.0	91.0	89.5
40.0	91.7	90.2	93.0	91.7	93.0	91.7	91.0	89.5
50.0	92.4	91.0	93.0	91.7	93.0	91.7	91.7	90.2
60.0	93.0	91.7	93.6	92.4	93.6	92.4	92.4	91.0
75.0	93.0	91.7	94.1	93.0	93.6	92.4	93.6	92.4
100.0	93.0	91.7	94.1	93.0	94.1	93.0	93.6	92.4
125.0	93.6	92.4	94.5	93.6	94.1	93.0	93.6	92.4
150.0	93.6	92.4	95.0	94.1	94.5	93.6	93.6	92.4
200.0	94.5	93.6	95.0	94.1	94.5	93.6	93.6	92.4
250.0	94.5	93.6	95.4	94.5	95.4	94.5	94.5	93.6
300.0	95.0	94.1	95.4	94.5	95.4	94.5	—	—
350.0	95.0	94.1	95.4	94.5	95.4	94.5	—	—
400.0	95.4	94.5	95.4	94.5	—	—	—	—
450.0	95.8	95.0	95.8	95.0	—	—	—	—
500.0	95.8	95.0	95.8	95.0	—	—	—	—
Enclosed motors								
1.0	75.5	72.0	82.5	80.0	80.0	77.0	74.0	70.0
1.5	82.5	80.0	84.0	81.5	85.5	82.5	77.0	74.0
2.0	84.0	81.5	84.0	81.5	86.5	84.0	82.5	80.0
3.0	85.5	82.5	87.5	85.5	87.5	85.5	84.0	81.5
5.0	87.5	85.5	87.5	85.5	87.5	85.5	85.5	82.5

Table 5-11—Full-load efficiencies of energy-efficient motors (*Continued*)

hp	2-pole		4-pole		6-pole		8-pole	
	Nominal efficiency	Minimum efficiency	Nominal efficiency	Minimum efficiency	Nominal efficiency	Minimum efficiency	Nominal efficiency	Minimum efficiency
Enclosed motors								
7.5	88.5	86.5	89.5	87.5	89.5	87.5	85.5	82.5
10.0	89.5	87.5	89.5	87.5	89.5	87.5	88.5	86.5
15.0	90.2	88.5	91.0	89.5	90.2	88.5	88.5	86.5
20.0	90.2	88.5	91.0	89.5	90.2	88.5	89.5	87.5
25.0	91.0	89.5	92.4	91.0	91.7	90.2	89.5	87.5
30.0	91.0	89.5	92.4	91.0	91.7	90.2	91.0	89.5
40.0	91.7	90.2	93.0	91.7	93.0	91.7	91.0	89.5
50.0	92.4	91.0	93.0	91.7	93.0	91.7	91.7	90.2
60.0	93.0	91.7	93.6	92.4	93.6	92.4	91.7	90.2
75.0	93.0	91.7	94.1	93.0	93.6	92.4	93.0	91.7
100.0	93.6	92.4	94.5	93.6	94.1	93.0	93.0	91.7
125.0	94.5	93.6	94.5	93.6	94.1	93.0	93.6	92.4
150.0	94.5	93.6	95.0	94.1	95.0	94.1	93.6	92.4
200.0	95.0	94.1	95.0	94.1	95.0	94.1	94.1	93.0
250.0	95.4	94.5	95.0	94.1	95.0	94.1	94.5	93.6
300.0	95.4	94.5	95.4	94.5	95.0	—	—	—
350.0	95.4	94.5	95.4	94.5	95.0	—	—	—
400.0	95.4	94.5	95.4	94.5	—	—	—	—
450.0	95.4	94.5	95.4	94.5	—	—	—	—
500.0	95.4	94.5	95.8	95.0	—	—	—	—

Source: NEMA MG 1-1993 [B164]

Table 5-12—Full-load efficiencies of Design E motors

hp	2-pole		4-pole		6-pole		8-pole	
	Nominal efficiency	Minimum efficiency	Nominal efficiency	Minimum efficiency	Nominal efficiency	Minimum efficiency	Nominal efficiency	Minimum efficiency
Open motors								
0.75	—	—	—	—	82.5	80.0	—	—
1.0	—	—	86.5	84.0	84.0	81.5	78.5	75.5
1.5	86.5	84.0	87.5	85.5	87.5	85.5	81.5	78.5
2.0	87.5	85.5	87.5	85.5	88.5	86.5	88.5	86.5
3.0	87.5	85.5	89.5	87.5	89.5	87.5	89.5	87.5

Table 5-12—Full-load efficiencies of Design E motors (*Continued*)

hp	2-pole		4-pole		6-pole		8-pole	
	Nominal efficiency	Minimum efficiency	Nominal efficiency	Minimum efficiency	Nominal efficiency	Minimum efficiency	Nominal efficiency	Minimum efficiency
Open motors								
5.0	88.5	86.5	90.2	88.5	90.2	88.5	90.2	88.5
7.5	90.2	88.5	91.0	89.5	91.0	89.5	91.0	89.5
10.0	91.0	89.5	91.7	90.2	92.4	91.0	91.7	90.2
15.0	91.0	89.5	92.4	91.0	92.4	91.0	91.7	90.2
20.0	92.4	91.0	93.0	91.7	92.4	91.0	92.4	91.0
25.0	93.0	91.7	93.6	92.4	93.6	92.4	92.4	91.0
30.0	93.0	91.7	94.1	93.0	93.6	92.4	93.0	91.7
40.0	93.6	92.4	94.5	93.6	94.5	93.6	93.0	91.7
50.0	93.6	92.4	95.4	94.5	94.5	93.6	93.6	92.4
60.0	93.6	92.4	95.4	94.5	95.0	94.1	94.1	93.0
75.0	94.5	93.6	95.4	94.5	95.4	94.5	95.0	94.1
100.0	95.0	94.1	95.4	94.5	95.4	94.5	95.0	94.1
125.0	95.4	94.5	95.4	94.5	95.4	94.5	95.0	94.1
150.0	95.8	95.0	95.8	95.0	95.8	95.0	95.4	94.5
200.0	95.8	95.0	95.8	95.0	95.8	95.0	95.4	94.5
250.0	95.8	95.0	96.2	95.4	96.2	95.4	95.8	95.0
300.0	96.2	95.4	96.2	95.4	96.2	95.4	—	—
350.0	96.2	95.4	96.5	95.8	96.5	95.8	—	—
400.0	96.5	95.8	96.5	95.8	—	—	—	—
450.0	96.5	95.8	96.8	96.2	—	—	—	—
500.0	96.8	96.2	96.8	96.2	—	—	—	—
Enclosed motors								
0.75	—	—	—	—	82.5	80.0	—	—
1.0	—	—	86.5	84.0	84.0	81.5	78.5	75.5
1.5	86.5	84.0	87.5	85.5	86.5	84.0	82.5	80.0
2.0	87.5	85.5	87.5	85.5	88.5	86.5	86.5	84.0
3.0	88.5	86.5	88.5	86.5	89.5	87.5	88.5	86.5
5.0	90.2	88.5	89.5	87.5	90.2	88.5	89.5	87.5
7.5	91.0	89.5	90.2	88.5	91.7	90.2	89.5	87.5
10.0	91.0	89.5	91.0	89.5	92.4	91.0	91.0	89.5
15.0	91.7	90.2	92.4	91.0	92.4	91.0	91.0	89.5
20.0	92.4	91.0	93.0	91.7	92.4	91.0	91.7	90.2
25.0	93.0	91.7	93.6	92.4	93.6	92.4	91.7	90.2
30.0	93.6	92.4	94.1	93.0	93.6	92.4	92.4	91.0
40.0	94.1	93.0	94.5	93.6	94.1	93.0	92.4	91.0
50.0	94.5	93.6	95.0	94.1	94.1	93.0	93.0	91.7
60.0	94.5	93.6	95.0	94.1	95.0	94.1	93.0	91.7

Table 5-12—Full-load efficiencies of Design E motors (*Continued*)

hp	2-pole		4-pole		6-pole		8-pole	
	Nominal efficiency	Minimum efficiency	Nominal efficiency	Minimum efficiency	Nominal efficiency	Minimum efficiency	Nominal efficiency	Minimum efficiency
Enclosed motors								
75.0	94.5	93.6	95.4	94.5	95.0	94.1	93.6	92.4
100.0	95.0	94.1	95.4	94.5	95.4	94.5	93.6	92.4
125.0	95.4	94.5	95.8	95.0	95.4	94.5	94.5	93.6
150.0	95.8	95.0	95.8	95.0	95.8	95.0	94.5	93.6
200.0	96.2	95.4	96.2	95.4	96.2	95.4	95.0	94.1
250.0	96.2	95.4	96.2	95.4	96.2	95.4	95.4	94.5
300.0	96.2	95.4	96.2	95.4	96.2	95.4	—	—
350.0	96.5	95.8	96.5	95.8	96.5	95.8	—	—
400.0	96.5	95.8	96.5	95.8	—	—	—	—
450.0	96.5	95.8	96.8	96.2	—	—	—	—
500.0	96.8	96.2	96.8	96.2	—	—	—	—

Source: NEMA MG 1-1993 [B164]

The minimum level of nominal full-load efficiency for more efficient Design E motors is shown in tables 5-11 and 5-12. This is the only NEMA design type with a minimum level of efficiency specified. Hence, this design type may not always be the most efficient motor for a specific application.

5.3.4 Single-phase motors

Single-phase motors in sizes 1/20 through 10 hp are commonly manufactured in six types:

a) Shaded-pole
b) Split-phase
c) Capacitor-start induction-run
d) Permanent-split capacitor
e) Capacitor-start capacitor-run
f) Universal

Refer to annex 5E for a summary of single-phase-motor efficiency characteristics.

5.3.5 Polyphase motors

Polyphase motors are manufactured as induction and synchronous motors.

Induction motors operate asynchronously due to slip. The synchronous speed of an induction motor, N_s, is fixed by the number of poles, p, and the supply frequency, f:

$$N_s = \frac{120f}{p}$$

Squirrel-cage induction motors, the workhorses of industry, are the standard induction motor.

Wound-rotor induction motors are constructed using a wound rotor accessible through slip rings. Wound-rotor motors are generally started with resistance in the rotor circuit. This resistance is sequentially reduced to permit the motor to come up to speed. Reduced speed is available down to about 50%, but the efficiency is low.

Synchronous motors are manufactured in two major types: nonexcited and dc excited. Nonexcited synchronous motors, ranging to 30 hp, employ a self-starting circuit and require no external excitation supply. DC-excited synchronous motors require dc supplied through slip rings for excitation.

The high-horsepower, industrial-sized synchronous motors are very efficient. Synchronous motors can operate at leading or unity power factor and thereby can provide power factor correction.

Squirrel-cage induction motors are divided into torque classifications by the specifications in NEMA MG 1-1993 [B164]: Motor performance according to torque classifications is described as follows (NEMA MG 10-1994 [B166]):

— *Design A*. These motors are usually designed for a specific use. They have a higher starting current and breakdown torque than Design B motors and have a slip of 5% or less.

— *Design B*. These are the general purpose motors. Slip is 5% or less.

— *Design C*. These motors have high starting torque with normal starting current and low slip. Slip is 5% or less.

— *Design D*. These motors have high slip, high starting torque, low starting current, and low full-load speed. Slip is about 5–8%.

— *Design E*. These motors are general purpose motors having speed-torque-current performance equivalent to those of international standards with a minimum level established for the nominal efficiency. The starting current can be higher than that for Design B motors. Slip is 3% or less.

The characteristics and application of polyphase induction motors are given in annex 5F.

5.3.5.1 Adjustable speed drives (ASDs)

Induction motors, either general or definite purpose, may be suitable for operation on inverter drives to achieve variation in the output delivered from the driven device by speed control rather than by other means of mechanical control of the flow. When the application requires variable or constant torque from low to base speed, or beyond, then an inverter drive may be appropriate.

5.3.5.2 Vector drive motors

Vector drive motors are motors suitable for operation on vector drives. When the application requires full torque from zero to base speed then a vector drive motor equipped with an encoder for feedback to the drive controller may be appropriate. Vector drives offer independent control of torque and speed.

5.3.5.3 Multispeed motors

Multispeed motors are available to operate at two, three, or four specific base speeds. This is accomplished by designing one or more sets of windings within one magnetic structure. Efficiency characteristics of multispeed motors may be lower than in the equivalent constant speed motor.

The multispeed motor classifications and typical applications are as follows:

— "Variable-torque" variable-horsepower: fans, pumps, blowers

— "Constant-horsepower" variable-torque: conveyors

— "Constant-torque" variable-horsepower: machine tools, reciprocating pumps, reciprocating air compressors

Multispeed motors operate at the following fixed speeds:

— Single-winding, two-speed: 900 and 1800 r/min
— Two-winding, three-speed: 900, 1200, and 1800 r/min
— Two-winding, four-speed: 600, 900, 1200, and 1800 r/min
— Two-winding, two-speed: 1800/1200, 1800/900, and 1800/600 r/min

5.3.5.4 Applicable standards

ANSI/ASHRAE/IESNA 90.1-1989 requires minimum motor efficiencies for all motors 1 hp and above operating 500 hours or more per year.

The ANSI/ASHRAE/IESNA 100.3-1985 [B18] recommends replacing motors for higher efficiency motors.

5.3.5.4.1 Incentives

Customer incentive programs and rebate programs for installing or retrofitting energy-efficient motors and ASDs are offered by many utilities. The staff of the Edison Electric Institute (EEI) can provide information on utilities' programs and their qualifications. It is important to consider these programs' requirements before designing, specifying, stocking, or repairing equipment in order to qualify for payments.

5.3.5.5 Duty-cycle applications

There are three duty cycles for sizing motors. "Duty cycle" refers to the energization/de-energization and load variations with respect to time for any application, as follows:

Duty cycle	Description
1	*Continuous duty.* This is a requirement of service that demands operation at an essentially constant load for an indefinite time. To size the motor, select proper horsepower based upon the continuous load.
2	*Intermittent duty.* This is a requirement of service that demands operation for alternate intervals of load and no-load; or load and rest; or load, no-load, and rest; each interval of which is definitely specified. Select a motor for these applications to match the horsepower requirements under the maximum loaded condition.
3	*Varying duty.* This is a requirement of service that demands operation at loads and at intervals of time that may be subject to wide variation. It is necessary to determine the peak horsepower required and a calculation of the rms horsepower to indicate the proper motor rating from a heating aspect.

An example of motor sizing for varying duty cycle is given in the following table. Column (4) is derived by squaring the horsepower and multiplying by the time for each part of the cycle.

(1) Part of cycle	(2) Time per part cycle, t, in seconds	(3) Horsepower required	(4) hp^2t
1	15	32	15 360
2	40	74	219 040
3	30	27	21 870
4	5	32	5 120
5	148	66	644 690
6	200	27	145 800
7	12	32	12 290
8	70	27	51 030

Using the above given and derived data,

Sum of the periods of time, column (2) = 520

Sum of the hp^2t, column (4) = 1 115 200

147

Therefore rms hp $= \left[\dfrac{1\ 115\ 200}{520} \right]^{1/2} = 46.3$ hp

The rms hp determines the motor thermal capability at constant speed to allow for plus or minus 10% voltage variation and the resulting additional motor heating. Particularly at peak loads at 90% voltage, a 10% allowance is added to the rms hp calculation:

required hp = rms hp · 1.10

required rms hp = 46.3 · 1.1 = 50.9 hp

The motor-usable horsepower is determined by nameplate hp · service factor, sf, and must be equal to or greater than the required hp. In this example, there is a choice of either a

    a)   50 hp motor with a 1.15 sf, or
    b)   60 hp motor with a 1.0 sf.

The motor must also be capable of carrying the peak hp (torque) value from the duty cycle at 90% voltage.

Since *motor breakdown torque* (BDT) is reduced by the voltage squared, the required BDT is determined from

$$\% \, BDT = \left[\frac{peak \ loak \ hp \cdot 100}{nameplate \ hp \cdot 0.9^2} \right] + 20$$

where "20" is a 20% margin added to prevent inefficiencies of operation too close to the actual breakdown torque.

Then, for the two motors:

50 hp motor with a 1.15 sf: $\% \, BDT \left[\dfrac{74}{50} \cdot 123 \right] + 20 = 202\%$

60 hp motor with a 1.0 sf: $\% \, BDT = \left[\dfrac{74}{60} \cdot 123 \right] + 20 = 172\%$

Since the breakdown torque for either a 50 hp or a 60 hp motor is 200% (NEMA MG 1-1993 [B164]), then the best choice is the 60 hp rating since only 172% of the torque is required at the peak-load horsepower.

5.3.6 Speed control and duty cycling

Energy can be saved by applying ASDs where constant-speed motors were previously used. Furthermore, older inefficient ASDs can be replaced with later-generation, more efficient drives. In some replacements, this change could eliminate a throttling valve on a pump or vanes on a fan inlet. Hence, greatly improved efficiency would be seen.

Available mechanical speed control equipment includes

— Gear motors: helical; worm gear; planetary gear
— Hydroviscous-type couplings
— Variable-speed belt drive
— Dry-fluid drives and couplings
— Wet-traction drives
— Silicone-fluid drives
— Ball variator

Available electrical speed control equipment includes

— Variable voltage input (VVI); chopper variable voltage input (CVVI)

— Current-source input (CSI)

— Pulse-width modulation (PWM)

— Digital vector

— Analog vector

— Brushless

— Two-speed pumping system with two electric motors coupled to the same pump, large horsepower motor at 1800 r/min and small horsepower motor at 1200 r/min

— Two-speed fan drive with two electric motors belted to the same fan

— Permanent magnet hysteresis drives

— Converter-fed variable-speed dc drive combining dc motor and thyristor converter in one unit

— Wound-rotor-motor variable-speed controller with slip power recovery

— Switched reluctance drive

— Eddy-current drive

5.3.7 Evaluating motor and drive decisions

5.3.7.1 Energy-efficient motor/drive simple payback analysis

The simple payback method gives the number of years required to recover the differential investment for energy-efficient motors. To determine the payback period, the premium for the energy-efficient motor is divided by the annual savings. First the savings must be determined using the following formula:

$$S = 0.746 \cdot \text{hp} \cdot L \cdot C \cdot N \left[\frac{100}{E_A} - \frac{100}{E_A} \right]$$

where

S	is the annual dollar savings
hp	is the horsepower
L	is the percentage load divided by 100
C	is the energy cost in dollars per kilowatthour
N	is the annual hours of operation
E_A	is the percent efficiency for standard motor
E_B	is the percent efficiency for energy-efficient motor

The efficiencies for the percent load at which the motor will be operated must be used, since motor efficiency varies with load.

Then the motor cost premium is divided by the annual savings to compute the payback period for the energy-efficient motor.

The following example makes a comparison between a 100 hp, 1800 r/min, standard TEFC motor with a cost of $2825 and a similar energy-efficient TEFC motor with a cost of $4529. Standard motor efficiency is 92.4% at full-load. Energy-efficient motor efficiency is 95.8% at full-load. The motor will operate at full-load 20 h/day, 300 days/year. Energy cost is $0.06/kWh.

$$S = 0.746 \text{ hp/kW} \cdot 100 \text{ hp} \cdot 1.00 \cdot \$0.06/\text{kWh} \cdot 20 \text{ h/day} \cdot 300 \text{ day/yr} \left[\frac{100}{92.4} - \frac{100}{95.8} \right]$$

$$S = \$1031.53 \text{ annual energy savings}$$

$$\text{Simple payback} = \frac{\$4529 - \$2825}{\$1031.53/\text{yr}} = 1.65 \text{ years payback}$$

5.3.7.2 Energy-efficient motor/drive present-worth life cycle analysis

For greater precision, the present-worth method of life cycle savings may be used. Refer to Chapter 3 for details.

5.3.7.3 Load cycling of motors

Applications where the electric motor is required to run at load followed by a period where no useful work is done by the driven machine should be reviewed to determine whether energy may be saved by switching off the motor and restarting it for the next load cycle.

Following is an example for a NEMA Design B motor from NEMA MG 10-1994 [B166]:

a) Assumptions:
    1) Motor rating: 10 hp @ 1800 r/min
    2) Motor full load efficiency: 89.2%
    3) Motor output with the drive machine running idle: 2.5 hp
    4) Motor efficiency at 2.5 hp = 86%
    5) Motor acceleration loss + system stored energy: 3750 watt seconds (W·s)

$$\text{System stored energy} = \frac{Wk^2 \, (\text{r/min})^2}{3.23 \cdot 10^6} \cdot 746 \, \text{W·s}$$

$$= \left(\frac{(0.7 + 0.7)\,(1800)^2}{3.23 \cdot 10^b} \cdot 746 \, \text{W·s} \right)$$

$$= 1047.63 \, \text{W·s}$$

$$\text{Motor acceleration loss} = \left(1 + \frac{r_1}{r_2} \right) \frac{Wk^2 \, (\text{r/min})^2}{3.23 \cdot 10^6} \cdot 746 \, \text{W·s}$$

where
r_1 is the stator/phase resistance = 0.615 Ω
r_2 is the rotor/phase resistance = 0.390 Ω

$$= 1 + \frac{0.615}{0.390} \, \frac{(0.7 + 0.7)\,(1800)^2}{3.23 \cdot 10^6} \cdot 746 \, \text{W·s}$$
$$= 2699.65$$

Total loss = 3747.28 W·s, or 3750 W·s

    6) Driven machine is fully loaded for 15 min followed by a 15 min idle period.
    7) Utilizations: 2000 hours annually
    8) Cost of energy: $0.06/kWh
    9) Motor inertia: 0.7 lbf/ft^2
    10) Driven machine inertia: 0.7 lbf/ft^2

b) Annual cost of energy to run the motor continuously:

    1) Cost to run at full load:

$$\$ = \frac{\text{running hours} \cdot \text{hp} \cdot 0.746 \, \text{hp/kW} \cdot \$/\text{kWh}}{\text{efficiency at full load}}$$

$$= \frac{1000 \, \text{h} \cdot 10 \, \text{hp} \cdot 0.746 \, \text{hp/kW} \cdot \$0.06/\text{kWh}}{0.892 \, \text{efficiency}}$$

$$= \$501.79$$

2) Cost to run the machine at idle:

$$= \frac{1000 \text{ h} \cdot 2.5 \text{ hp} \cdot 0.746 \text{ hp/kW} \cdot \$0.06\text{/kWh}}{0.86 \text{ efficiency}} = \$130.12$$

Total annual cost = \$501.79 + \$130.12 = \$631.79

c) Annual cost of energy if motor is switched off after each load cycle:

1) Cost to run at full load, same as before = \$501.79
2) Cost to restore system energy and supply motor acceleration losses:

$$\$ = \frac{N_s \cdot J \cdot \$\text{/kWh}}{1000 \ (\text{W/kW} \cdot 60 \cdot \text{min/h} \cdot 60 \text{ s/min})}$$

where

N_s is the number of starts annually
J is the motor acceleration loss plus system stored energy in W·s

$$\$ = \frac{4000 \text{ starts/yr} \cdot 3750 \text{ W·s} \cdot \$0.06\text{/kWh}}{1000 \ (\text{W/kW} \cdot 60 \text{ min/h} \cdot 60 \text{ s/min})} \quad \$ = \$0.25$$

3) Total annual cost = \$501.79 + 0.25 = \$502.04
4) Annual energy savings due to switching off the motor between load cycle:
\$ = 130.12 − 0.25 = \$129.87

Note that in some load-cycling applications a NEMA Design D motor should be evaluated with the NEMA Design B motor due to the differences in starting losses. That is, a NEMA Design D motor may have lower starting losses and higher running losses. Depending upon the cycle times, the NEMA Design D motor may have lower overall costs.

5.3.7.4 Allowable number of starts and minimum time between starts for Design A and Design B motors (NEMA MG 1-1993 [B164] and NEMA MG 10-1994 [B166])

NEMA MG 1-1993, 12.54, provides guidance on the number of successive starts, that is, two starts from ambient or one start from rated load operating temperature; however, the information from MG 1-1993, 12.54, is not applicable to repetitive start-run-stop-rest-cycles.

Table 5-13 provides guidance on (1) the minimum off time required to allow the motor to cool sufficiently to permit another start; (2) the maximum number of starts per hour, irrespective of load inertia (Wk^2) to minimize the winding stress; (3) a means of adjusting the starts per hour as a function of load inertia.

WARNING: *When using the upper range of starting duty there will be a reduction in motor life expectancy.*

**Table 5-13—Allowable number of starts and minimum time between
starts for Design A and Design B motors**

hp	2-pole			4-pole			6-pole		
	A	*B*	*C*	*A*	*B*	*C*	*A*	*B*	*C*
1	15	1.2	75	30	5.8	38	34	15	33
1.5	12.9	1.8	76	25.7	8.6	38	29.1	23	34
2	11.5	2.4	77	23	11	39	26.1	30	35
3	9.9	3.5	80	19.8	17	40	22.4	44	36
5	8.1	5.7	83	16.3	27	42	18.4	71	37
7.5	7.0	8.3	88	13.9	39	44	15.8	104	39
10	6.2	11	92	12.5	51	46	14.2	137	41
15	5.4	16	100	10.7	75	50	12.1	200	44
20	4.8	21	110	9.6	99	55	10.9	262	48
25	4.4	26	115	8.8	122	58	10.0	324	51
30	4.1	31	120	8.2	144	60	9.3	384	53
40	3.7	40	130	7.4	189	65	8.4	503	57
50	3.4	49	145	6.8	232	72	7.7	620	64
60	3.2	58	170	6.3	275	85	7.2	735	75
75	2.9	71	180	5.8	338	90	6.6	904	79
100	2.6	92	220	5.2	441	110	5.9	1181	97
125	2.4	113	275	4.8	542	140	5.4	1452	120
150	2.2	133	320	4.5	640	160	5.1	1719	140
200	2.0	172	600	4.0	831	300	4.5	2238	265
250	1.8	210	1000	3.7	1017	500	4.2	2744	440

Source: NEMA MG 10-1994 [B166].

NOTES

Where *A*   is the maximum number of starts/h

   B   is the maximum product of starts/h $\cdot$ load Wk^2

   C   is the minimum rest or off time in seconds between starts

The allowable starts/h is the less of (1)*A* or (2) *B* divided by the load Wk^2, i.e.:

$$\text{Starts/h} \leq A \leq \frac{B}{\text{load } Wk^2}$$

This table is based on the following conditions:

a)   Applied voltage and frequency in accordance with NEMA MG 1-1993 [B164], 12.44.

b)   During the acceleration period, the connected load torque is equal to or less than a torque
that varies as the square of the speed and is equal to 100% of rated torque at rated speed.

c)   External load Wk^2 equal to or less than the values listed in NEMA MG 1-1993 [B164], 12.54.

For all other conditions, the manufacturer should be consulted.

To calculate the number of allowable starts per hours, consider the following examples. Note that Example 1 is a direct connected load while Example 2 is a belt connected load.

a) Example 1 (from NEMA MG 10-1994 [B166]), Direct connected load: 50 hp, 4-pole, Design B motor directly connected to a pump with a moment of inertia (Wk^2) of 20 lbf/ft^2.

From table 5-13:

$A = 6.8 =$ maximum number of starts
$B = 232 =$ maximum product of starts times load inertia Wk^2
$C = 725 =$ minimum off time
therefore,

$$\frac{B}{\text{load } Wk^2} = \frac{232}{20} = 11.6 \text{ starts/h}$$

since 11.6 exceeds the 6.8 maximum number of starts per hour then the motor must be limited to 6.8 starts/h with a minimum off time between starts of 72 s.

b) Example 2 (from NEMA MG 10-1994 [B166]): Belt connected load: 25 hp, 2-pole, 3550 r/min, Design B motor, belt connected to a 5000 r/min blower with a moment of inertia (Wk^2) of 3.7 lbf/ft^2.

Note, since the load is at a different speed than the motor, it is necessary to calculate the inertia referred to the motor shaft.

Load Wk^2 referred to motor shaft $= \left[\dfrac{5000}{3550}\right]^2 \cdot 3.7 \text{ lbf/ft}^2 = 7.34 \text{ lbf/ft}^2$

From table 5-13:

$A = 4.4 =$ Maximum number of starts per hour
$B = 26 =$ Maximum product of starts per hour times load Wk^2
$C = 115 =$ Minimum rest or off time

and, $\dfrac{B}{Wk^2} = \dfrac{26}{7.34} = 3.5 \text{ starts/h}$

Therefore, the motor must be limited to 3.5 starts/h with a minimum off time between starts of 115 s.

Refer to NEMA MG 1-1993 [B164], 20.43, for the number of starts for large induction motors and to 21.43 for the number of starts for large synchronous motors.

5.3.7.5 Applications involving extended periods of light load operation

At less than full-load, the squirrel-cage induction motor power factor will be less than optimum and the magnetization losses become a higher percentage of the total losses.

Power factor controllers, included within an adjustable speed drive or installed as separate equipment have, can be used to reduce these losses at light load periods in certain horsepower ratings of single-phase and three-phase motors, especially motors 5 hp and less.

The power factor controller senses motor voltage and current, and then calculates the phase angle between voltage and current. This phase angle is called the displacement power factor. The ideal power factor is when the ratio of watts to voltamperes approaches unity. The controller compares the displacement power factor to the ideal power factor, and then generates a feedback signal that is used to command adjustments in the input voltage to the motor. This action reduces the current and voltage under light load conditions. Careful analysis should determine the savings based upon the duty cycle, the losses in the power factor controller, and the actual motor losses at the various low-load operating points. The magnetization losses that can be saved may not equal the losses added by the controller plus the additional motor losses due to the distorted waveform caused by the controller.

CAUTION: *Sudden application of load may stall the motor due to lack of available torque.*

5.3.7.6 Adjustable speed loads

Load characteristics are necessary to be defined in the application of ASDs. There are the following basic types:

— *Constant torque.* Where the power (i.e., product of torque and speed) requirement varies linearly with speed. Examples are conveyor belt loads and positive displacement pumps.

— *Constant power.* Where the torque requirements decrease as the speed increases. Examples are machine tools and vehicle drives.

— *Variable torque.* Where the torque varies in any other way than as described for constant power and constant torque.

In addition, centrifugal loads are also referred to as exponential loads. There are two categories:

— *Cubed exponential loads.* In cubed exponential loads the torque varies as the square of the speed. The horsepower varies as the cube of the speed. All centrifugal pumps and compressors and some fans and blowers are of this type of load.[3]

— *Squared exponential loads.* In squared exponential loads the torque varies directly as the speed. The horsepower varies as the square of the speed. Reciprocating pumps and some extruders are representative of such loads.

---

[3]It is important to realize that the theoretical "cubed" relationship between fan speed and input horsepower excludes the total system interaction. The effect of static back pressure in the system duct work or the static head in a centrifugal pump application are excluded from this ideal situation. Hence, the actual curve using ASD equipment in centrifugal fan and pump systems will often show something closer to a "square" relationship between speed and input horsepower.

The drive could be mechanical or an electrical adjustable frequency controller. The principle of speed control for adjustable frequency drives is based on the following fundamental formula for a standard ac motor:

$$N_s = \frac{120 \cdot f}{p}$$

where

N_s is the synchronous speed, r/min
f is the frequency, Hz
p is the number of poles

The number of poles is established during the motor's design and manufacture.

The adjustable frequency system controls the frequency, f, applied to the motor. The speed, N_s, of the motor is then proportional to the applied frequency. Control frequency is adjusted by means of a potentiometer or external signal. The frequency output of the control is infinitely adjustable over the speed range and therefore the speed of the motor is infinitely adjustable.

The controller automatically maintains the required volts/hertz (V/Hz) ratio to the motor at any speed. This provides maximum motor capability throughout the speed range (Emerson Electric Co. Publication ADG-044A [B252]).

The input hp shall be the drive hp divided by the drive efficiency, as follows:

$$\text{input hp} = \frac{\text{drive hp input to load}}{\text{efficiency of motor} \cdot \text{efficiency of inverter at actual 1}}$$

For electrical power input kW:

$$\text{kW input} = \text{hp input} \cdot 0.746 \text{ kW/hp}$$

The drive in the equation could be mechanical or an electrical adjustable frequency controller. Inverter is used here as an example only. The efficiency of a pulse width modulated (PWM) inverter, for example, is calculated as follows: 2% of load plus 500 W to 1 kW of no-load losses, depending upon inverter size.

Refer to annex 5G for illustrations of the efficiency of a typical 300 hp adjustable frequency drive. Annex 5G illustrates the variation of efficiency with frequency for the motor, the adjustable frequency controller, and the complete drive. Annex 5H illustrates the variation in motor output kilowatts power, the controller output kilowatts, and the drive input kilowatts as a function of frequency.

Refer to 5.6 for sources of personal computer programs to use in evaluating motor and drive application economics.

The basic equations used in evaluating variable speed applications are as follows:

$$\text{Rotating objects: hp} = \frac{T \cdot n}{5252} = \frac{2\pi n T}{33\,000} \qquad (5\text{-}1)$$

$$\text{Horizontal conveyors: hp} = \frac{F \cdot V \cdot \text{coefficient of friction}}{33\,000} \qquad (5\text{-}2)$$

$$\text{Vertical conveyors: hp} = \frac{F \cdot V}{33\,000} \qquad (5\text{-}3)$$

$$\text{Pumps: hp} = \frac{(\text{gal/min}) \cdot \text{head} \cdot \text{specific gravity}}{3960 \cdot \text{efficiency of pump}} \qquad (5\text{-}4)$$

or

$$\frac{(\text{gal/min}) \cdot (\text{lbf/in}^2) \cdot \text{specific gravity}}{1713 \cdot \text{efficiency of pump}} \qquad (5\text{-}5)$$

$$\text{Fans and blowers: hp} = \frac{\text{ft}^3/\text{min} \cdot \text{lbf/ft}^2}{33\,000 \cdot \text{efficiency of fan}} \qquad (5\text{-}6)$$

or

$$\text{hp} = \frac{(\text{ft}^3/\text{min}) \cdot (\text{inH}_2\text{O})}{6356 \cdot \text{efficiency of fan}} \qquad (5\text{-}7)$$

$$\text{Accelerating torque: } T_a = \frac{Wk^2 \cdot \text{change in speed}}{308 \cdot t} \qquad (5\text{-}8)$$

where

t	is time, s
T	is torque, lbf·ft
n	is speed, r/min
F	is force, lbf
V	is velocity, ft/min
gal/min	is gallons per minute
head	is height of water, ft
efficiency	is percent per 100

157

lbf/in^2 is pound-force per square inch

inH_2O is pressure, inches of water gauge

lbf/ft^2 is pound-force per square foot

ft^3/min is cubic foot per minute

hp   is horsepower

Wk^2 is moment of inertia

Torque: $1\ lbf \cdot ft = 1.355\ 818\ N \cdot m$

Speed: $1\ r/min = 0.104\ 719\ 8\ rad/s$

Force: $1\ lbf = 4.448\ 222\ Newton\ (N)$

Velocity: $1\ ft/min = 0.304\ 8\ m/s$

Volume per time (includes flow): $1\ gal/min = 0.063\ 090\ 20\ L/s$

Volume per time (includes flow): $1\ ft^3/min = 4.719\ 474 \times 10^{-4}\ m^3/s$

Volume (includes capacity): $1\ gal\ (U.S.) = 3\ 785\ 412\ L$

Volume (includes capacity): $1\ L = 0.001\ m^3$

Pressure (force per area): $1\ in\ water\ (inH_2O) = 249.089\ Pascals\ (Pa)$

Head: pressure (force per area): $1\ ft\ water\ (ftH_2O) = 2.989\ 07\ kiloPascals\ (kPa)$

Pressure (force per area): $1\ lbf/in^2 = 6.894\ 757\ kiloPascals\ (kPa)$

Power: $1\ hp = 746\ W$

Length: $1\ in = 0.025\ 4\ mm$

Length: $1\ ft = 0.304\ 8\ m$

5.3.7.6.1 Centrifugal pump application example (see figure 5-3) (Reliance Technical Paper D-7100-1 [B189])

a)   First, obtain the pump performance curve and efficiency data from the pump manufacturer. In this graph, the head is the ordinate and the flow is the abscissa. The pump curve is plotted from the total head out to maximum flow. On the pump curve the pump efficiencies are marked at discrete points.

b)   Second, determine the lift required by the system. This lift (or head) is the difference in elevation between input tank and output tank or, if there are no tanks, then the pressure of the fluid if a straight pipe replaced the pump. Mark this pressure or head on the ordinate.

c)   Third, measure the flow required with the valve wide open to find the 100% flow operating point. Mark this point on the pump curve.

d)   Fourth, interpolate a system curve between the ordinate, at the required lift point found in the second step, and the 100% flow operating point marked on the pump curve. Assume a squared relationship between the flow and head such that at half speed, one quarter of the friction head, above the required lift, is required across the pump.

e)   Fifth, establish a representative series of duty cycle operating points with a weighted average proportion of the operating time assigned to each. For example:

—   40% of the time the flow is 700 gal/min

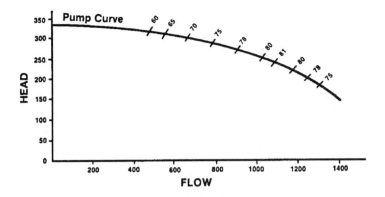

(a) Pump curve and efficiency data

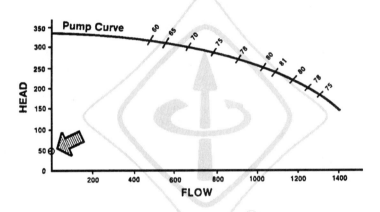

(b) Measure required lift

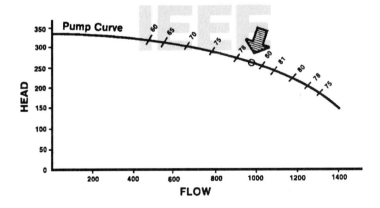

(c) Measure flow with valve fully open

Figure 5-3—Evaluating centrifugal pump application

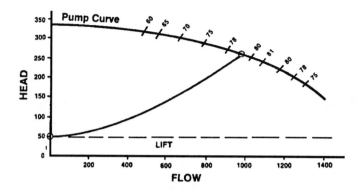

(d) Interpolate system curve

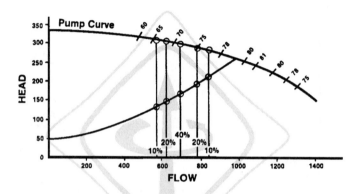

(e) Establish duty cycle

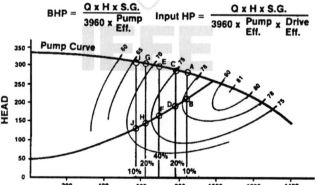

(f) Plot pump efficiencies

Source: All parts of this figure from Reliance Technical Paper D-7100 [B189]

Figure 5-3—Evaluating centrifugal pump application *(Continued)*

— 20% of the time the flow is 630 gal/min
— 20% of the time the flow is 770 gal/min
— 10% of the time the flow is 830 gal/min
— 10% of the time the flow is 570 gal/min

f) Sixth, plot the pump efficiency, obtained from the pump manufacturer, on the graph. These curves are drawn from an efficiency value at a low flow on the pump curve to the same efficiency value at a higher flow on the pump curve and resemble arcs.

Where the required lift or static head is not available, the system may be defined as high lift or low lift. Assume 20% of the natural operating point pressure for a low lift system and 50% for a high lift system.

Where the duty cycle is unknown, a leading manufacturer of drives recommends the following: assume 50% of the operating time to the average flow, a 5% operating time to the minimum and maximum flows, and a 20% operating time to the points calculated between the average and minimum and average and maximum flow points using a standard normal distribution.

g) Seventh, calculate the energy savings. The head, flow, specific gravity, and pump efficiency are listed for each point. From this information the brake horsepower is calculated. The drive manufacturer, supplier of inverter and motor, then provides the drive efficiency, product of motor and inverter efficiency, based upon this brake horsepower. Input horsepower then can be calculated. The horsepower savings between valve control and adjustable speed control is then calculated by subtraction. The weighted average horsepower savings can then be added to derive the average horsepower saved throughout the duty cycle, in this example 29.2 hp. The annual energy savings is then the product of the weighted horsepower and the total annual operating hours. (See table 5-14.)

5.3.7.6.2 Centrifugal fan application example (see figure 5-4) (Reliance Technical Paper D-7102 [B190])

a) Obtain the fan performance curve and efficiency data from the fan manufacturer. In this graph, static pressure is the ordinate and static pressure volume is the abscissa. The fan curve is plotted from the total static pressure out to maximum volume.

b) Plot the operating point on the fan curve where the output of the fan is constricted and therefore significant pressure must be developed to produce substantial air volume. This is called point A. Plot a second point on the fan curve for the operating point B where the fan is applied to a less restrictive duct system (less static pressure).

c) Plot system curves from the origin to point A and to point B. These system curves follow an exponential relationship. These system curves show that for this particular fan, at constant speed it will operate at point A on System 1 and at point B on System 2.

d) To decrease volume, System 2 could be modified by the addition of a partially closed discharge damper to increase the resistance in the output duct work and therefore produce reduced volume. The modified system, with partially closed damper, is then

Table 5-14—Centrifugal pump example of an adjustable speed drive application

Point on pump curve and system curve	Head (ft)	Flow, (gal/min)	Specific gravity	Pump efficiency	Brake (hp)	Drive efficiency[a]	Input (hp)	Point-to-point savings (hp)	Percent of duty cycle	Weighted average hp
A	285	830	1.0	0.76	78.6	0.92	85.4	16.6	10%	1.7
B	200	830	1.0	0.78	53.7	0.79	68.0			
C	295	770	1.0	0.75	76.5	0.91	84.1	24.2	20%	4.8
D	180	770	1.0	0.77	45.5	0.76	59.9			
E	300	700	1.0	0.71	74.7	0.90	83.0	30.7	40%	12.3
F	155	700	1.0	0.76	36.1	0.69	52.3			
G	305	630	1.0	0.67	72.4	0.89	81.3	34.4	20%	6.9
H	140	630	1.0	0.73	30.5	0.65	46.9			
I	310	570	1.0	0.64	69.7	0.86	81.0	35.1	10%	3.5
J	130	570	1.0	0.69	27.1	0.59	45.9			
									Total =	29.2

Source: Reliance Technical Paper D-7100-1 [B189]

NOTES—Motor efficiency is decreased on pulse width modulation (PWM) power. Losses go up by 20%; that is, when the motor efficiency is published as 92% on sine wave power, then the losses are 8%. Therefore, on PWM power, 8% · 0.2 = 1.6% more losses on adjustable speed drive, and the new motor efficiency is 92% − 1.6% = 90.4%.

Annual energy savings = 29.2 hp · annual operating hours · 0.746 kW/hp; for 4000 h, this is 87 132.8 kWh.

[a] Includes efficiency of motor and inverter. Drive could be mechanical or electrical speed controllers. Inverter is used here as an example only.

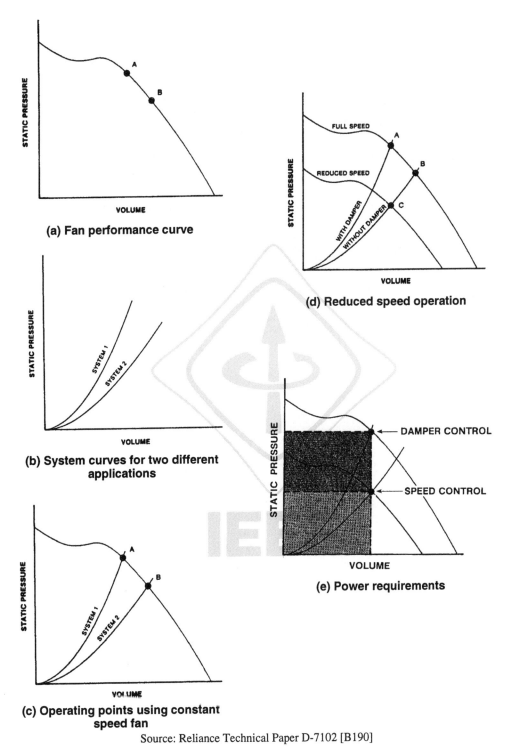

(a) Fan performance curve

(b) System curves for two different applications

(c) Operating points using constant speed fan

(d) Reduced speed operation

(e) Power requirements

Source: Reliance Technical Paper D-7102 [B190]

Figure 5-4—Evaluating centrifugal fan application

163

described by the system curve labeled System 1. The fan will then produce less volume by operating at point A rather than point B.

This type of operation causes the need for the fan to operate against greater pressure and waste energy.

The volume can also be decreased by reducing the speed of the fan.

e)   Plot the fan curve at other speeds representing operation producing less volume.

As a result of reducing the fan speed the system curve changes and the same reduced volume can be produced with less pressure.

The variation of shaft hp and torque with output speed is shown in table 5-15.

Table 5-15—Centrifugal fan example of variation of shaft horsepower and torque with rated speed

Speed (%)	Torque (%)	Percent horsepower (hp)
100	100	100
90	81	72.9
80	64	51.2
70	49	34.3
60	36	21.6
50	25	12.5

Since, in centrifugal fans volume is directly proportional to speed change:

$$\frac{(\text{ft}^3/\text{min})_2}{(\text{ft}^3/\text{min})_1} = \frac{(\text{r/min})_2}{(\text{r/min})_1}$$

Fan horsepower, however, is directly proportional to the cube of the speed change:

$$\frac{\text{hp}_2}{\text{hp}_1} = \left(\frac{\text{r/min}_2}{\text{r/min}_1}\right)^3$$

It is very important to realize that this theoretical relationship between fan speed and horsepower does not take into account the effect of total system interactions, and that it does not include the effect of static back pressure in the system duct work (or the effect of static head in a centrifugal pump application). The magnitude of the static back pressure will have an adverse effect on the theoretical power and energy savings.

The actual system curve using ASD equipment in centrifugal fan and pump systems will approximate a square function. The prudent approach is to take this into consideration when developing expected savings using ASD equipment.

   f)   Calculate the energy savings as shown in table 5-16.

Table 5-16—Worksheet for ASD operation, weighted average calculation

Point on fan curve and system curve	Static pressure (lbf/in²)	Volume (ft³/min)	Fan efficiency	Brake hp	Drive efficiency	Input hp	Point to point hp savings	Percent of duty cycle	Weighted average hp
A									
B									
C									
D									
E									
F									
G									
H									
I									
J									
								Total =	

NOTE—The annual energy savings is then the weighted average horsepower times the annual operating hours.

Table 5-17 shows an example of centrifugal fan savings using figure 5-5.

5.3.7.6.3 Application involving overhauling loads

Energy can be returned to the supply by installing adjustable speed controllers with regeneration capability when the application is an overhauling load, for example, elevators, downhill conveyors, escalators, absorption test stands, web process stands, etc.

Regeneration capability may be available with mechanical speed drives or with ac variable voltage or eddy current clutch drives. Regeneration capability should be in the drive specification, when the application warrants:

— Regeneration is approximately 10% as standard on PWM ac drives.

— Additional regeneration is available by snubbers or synchronous rectifiers at extra charge.

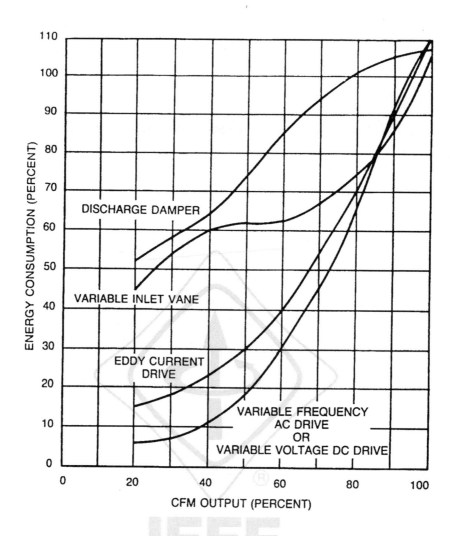

Source: NEMA MG 10-1994 [B166]

Figure 5-5—Typical energy consumption of a centrifugal fan system with discharge damper control, variable inlet vane control, and variable speed control with eddy current and variable frequency drives

5.3.8 Increasing efficiency

The objectives in increasing motor efficiency can involve any of the following:

a) Reduce energy consumption with existing equipment by using some form of demand control or power factor improvement or both.

b) Investigate the savings possible by conversion to motors with higher efficiency, either synchronous motors or premium efficiency squirrel-cage induction motors.

Table 5-17—Energy consumption comparison using figure 5-5

Volume percent of full ft^3/min	Discharge damper control	Variable speed drive	Difference	Utilization annual hours	Savings, weighted full load hours
100	107%	105%	2%	1000	20
75	97%	55%	42%	3000	1260
50	75%	18%	57%	2000	1140
Total					2420

NOTE—Annual energy savings = 2420 h · drive input hp
Assuming drive input = 100 hp
Then annual energy savings = 242 000 hp h
or, 242 000 hp h · 0.746 kW/hp = 180 532 kWh

c) Investigate the savings possible by improvement of the efficiency of the mechanical power transmission between the motor and the load, such as

1) Changing from worm to helical gearing

2) Changing V-belts from ordinary wrapped belts to V-belts with raw edge (coverless) construction and a rounded notch profile

d) Evaluate savings possible by changing the motor process, for example, converting a process from constant speed to adjustable speed.

5.3.8.1 Reducing consumption

Three techniques can be used without changing the motor-load power train:

a) Turn off load equipment non-essential to process continuity or human comfort.

b) Install automatic demand control equipment to shed predetermined loads to prevent exceeding a pre-set limit demand.

c) Improve the system power factor by installing power factor correction equipment at the load so as to reduce I^2R losses in the transformers and conductors supplying the load.

Perform an energy balance calculation to compare the energy consumed in starting the load against energy consumed in running the equipment lightly loaded, as part of the decision to shed unused load equipment and reduce total kWh. Smaller motors, for example, 1–5 hp, with low inertia loads, such as centrifugal pumps, may be able to save energy with off times as short as 15–30 s provided the additional starts do not damage the motor. As motor hp ratings and load inertias increase, the off times necessary to save energy increase. Perform the calculations described earlier from NEMA MG 10-1994 [B166] (see 5.3.7.4) to determine the allowable rest periods. Some process machinery, such as chillers, have complex start and shutdown cycles and the equipment manufacturer should be consulted before establishing programs to automatically shed and restart these loads.

Motor power factor improvement is beneficial since it eliminates the wiring and transformer energy losses due to providing excessive magnetizing current at the motor. An energy balance should be performed to make sure that the kWh saved through reduction in wiring and transformer energy losses are greater then the losses within the power factor correction capacitors. Where a utility has a rate structure encouraging high power factor, there are benefits through savings in the power factor penalty. Refer also to clauses 5.5.2 and 5.5.3.

CAUTION: *For safety reasons, it is generally better to improve power factor for multiple loads as a part of the plant distribution system.*

The rules for power factor improvement of polyphase squirrel-cage induction motors are as follows:

a) Rules for power factor improvement are published in NEMA MG 1-1993 [B164], Section II, Part 14.43.

b) **WARNING:** *The use of power capacitors for power factor correction on the load side of an electronic power supply connected to an induction motor is not recommended!*

c) **WARNING:** *In no case should power factor improvement capacitors be applied in ratings exceeding the maximum safe value specified by the motor manufacturer. Excessive improvement may cause over-excitation resulting in high transient voltages, currents, and torques that can increase safety hazards to personnel and cause possible damage to the motor or to the driven equipment.*

d) Tables of capacitive kvar for power factor improvement of unspecified manufacturers' squirrel-cage induction motors—versus horsepower, speed, and voltage—should be only used as a general guide. Design characteristics vary between manufacturers, even for motors of identical rating. Motor manufacturers will provide the maximum capacitor rating that can be switched with a specific motor.

e) Where capacitors are not switched by the motor starters, the proper application of power capacitors to a bus with harmonic currents requires an analysis of the power system to avoid potential harmonic resonance of the power capacitors in combination with transformer and circuit inductance.

f) Induction motor-and-starter combinations where power factor improvement capacitors may be switched by the starter are as follows:

Motor type	Starter type	Starter configuration	Comments
Single-speed motors	Full-voltage	Non-reversing	—
Squirrel-cage induction motors	Reduced-voltage	Primary resistor; primary reactor; autotransformer with closed transition	Switch capacitors with "run" contacts

g) Induction motor-and-starter combinations, where power factor improvement capacitors may not be switched by the starter, are as follows:

Motor type	Starter type	Starter configuration	Comments
Squirrel-cage induction motors	Part-winding	1/2—1/2 and 1/3—2/3 starters	Obtain maximum kvar value form motor manufacturer. Do not use standard kvar vs.hp tables
Single-speed motor	Full-voltage	Reversing	—
Squirrel-cage induction motors	Reduced-voltage	Autotransformer open transition including manual autotransformer starter	—
Squirrel-cage induction motors	Wye-delta	Either open or closed transitions	No exceptions!
Multispeed motors	Multispeed	Two or more speeds, any torque configuration	—

5.3.8.2 Large energy-efficient motors

There are opportunities to conserve energy through the reduction in motor losses with energy-efficient large-motor designs in the 500–10 000 hp range as well as in the smaller frame sizes, especially for large motors running continuously at or near full load.

For example, an energy-efficient 1000 hp 1800 r/min motor may have an efficiency at full-load of 94.8%. The motors total losses would be 40.9 kW. A standard motor may have losses of 49.1 kW at full load or a full-load efficiency of 93.8. The 8.2 kW energy saving amounts to 71832 kWh annually in favor of the energy-efficient motor.

Synchronous motors have full-load efficiencies of 1–2 points better than squirrel-cage induction motors of similar ratings. Synchronous motors and their starting equipment are complex. Large, horizontal, slow-speed pump and fan loads, particularly where the driven equipment can be unloaded during starting, may be considered for synchronous motors, in consultation with the motor manufacturer.

5.3.8.3 Transmission efficiency

Motor speeds are reduced for most loads in the 1–125 hp range, excluding pumps and compressors. The transmission efficiency of the equipment transferring power from the motor shaft to the load is just as important as the efficiency of the motor.

The motor output shaft may be driving a gear train, a pulley, a sheave, or a sprocket.

5.3.8.3.1 Gear drives

Worm gearing is commonly used in ratings below 20 hp. Worm gearing provides large values of speed reduction per stage. A typical mechanical drive would be a right-angle worm gear integral with a C-face gear motor, as specified by NEMA MG 1-1993 [B164]. Worm gearing efficiency decreases as the gear ratio (amount of speed reduction) increases as shown in figure 5-6.

Worm gear losses are approximately equal to half the gear ratio plus 10%. For example, a worm gear having a gear ratio of 50 : 1 driven by a 1750 r/min motor would have an output of 35 r/min. The losses would then be equal to the following:

$0.5 \cdot 50$ gear ratio $+ 10\% = 25\% + 10\% = 35\%$

The transmission efficiency would then be equal to the following:

$100\% - 35\% = 65\%$

Parallel shaft, helical, and spur gearing have rolling contact between mating teeth, compared with sliding contact between mating teeth of the worm gear.

Helical gears are approximately 98.5% efficient per stage or 95% efficient for a 3-stage unit (SEW-Eurodrive, Inc.).

For the 50 : 1 gear ratio required in the example above three stages of gearing would be required. Therefore, the helical gear reducer efficiency would be

$98.5 \cdot 98.5 \cdot 98.5 = 95.5\%$, or 95%

The impact of the efficiencies on motor hp is shown in this example:

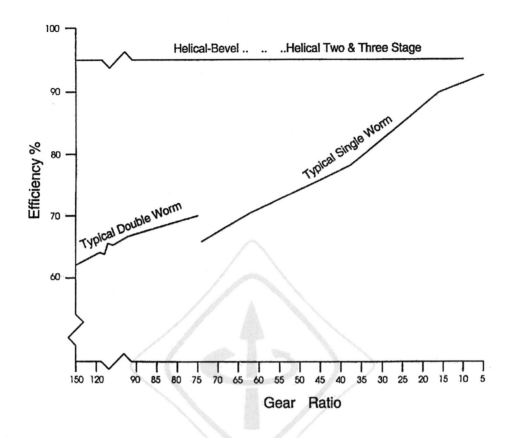

Source: Courtesy of SEW-Eurodrive, Inc.

Figure 5-6—Efficiency comparison of gear drives

Required: 9 hp delivered to a load rotating at 35 r/min

Helical gear reducer: $\dfrac{9\ \text{hp}}{95\%\ \text{efficiency}}$ = 9.47 hp input required. Therefore, use a 10 hp motor.

Worm gear reducer: $\dfrac{9\ \text{hp}}{65\%\ \text{efficiency}}$ = 13.8 hp input required. Therefore, use a 15 hp motor.

The purpose of this example is to point out the need for transmission efficiency to be evaluated. Worm gearing has many good applications; however, in this example the helical gear reducer is the more efficient alternative.

5.3.8.3.2 V-belt transmissions

Belt-drive efficiency is affected by these seven prime factors (Eaton Corp. Technical Bulletin No. 700-4006 [B82]):

a) *Belt type.* Conventional, wedge, and positive drive belts each have efficiency characteristics to be considered. There is a relationship between belt thickness, sheave diameter, and efficiency. If there is a choice, the thin cross-section belts yield greater efficiency. If thicker cross sections must be employed, efficiency can be maximized by using a driver sheave in the larger diameter range.

b) *Belt style.* Machined edge-molded cog belts are energy savers; however, envelope belts and banded belts can be equally efficient when properly applied. Standard envelope belts can be applied with efficiencies approaching the cog style belt if the drive utilizes larger sheaves above recommended minimums. Banded belts can be more efficient than individual belts where belt turnover and vibration are problems.

c) *Drive maintenance.* Proper tensioning is a critical element of V-belt efficiency. V-belt drives should have an efficiency in the 95–98% range when properly applied. If incorrect tensioning allows a 4% slip, the efficiency of a drive normally at 97% will drop to 92%. As the slip drops to 6%, an excess amount of heat is generated and the drive is now at 88% efficiency. Approaching 8% slip, the efficiency drops to 80%. At 8% slip, the drive will not function.

d) *Drive design.* The number of belts on a drive impacts the efficiency: overbelting (too many belts on a drive) can be as detrimental as underbelting (too few belts on a drive). Overdesign causes individual belts to run at less than rated capacity. For a drive running at one half or 50% of rated load capacity, a decrease in efficiency of nearly 4% will result.

e) *Coefficient of friction.* When the drive is located in a very dusty or very wet environment a 4% slippage is common and the drive efficiency will drop to 92%. If the slippage increases, then proper belt tightening is required.

f) *Sheave diameter.* The minimum recommended sheave diameter is a key point to consider on any V-belt drive. When the small sheave utilized on a drive falls below the minimum recommended diameter, the efficiency drops off drastically.

g) *Belt manufacturer.* On smaller diameter sheaves, there may be as much as a 6% efficiency advantage between V-belts of different manufacturers.

5.3.8.4 Voltage and harmonics

The significant factors related to motor voltage that have an effect on efficiency are design voltage, operating voltage, and harmonics:

a) *Design voltage.* High-voltage motors will have a lower efficiency than medium-voltage motors due to the increase in ground and turn insulation required on higher voltage machines. The increased insulation causes a proportional decrease in available space for the copper in the slot, hence, higher I^2R losses. Energy conservation considerations suggest that, where the motor manufacturer's performance data indicates an efficiency rating in favor of the lower voltage motor, there is an obligation for the engineer to analyze the economics of each proposed installation. One option may be to obtain the higher voltage motor with a premium efficiency rating, for example:

b) *Operating at other than normal voltage conditions.*

1) *Operating at undervoltage.* When a fully loaded motor is operated at reduced voltage (even within the allowable 10% limit) the motor will draw increased current in order to produce the required torque. This will cause an increase in the I^2R loss of the stator and rotor. Refer to figure 5-7. However, a lightly loaded motor will operate more efficiently at lower voltage.

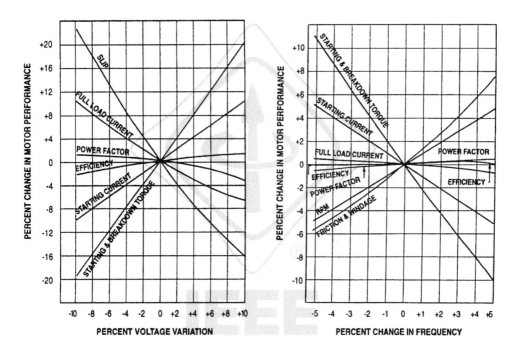

Source: General Electric Co. GET-6812B [B104]

Figure 5-7—Motor characteristics and losses versus applied voltage and frequency

2) *Operating at overvoltage.* As the voltage increases the magnetizing current increases by at least a square function. At some point saturation will occur. At about 10–15% overvoltage, the efficiency will decrease. Both the stator and core losses will increase. There will also be a significant decrease in power factor. Refer to figure 5-7.

3) *Operating at unbalanced voltage.* This condition will produce negative sequence currents that cause extraneous losses and increased heating in the winding. Refer to figure 5-8.

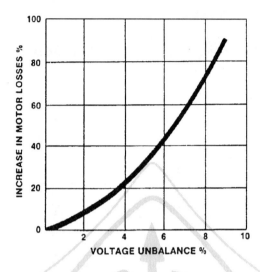

Source: Reliance Electric Co. Bulletin B-2615 [B187]

Figure 5-8—Motor voltage unbalance versus increase in motor losses

4) Since motor operating voltage is important for motor life and efficiency, there are several techniques to be implemented in the initial design:

i) Install protective relaying to monitor the operating voltage and to advise the staff when conditions are abnormal.

ii) Install protective relaying to shutdown the motor for unbalanced voltage conditions.

iii) Specify a regulated supply voltage so that the motor terminal voltage is kept within rated conditions. Select the appropriate voltage tap on the transformer.

iv) Consider the use of power factor correction capacitors for voltage control at the motor.

c) *Operating constant-speed motors on a sinusoidal bus with harmonic content and general-purpose motors used with variable-voltage or variable-frequency controls or both.* NEMA MG 1-1993 [B164] reports that efficiency will be reduced when a motor is operated on a bus with harmonic content: "The rated horsepower of the motor should be multiplied by the factor shown in figure 5-9 to reduce the possibility of damage to the motor" when the motor is operated at its rated conditions and the voltage applied to the motor consists of components at other than the nominal frequency. Refer to figure 5-9 and the example of derating.

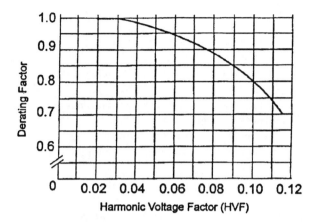

Source: NEMA MG 1-1993 [B164].

NOTE—The harmonic voltage factor (HVF) is defined as follows:

$$\sqrt{\sum_{n=5}^{n=\infty} \frac{(V_n)^2}{n}}$$

where

n is the order of odd harmonic, not including those divisible by three
V_n is the per-unit magnitude of the voltage at the nth harmonic frequency

Example: With per-unit voltages of 0.10, 0.07, 0.045, and 0.036 occurring at the 5, 7, 11, and 13th harmonics, respectively, the value of the HVF is

$$\sqrt{\frac{0.10^2}{5} + \frac{0.07^2}{7} + \frac{0.45^2}{11} + \frac{0.036^2}{13}} = 0.0546$$

Figure 5-9—Proposed derating curve for harmonic voltage

5.3.8.5 Motor size

Larger motors are far more efficient than smaller motors, especially in the 1–100 hp range. Serious consideration should be given to designing systems that use a few large motors in place of a quantity of small motors. In some applications it may be better to use several smaller motors in place of a few larger units, such as where wide variations in load demand exist.

5.3.8.6 Motor speed

There may be significant efficiency advantages to selecting higher speed motors rather than lower speed motors. The efficiency difference may be as much as 4% at 2 hp between a 3600 r/min and a 1200 r/min motor. At 1000 hp, the efficiency difference may be 1% between a 3600 r/min and an 1800 r/min motor. Referring to figure 5-10, the efficiency difference between a 200 hp, 3600 r/min motor and a 900 r/min motor may be as much as 12%. Referring to figure 5-11, note that the power factor of high-speed motors is generally better than low-speed motors. Careful review of motor performance data is required since as speed increases, the windage and friction losses increase. The optimum speed is usually 1800 r/min.

5.3.8.7 Loading

Maximum motor efficiency does not always occur at full load. For larger motors as long as the motor is operating beyond 50% rated load the efficiency does not vary severely. However, beyond that range the efficiency drops severely. Referring to figure 5-12, for the 10 hp motor, the efficiency is relatively constant between 50 and 100% rated load. Large motors can be designed for peak efficiency at low load for situations where they are operated most of the time at low loads.

5.3.8.8 Bearings

Friction loss is affected by the bearing selection. Ball bearings will generally have lower losses than sleeve bearings on horizontal motors. Sleeve bearings are not suitable for belted applications. On vertical motors with low or normal thrust loads the spherical roller or the angular-contact, ball-type, thrust bearings will have lower losses than the plate-type bearing. Where higher thrust loads exist, the plate-type bearing is suitable. Roller bearings are used, when specified, on the drive-end to handle heavy or overhung (radial) shaft loads. Motor evaluations should consider the L_{10} bearing life. That is, the motors being compared should seek to maximize the L_{10} bearing life for comparable bearing systems.

5.3.8.9 Motor space heaters

Consult the motor manufacturer during the preparation of the motor specification to determine the appropriate type of space heater, indirect heating or direct heating, and controls for prevention of condensation in idle motors.

5.3.8.10 Starting conditions (large motors)

a) *Starting load.* By reducing the starting demand, the motor designer can frequently design a motor with improved efficiency.

b) *Reduced-voltage starting.* Specification of a lower inrush for direct on-line starting will cause the motor designer to compromise efficiency and power factor. An alternative is to use a reduced-voltage starter, such as an autotransformer starter, with a higher efficiency motor.

c) *VVVF drives* offer a distinct advantage in starting as they allow slow acceleration without high inrush currents.

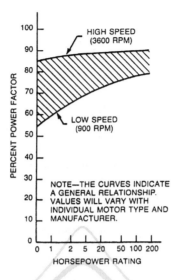

Source: NEMA MG 10-1994 [B166]

Figure 5-10—Typical power factor versus load curves for 1800 r/min, three-phase, 60 Hz, Design B, squirrel-cage induction motors

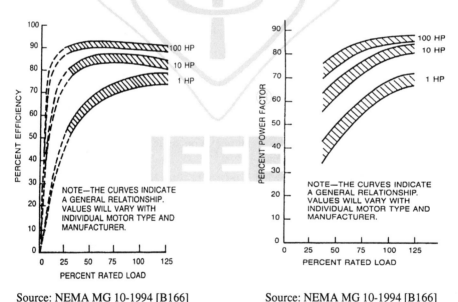

Source: NEMA MG 10-1994 [B166]   Source: NEMA MG 10-1994 [B166]

Figure 5-11—Typical full-load power factor versus horsepower rating curves for three-phase, 60 Hz, Design B, squirrel-cage induction motors

Figure 5-12—Typical efficiency versus load curves for 1800 r/min, three-phase, 60 Hz, Design B squirrel-cage induction motors

5.3.8.11 Vertical motors

Vertical motors, such as used for pumping applications, are available in energy-efficient designs in the range of 1.5–1000 hp, Design B (General Electric GEP-1839C [B112]). Efficiency evaluations should be considered for vertical motors as well as for horizontal motors.

5.3.8.12 Rewinding motors

There are precautions to be taken in the rewind shop to avoid damage to the motor efficiency (General Electric GEA-10951C [B106]): "When stripping out the old winding, and this applies to normal as well as high-efficiency motors, it is essential to keep the stator core below 700 °F. If the stator gets too hot, the insulation between the stator laminations will break down, the eddy current losses will rise and the efficiency will drop."

The increased operating costs due to receiving a poor-quality rewound motor will be equal to or greater than the reduction in operating costs achieved by replacing the damaged motor with a new energy-efficient motor.

5.3.8.13 Motor location

Energy can be reduced in the operation of the ventilating or air-conditioning plant by proper location of the motor and driven equipment with respect to the conditioned space. The ASHRAE 1993 Handbook [B17] reports the following:

The heat equivalent of equipment operated by electric motors within a conditioned space is given by the following equation:

$$qem = (P/E_M)\,(F_{UM})\,(F_{LM})$$

where

qem	is the heat equivalent of equipment operation, in W
P	is the motor horsepower rating, in kW
E_M	is the motor efficiency, as decimal fraction < 1.0
F_{LM}	is the motor load factor, 1.0 or decimal fraction < 1.0
F_{UM}	is the motor use factor, 1.0 or decimal fraction < 1.0

The above equation is used when both the motor and the driven equipment are in the conditioned space. Refer to table 5-18 for the results of the equation when F_{LM} and F_{UM} are both equal to 1.

When the motor is outside the conditioned space or airstream, then the heat equivalent is given by

$$qem = P = (F_{UM})\,(F_{LM})$$

The results of the above equation are in Location B of table 5-18.

When the motor is inside the conditioned space or airstream but the driven machine is outside, then the heat equivalent is given by

$$qem = P\,[(1.0 - E_M)/E_M]\,(F_{LM})\,(F_{UM})$$

Refer to Location C of table 5-18.

This equation also applies to a fan or pump in the conditioned space that exhausts air or pumps fluid outside that space.

Table 5-18—Heat gain from typical electric motors

Motor nameplate or rated horsepower	kW	Motor type	Nominal r/min	Full load motor efficiency (%)	Location of motor and driven equipment with respect to conditioned space or airstream		
					A	B	C
					Motor in, driven equipment in (W)	Motor out, driven equipment in (W)	Motor in, driven equipment out (W)
0.05	0.04	Shaded-pole	1500	35	105	130	70
0.08	0.06	Shaded-pole	1500	35	170	59	110
0.125	0.09	Shaded-pole	1500	35	264	94	173
0.16	0.12	Shaded-pole	1500	35	340	117	223
0.25	0.19	Split-phase	1750	54	346	188	158
0.33	0.25	Split-phase	1750	56	439	246	194
0.50	0.37	Split-phase	1750	60	621	372	249
0.75	0.56	3-phase	1750	72	776	557	217
1	0.75	3-phase	1750	75	993	747	249
1.5	1.1	3-phase	1750	77	1453	1119	334
2	1.5	3-phase	1750	79	1887	1491	396
3	2.2	3-phase	1750	81	2673	2238	525
5	3.7	3-phase	1750	82	4541	3721	817
7.5	5.6	3-phase	1750	84	6651	5596	1066
10	7.5	3-phase	1750	85	8760	7178	1315
15	11.2	3-phase	1750	86	13 009	11 192	1820
20	14.9	3-phase	1750	87	17 140	14 193	2230

Table 5-18—Heat gain from typical electric motors (*Continued*)

Motor nameplate or rated horsepower	kW	Motor type	Nominal r/min	Full load motor efficiency (%)	Location of motor and driven equipment with respect to conditioned space or airstream		
					A	B	C
					Motor in, driven equipment in (W)	Motor out, driven equipment in (W)	Motor in, driven equipment out (W)
25	18.6	3-phase	1750	88	21 184	18 635	2545
30	22.4	3-phase	1750	89	25 110	22 370	2765
40	30	3-phase	1750	89	33 401	29 885	3690
50	37	3-phase	1750	89	41 900	37 210	4600
60	45	3-phase	1750	89	50 395	44 829	5538
75	56	3-phase	1750	90	62 115	55 962	6210
100	75	3-phase	1750	90	82 918	74 719	8290
125	93	3-phase	1750	90	103 430	93 172	10 342
150	110	3-phase	1750	91	123 060	111 925	11 075
200	150	3-phase	1750	91	163 785	149 135	14 738
250	190	3-phase	1750	91	204 805	186 346	18 430

Source: ASHRAE 1993 Handbook [B17]

5.4 Transformers and reactors

5.4.1 General

Transformers in the U.S. are manufactured to standards of ANSI and UL as follows:

- IEEE Std C57.12.00-1993 [B132]
- IEEE Std C57.12.01-1989 [B133]
- NEMA ST 20-1992 [B168]
- UL 1561-1994 [B225]
- UL 1562-1994 [B226]

The transformer standards do not require efficiency targets in the transformer design. The goal of the standards is safety, convenience, compatibility, security, reliability, noise control, and other engineering and environmental parameters.

Energy savings are achievable by

- Specifying and purchasing efficient transformers
- Operating the transformers efficiently

Losses in transformers occur within three distinct circuits, as follows:

a) Electric circuit losses:
   1) I^2R loss due to load currents
   2) I^2R loss due to no-load currents
   3) Eddy current loss in conductors due to leakage fields
   4) I^2R loss due to the loss currents
b) Magnetic circuit losses:
   1) Hysteresis loss in core laminations
   2) Eddy current loss in core laminations
   3) Stray eddy current loss in core clamps, bolts, etc.
c) Dielectric circuit loss: Up to 50 kV this is a small loss and is usually included in the no-load losses.

The above losses are usually categorized as either load losses or no-load losses, with the load loss category containing a1), a3), and a4), and the no-load loss category containing a2), b1), b2), b3), and c).

No-load losses are measured at rated frequency and rated (nameplate) secondary voltage. No-load losses are assumed to be independent of load.

Load losses are measured at rated frequency and rated secondary current but with the secondary short-circuited and reduced voltage applied to the primary. Load losses are assumed to vary as the square of the current and are corrected to equivalent values at the operating temperature.

5.4.1.1 Efficiency

Transformer efficiency is the ratio of power output to power input. Where the losses are the total losses, the sum of the no-load and load losses:

$$\text{Efficiency} = \frac{\text{output}}{\text{input}} = 1 - \frac{\text{losses}}{\text{input}}$$

5.4.1.2 Loss ratio

The ratio of losses, R, is the ratio of the load loss to the no-load loss:

$$R = \frac{\text{load loss (kW)}}{\text{no-load loss (kW)}}$$

With a given loss ratio, the maximum efficiency of a transformer is determined by the following equation:

$$L = \frac{1}{\sqrt{R}}$$

where

L is the per unit kVA load at which the transformer operates most efficiently
R is the loss ratio

The relationship between the transformer loss ratio and the most efficient loading is as follows:

Transformer loss ratio, R	Transformer most efficient loading, L (per unit kVA load)
1.0	1.000
1.5	0.816
2.0	0.707
2.5	0.632
3.0	0.577
3.5	0.534
4.0	0.500
4.5	0.471
5.0	0.447
5.5	0.426
6.0	0.408

For example, it may be observed that a transformer with a loss ratio of 4 would most be efficiently operated at 50% load.

The specifier has some latitude in specifying the desired loss ratio R; however, this is not economical for distribution transformers because the cost of designing and constructing a distribution transformer with a loss ratio R different from the mass-produced design would outweigh any potential savings from loss reduction.

The loss ratio R for standardized distribution transformers generally falls in the range of 2.5 to 5, depending upon the size, manufacturer, and voltage rating of the transformer. The limits of the band and the distribution of units within the band vary among manufacturers.

The specifier's intentions to purchase an efficient transformer are best served by including a loss evaluation procedure in the request for proposal to the transformer manufacturer as shown in the following example.

The purchaser shall evaluate the transformer losses submitted in the supplier's quotation (or bid) at the top (or maximum) nameplate rating of the transformer, as follows:

— No-load losses to be evaluated at __$/kW (where the value of a no-load kilowatt can range from $500 to $10 000).

— Load losses to be evaluated at __$/kW (where the value of a load loss kilowatt can range from $300 to $8000).

— Auxiliary equipment losses to be evaluated at ___$/kW (where the value of running the auxiliary equipment, fans and pumps, on the transformer will be somewhat less than the load-loss kilowatt cost).

Evaluated losses shall be calculated for each transformer by multiplying the appropriate dollars per kilowatt values listed in the above table by the guaranteed load losses at the transformer rated winding temperature rise and no-load losses at 100% voltage, as stated in the appropriate spaces, with the products added to the bid price for evaluation.

Procedures for calculating the loss evaluation values are found in IEEE Std C57.120-1991 [B134]; chapter 10, section 152, of the Standard Handbook for Electrical Engineers [B241]; USDA REA Bulletin No. 65-2 [B227]; and EEI 1981 [B93].

ANSI/ASHRAE/IESNA 90.1-1989 now requires the completion of "Transformer Loss Calculation Estimate" Form 5-1 for each transformer 300 kVA and above. The purpose, as stated therein, is "to compare among types of transformers and configurations available to the designer in order to balance energy costs with necessary operating flexibility, reliability (redundancy), and safety."

5.4.2 Liquid-immersed transformers

Liquid-immersed power transformers are designed to operate at standard self-cooled kVA ratings with a 55 °C or a 65 °C winding temperature rise over average ambient temperature. (Some manufacturers may also offer a 115 °C rise liquid-filled transformer.)

A dual 55/65 °C rise insulation rating increases the transformers available capacity by 12%.

Forced-air fan-cooling equipment installed on the transformer provides additional capacity increases, as follows:

Self-cooled rating (kVA)		kVA increases above self-cooled rating when equipped with cooling fans	
Three-phase	Single-phase	First stage of cooling	Second stage of cooling
750–2000	833–1667	15%	Not available
2500–10 000	2500–8333	25%	Not available
12 000–60 000	Not listed	33-1/3%	66-2/3%

The operation of the forced-air fan-cooling stages may be initiated by sensing oil temperature or by sensing winding temperature.

Liquid-immersed transformers may also be cooled by the forced circulation of the insulating liquid through external heat exchangers or radiators.

The various classes of liquid-immersed transformer cooling systems are as follows:

Type letters[a]	Cooling method
OA	Oil-immersed, self-cooled
OW	Oil-immersed, water-cooled
OW/A	Oil-immersed, water-cooled/self-cooled
OA/FA	Oil-immersed, self-cooled/forced-air-cooled
OA/FA/FA	Oil-immersed, self-cooled/forced-air-cooled/forced-air-cooled
OA/FA/FOA	Oil-immersed, self-cooled/forced-air-cooled/forced-air-forced-oil-cooled
OA/FOA/FOA	Oil-immersed, self-cooled/forced-air-forced-oil-cooled/forced-air-forced-oil-cooled
FOA	Oil-immersed, forced-oil-cooled with forced-air-cooler
FOW	Oil-immersed, forced-oil-cooled with forced-water-cooler

[a]Where a liquid other than mineral oil is used as the insulating fluid, then the letter "O" is replaced by another letter, e.g., "L."

For three-phase substation transformers, 1500–50 000 kVA, the relative losses are as shown:

Cooling class	No-load loss	Load loss	Loss ratio	Percentage load for maximum efficiency
OA	100%	100%	1:1	100%
OW	80%	144%	1.8:1	74.5%
OA/FA	80%	144%	1.8:1	74.5%
OA/FA/FA	68%	182%	2.68:1	61.1%
OA/FOA	68%	182%	2.68:1	61.1%
OA/FOW	68%	182%	2.68:1	61.1%

NOTE—No-load loss and load loss data from one manufacturer's publication. Always consult transformer manufacturer's for latest product design and performance data.

5.4.3 Dry-type and solid cast resin transformers

Dry-type and solid cast resin transformers may have an additional capacity by fan-forced cooling. The classes of cooling systems for dry-type transformers are as follows:

Type letters	Cooling system class
AA	Dry-type, self-cooled
AFA	Dry-type, forced-air-cooled
AA/FA	Dry-type, self-cooled/forced-air-cooled
AA/AA/FA	Dry-type, self-cooled at 80 °C rise/self-cooled at 150 °C rise/forced-air-cooled at 150 °C rise
NA	Sealed dry-type, nitrogen-filled, self-cooled
GA	Sealed dry-type, gas-filled (C_2F_6, for example), self-cooled

The transformer cooling class AA/AA/FA means the transformer has a 220 °C insulation system and that the transformer may operate to 80 °C and to 150 °C rise without fans, and with fans running, would provide increased capacity.

The increase in capacity due to forced-air cooling is 33-1/3% for dry-type transformers. Solid cast resin transformers of some manufacturers allow 33-1/3% capacity increase through 750 kVA and 50% capacity increase for 1000 kVA and above with forced-air cooling.

Custom-designed dry-type transformers, usually transformers with a rating over 600 V, may be specified with a loss ratio that results in the most efficient operation for the load. Loss evaluation dollars per kilowatt for no-load loss and load-loss should be included with the request for proposal sent to the manufacturer.

The relationship between transformer materials and efficiency at various loads is shown in annex 5I for two winding temperature rises.

Mass-produced dry-type transformers, usually transformers with a rating less than 600 V, need to be evaluated on the basis of the no-load losses and load-losses inherent in the manufacturer's designs at the various insulation classes and winding temperature rises. The losses may vary between designs of different manufacturers even for the same kVA rating, insulation class, and winding temperature rise. After selecting the rating, insulation class, and temperature rise, the specifier should also include in the request for quotation sent to the manufacturers, the loss evaluation dollars per kilowatt for no-load loss and load-loss. This usually signals the manufacturer to take account of the loss evaluation penalty in setting the selling price.

Dry-type transformers in the 600 V class can be evaluated based upon transformer loading, losses, and energy costs, as shown in the following example (example data courtesy of Acme Electric Co.):

Evaluate a 75 kVA transformer operating 24 h/day: 8 h at 100% load, 8 h at 66% full load, and 8 h at zero load. The electric rate is 5 cents per kWh ($0.05/kWh). Using the loss data in table 5-20, calculate the annual costs of operating a 150 °C rise and an 80 °C rise transformer, and the simple payback period. (See tables 5-19 through 5-22.)

Table 5-19—Example of a transformer loading schedule for the sample calculation of determining the cost of transformers losses

kVA	Daily hours in use	Full load (%)	No-load loss (%)
75	8	100	100
	8	66	100
	8	0	100
Total	24		

Table 5-20—Dry-type transformer loss data

Winding temperature rise (°C)	Three-phase kVA	No-load loss (%)	Load loss (%)	Total loss (%)	Loss ratio, R[a]	Efficiency (%)	Maximum efficiency loading, L[b] (%)
(1)	(2)	(3)	(4)	(5)	(6)	(7)	(8)
80	30–75	0.31	2.73	3.04	8.806	97.05	34%
	112.5– 225	0.39	1.81	2.20	4.641	97.85	46%
	300–500	0.37	1.53	1.90	4.135	98.14	49%
115	30–75	0.27	2.87	3.13	10.630	97.10	31%
	112.5– 225	0.38	2.06	2.44	5.421	97.62	43%
	300–500	0.34	1.13	1.47	3.324	98.50	55%
150	30–75	0.53	3.84	4.37	7.245	95.82	37%
	112.5– 225	0.43	2.28	2.71	5.302	97.86	43%
	300–500	0.30	2.09	2.39	6.966	97.30	38%

Source: Data for columns 1–5 courtesy of Acme Electric Co.
[a]Column 6 = column 4 divided by column 3.
[b]Column 8 = 1 divided by the square root of column 6, expressed as a percent.

Table 5-21—Cost of losses example for a 75 kVA transformer rated 150 °C rise

Core losses (no-load losses)					Total loss = no-load loss + load loss ($)
Full-load (%)	Percent loss	kW loss	kWh loss	Loss $	
100	0.53	0.3975	1160.7	58.03	
66	0.53	0.3975	1160.7	58.03	
0	0.53	0.3975	1160.7	58.03	
			Total no-load:	$174.09	
Winding losses (load losses)					
100	3.84	2.88	8410	420.50	478.53
66	$3.84 \cdot (0.66)^2 = 1.67$	1.253	3659	182.95	240.98
0	$3.84 \cdot (0)^2 = 0$	0	0	0	58.03
			Total load:	$603.45	$777.54

NOTE—The energy cost is based on $0.5/kWh.

Table 5-22—Cost of losses example for a 75 kVA transformer rated 80 °C rise

Winding losses (load losses)					Total loss = no-load loss + load loss ($)
Full-load (%)	Percent loss	kW loss	kWh loss	Loss $	
100	0.31	0.2325	678.9	33.95	
66	0.31	0.2325	678.9	33.95	
0	0.31	0.2325	678.9	33.95	
			Total no-load:	$101.85	
Core losses (no-load losses)					
100	2.73	2.05	5986	299.30	333.25
66	$2.73 \cdot (0.66)^2 = 1.89$	0.892	2605	130.25	164.20
0	$2.73 \cdot (0)^2 = 0$	0	0	0	33.95
			Total load:	$429.55	$531.40

NOTE—The energy cost is based on $0.5/kWh.

The amount of savings using 80 °C rise low-loss transformers is summarized as follows:

Operating cost of 150 °C rise = $777.54
Operating cost of 80 °C rise = $531.40
Operating difference = $246.14
Difference in purchase price = $765.00

Simple payback period $= \dfrac{\$765.00}{\$246.14/\text{yr}} = 3.1$ years in favor of buying the 80 °C rise transformer.

5.4.4 Transformer energy saving recommendations

a)   Requests for quotations for new transformers should include the following information:

1)   Loss evaluation dollars per kilowatt for no-load losses.

2)   Loss evaluation dollars per kilowatt for load-losses.

3)   Loss evaluation dollars per kilowatt for auxiliary load loss.

4)   The transformer loss ratio indicating the percentage of full load that the transformer shall be operating for maximum efficiency.

5)   The distribution system load factor indicating the ratio of average annual load to peak annual load.

6)   Load power quality, sinusoidal or type of nonlinear load.

7)   For liquid-immersed transformers: alternate quotations for 55 °C, 65 °C, and dual-rated 55/65 °C rise.

8)   For dry-type transformers: alternate quotations for 80 °C, 115 °C, and 150 °C rise.

b)   Maintenance schedules should assure that transformer ventilating louvers, radiators, fans, are clean and unobstructed and that transformer surfaces are protectively coated with appropriate coatings.

c)   Transformer vaults should provide sufficient cooling air at a rate per the NEC Article 450 and not less than one square foot (0.0929 m^2) of inlet and outlet per 100 kVA of rated transformer capacity after deducting the area occupied by screens, gratings, and louvers.

d)   When available, dry-type transformers should be supplied with temperature-sensing bulbs for each phase with instrumentation capable of reading each phase on request and indicating maximum temperature. Fan forced air cooling should be controlled from this instrumentation.

e)   Operate the transformer at the correct voltage tap setting. Operating a transformer at a higher primary voltage setting adds more impedance in the primary winding and increases the losses. Therefore carefully select the primary voltage tap that produces the appropriate nominal secondary voltage at no-load condition, for example 480 V.

5.4.5 The minimum total loss loading of substation transformers operating in parallel (Franklin and Franklin 1983 [B242])

Transformer substations that are attended or that are provided with remote switching can be economically operated by reducing the total transformation losses to a minimum under all loading conditions.

Transformers operating in parallel should have these three characteristics identical:

    a)   Voltage ratios
    b)   Percentage impedances
    c)   Ratios of resistance to reactance voltages

The economic loading analysis determines where the transformer(s) are to be switched into or out of the circuit so that the minimum total losses, i.e., core plus coil losses, are obtained.

An example of minimum total loss loading is as follows:

Consider two 500 kVA transformers each with a core loss of 2.7 kW and a full-load coil loss at 60 °C of 5.7 kW.

The load at which one 500 kVA transformer operates with the minimum percentage losses is given by

$$K_1 = WK$$

where

$$W = \left(\frac{P_f}{P_c}\right)^{1/2}$$

    K is the rated kVA
    P_f is the core loss (in W) or as a percentage of the normal full-load rating
    P_c is the coil loss (in W) or as a percentage of the normal full-load rating

$$K_1 = \left[\frac{2700 \text{ W}}{5700 \text{ W}}\right]^{1/2} \cdot 500 \text{ kVA} = 344 \text{ kVA}$$

Thus, 344 kVA is the most efficient loading of a single transformer.

The critical output where the total losses in a single transformer are the same as the sum total losses of two transformers in parallel determines the substation output when the second transformer is switched into or out of the circuit. At the critical output each of the transformers in parallel would be carrying one-half the load of one transformer operating singly. This is expressed by the equation:

$$P_f + \left[\frac{K_x}{K}\right]^2 P_c = 2\left[P_f + \left[\frac{K_x}{2K}\right]^2 P_c\right]$$

For two identical transformers, the equation for the kVA value of the critical loading where the coneflower from one to two transformers should occur becomes

$$K_x = K\left(\frac{2P_f}{P_c}\right)^{1/2}$$

For the two 500 kVA transformers in this example the critical output value is

$$K_x = 500 \text{ kVA} \left[2 \cdot \frac{2700 \text{ W}}{5700 \text{ W}}\right]^{1/2} = 486.7 \text{ kVA}$$

Therefore, when the load increases on a single transformer to 486 kVA, then switching in a second transformer for parallel operation results in the substation total losses being minimized as load increases. Similarly when the substation load decreases to 486 kVA then the second transformer may be switched out of the circuit for most efficient operation.

Refer to Franklin and Franklin 1983 [B242] for techniques to analyze substations with more than two transformers and for substations with dissimilar transformers.

5.4.6 Reactors

5.4.6.1 General

Reactors are wound devices similar in many respects to transformers; they are used in the following applications:

a)   Motor-starting reactors for reduced-voltage starting
b)   Current-limiting reactors such as for motor-control centers
c)   Substation bus-tie current-limiting reactors
d)   Neutral grounding reactors
e)   Shunt reactors for compensation in long underground cable circuits
f)   Reactors as used in harmonic filters for power-factor improvement
g)   Generator mains reactors are used on generator synchronizing buses

The industry standards for reactors are as follows:

—   IEEE Std C57.21-1990 [B130]
—   NEMA ST 20-1992 [B168]

5.4.6.2 Energy-saving opportunities

Reactors are usually "specified" devices requiring the specifier to state the necessary imped-ance to perform the current-limiting compensation or filtering.

Smaller reactors for reduced-voltage starting and motor-control-center current-limiting appli-cations are available already designed; therefore, it usually is not economical to custom design smaller reactors solely for energy conservation.

Where the opportunity does exist for energy conservation in the larger reactors, the specifier can advise the manufacturer about the loss evaluation dollars per kilowatt as follows:

— *Shunt reactor.* A shunt reactor acts as a constant load at a given voltage. Its total loss cost evaluation (even though consisting mainly of I^2R losses) is calculated using only the no-load formula. Its losses increase as the impressed voltage increases.

— *Series reactors.* A series reactor experiences varying load. Because it does not have a no-load loss, its total power loss cost evaluation is calculated using only the load loss formula.

5.5 Capacitors and synchronous machines

5.5.1 Quality of power

Energy savings can be achieved by improving the quality of the power supply to the utiliza-tion equipment.

The quality of the power supply is defined by five independent parameters:

a) *Nominal supply quality.* The magnitude, harmonic content, and balance of the supply voltage waveform under normal conditions.

b) *Long-term reliability.* The frequency and extent of long-term power outages (greater than 10 s) such as caused by a permanent fault.

c) *Short-term interruption.* The frequency of short-term total supply interruptions (less than 10 s) such as an auto recloser operation.

d) *Short-term disturbances.* The frequency and magnitude of short-term disturbances (less than 10 s), including voltage drops, such as caused by reflected faults from adja-cent feeders.

e) *Transient disturbances.* The frequency and magnitude of transient disturbances (often less than one cycle duration and no more than several seconds), including impulses, voltage surges (spikes), and voltage dips (sags).

The nominal voltage to the service point may vary according to ANSI C84.1-1989 [B26] within the limits of Range A or 105% to 95% of the nominal system voltage on a regulated power distribution system. On an unregulated power distribution system the voltage may vary within Range B or from 105.8% to 91.7% of the nominal system voltage. Refer to IEEE Std 141-1993 for a further discussion.

Knowing the allowable voltage range at the service, the responsible industrial facility staff can set the transformer taps, or the automatic tap changing control, to provide the optimum voltage within the facility for most efficient running of motors and other equipment.

The system frequency at the supply service may vary within the range of 60 Hz plus or minus 0.1 Hz. This is outside the control of the responsible industrial facility staff unless frequency converters are installed.

The power supply system designed by the utility should be capable of keeping the maximum system unbalance to below 3%.

5.5.1.1 Harmonic limits

5.5.1.1.1 Definitions

distortion factor (harmonic factor). The ratio of the root-mean-square of the harmonic content to the root-mean-square value of the fundamental quantity, expressed as a percent of the fundamental.

$$DF = \left[\frac{\text{sum of the squares of amplitudes of all harmonics}}{\text{square of amplitude of fundamental}} \right]^{1/2} \cdot 100\%$$

filter. A generic term used to describe those types of equipment whose purpose is to reduce the harmonic current or voltage flowing in or being impressed upon specific parts of an electrical power system, or both.

harmonic. Sinusoidal quantities, the frequencies of which are whole multiples of the 60 Hz fundamental frequency.

harmonic analyzer. An instrument for recording and displaying the waveform of the electrical circuit into its component frequencies and amplitudes.

harmonic factor. The ratio of the root-sum-square (rss) value of all the harmonics to the root-mean-square (rms) value of the fundamental use:

$$\text{harmonic factor (for current)} = \frac{\left(I_3^2 + I_5^2 + I_7^2 \right)^{1/2}}{I_1}$$

$$\text{harmonic factor (for voltage)} = \frac{\left(E_3^2 + E_5^2 + E_7^2 + \ldots \right)^{1/2}}{E_1}$$

harmonic order (*n*). The ratio between the harmonic frequency and the fundamental frequency.

nonlinear load. A load that draws a nonsinusoidal current wave when supplied by a sinusoidal voltage source.

point of common coupling (PCC). Point at which the customers supply tees off the general utility system supplying other customers.

total demand distortion (TDD). The total root-sum-square harmonic current distortion in percent of the maximum demand load current (15 min or 30 min demand).

total harmonic distortion (THD). This term has come into common usage to define voltage or current "distortion factor."

triplen harmonics. Harmonics having a frequency exactly divisible by three times the fundamental frequency.

voltage unbalance. The voltage unbalance, in percent, is 100 times the maximum voltage deviation from the average voltage, divided by the average voltage:

$$\text{Voltage: unbalance (\%) } 100 \cdot \frac{\text{maximum voltage deviation from average voltage}}{\text{average voltage}}$$

5.5.1.1.2 General

The problem of loads with nonlinear characteristics depends upon their connection within the building's electrical distribution system.

There are two sets of standards for harmonic limits as described:

a) *Utility harmonic limits (harmonic limits for suppliers).* Electric power provided to the building is required to have a quality defined by the voltage distortion content as measured at the *point of common coupling* (PCC).

Bus voltage at PCC	Individual voltage distortion (%)	Total voltage distortion (THD%)
69 000 V and below	3.0	5.0

b) *Consumer harmonic limits (harmonic limits for buildings/industrial facilities).* Consumers (the building owners/operators) are required to limit the harmonic current injection from their equipment into the utility system as measured at the PCC. The harmonic current limits are based on the size of the power system.

The following table shows current distortion limits for general distribution systems (120 V–69 000 V). The maximum harmonic current distortion is in percent of maximum demand load current individual harmonic order (odd harmonic).

I_{sc}/I_L	$h{<}11$	$11{\leq}h{<}17$	$17{\leq}h{<}23$	$23{\leq}h{<}35$	$35{\leq}h$	TDD
<20	4.0	2.0	1.5	0.6	0.3	5.0
20<50	7.0	3.5	2.5	1.0	0.5	8.0

where

I_{sc} is the utility maximum short-circuit current at the PCC

I_L is the maximum demand load current (60 Hz) at the PCC

NOTE—The above table, from table 10.4 of IEEE Std 519-1992 [B127], continues for other power systems. The usual consumers have power systems in the range of I_{sc}/I_L less than 20.

c)   *Equipment exceeding harmonic limits.* When it is found by the measurement survey that the harmonic limits are exceeded, then some or all of the following procedures may be followed to reduce the distortion to acceptable levels:

1)   Rearrangement of the incoming power supply conductors to a different utility distribution line or to a larger capacity utility distribution line. This may avoid poor power quality caused by another utility customer.

2)   Equipment with three-phase electronic *uninterruptible power supply* (UPS) systems, for power conditioning and supply during utility outages, should use 12-pulse rectifiers to minimize distortion on the input feeders and 12-step inverters to eliminate all harmonics below the 11th.

3)   Equipment with three-phase adjustable speed drives, for motor speed control, should use harmonic filters to reduce or eliminate the 5th, 7th, and 11th harmonics or others.

4)   Run a dedicated neutral conductor with each phase conductor when supplying single-phase loads with harmonic content.

5)   Run a full-size or oversized neutral conductor on 4-wire, 3-phase circuits where the major portion of the load consists of nonlinear loads, such as electric-discharge lighting (fluorescent and high-intensity discharge lighting), electronic computers (data processing), or similar equipment, where there are harmonic currents in the neutral conductor.

6)   In panelboards, switchboards, and motor control centers where there may be single-phase loads, rearrange the circuits among the phases to balance the currents so that the current in the neutral due to phase unbalance is reduced or eliminated.

7)   Use isolation transformers in branch circuits with harmonic distortion problems. For example, a delta-wye isolation transformer may improve the power factor,

block neutral currents from transferring further up the line, lower the current distortion on the branch circuit, and act to reground the secondary neutral to lower the neutral-to-ground voltage problems.

8) Use active power line conditioners (APLCs) to cancel harmonics. An active power line conditioner is a compact, integrated power conditioning device that helps to solve a variety of power quality problems. APLCs combine adaptive, active harmonic filtering, effective line voltage regulation, and transient voltage surge suppression to protect sensitive electronic equipment. APLCs are available in single-phase equipment at 5 kVA and in three-phase equipment up to 150 kVA for service entrance and non-service entrance applications.

9) Use "zero sequence filter traps," also called "third harmonic traps," to reduce the 3rd harmonic current. These traps are a version of a zig-zag transformer that has been designed to reduce high neutral currents. The devices are selected based upon measurements of the true rms current in the neutral conductors(s). The equipment is compact and typically available in current ratings of 100, 150, 300, and 450 A maximum neutral current. Typically the trap is installed on the line side of a 208Y/120 V panelboard.

10) Consult with specialists on replacing automatically switched power-factor correction capacitor bank controllers. Alternatives include controllers that switch only at the voltage waveform.

11) Consult with specialists on adding filters to unprotected capacitor banks, or other measures, to avoid resonance due to harmonics.

12) On new installations where power-factor correction equipment is required to avoid utility penalties, consider the application of synchronous motors. Synchronous motors can drive a compressor or other large load while improving the power, and there are no harmonic problems associated with their factor use.

13) Similar to item 12), consider synchronous condensers where there is no motor load but power-factor improvement is still required.

5.5.1.2 Transient overvoltage concerns

Each time a capacitor is energized, a transient voltage oscillation occurs between the capacitor and the power system inductance. The result is a transient overvoltage that can be as high as 2.0 V/unit at the capacitor location. These overvoltages are usually not a utility concern; however, these transients, caused by utility switching of capacitor banks, can be magnified at the customer facility if the customer has low-voltage capacitor banks for power-factor correction.

Adjustable speed drives (ASDs) are particularly susceptible to these transients because of the relatively low *peak inverse voltage* (PIV) ratings of semiconductor switches and the low energy ratings of the *metal oxide varistors* (MOVs) used to protect the power electronics. Refer to EPRI [B98] for more information.

The magnified transient overvoltages can be controlled by the following:

— Using vacuum switches with synchronous closing control to energize the capacitor bank.

— Installing high energy MOV protection on the 480 V buses to limit the transient magnitudes to about 1.8 per unit. The energy capability of these surge arresters should be at least 1 kJ.

— Using tuned filters for power-factor correction instead of just shunt capacitor banks. These filters change the response of the circuit and usually prevent magnifications from being a problem.

The quality of power parameters that can affect the losses and efficiency of the plant's electrical power distribution system are as follows:

Symptom	Possible cause	Probable solution
Motors overheat when powered by *variable frequency drives* (VFDs)	High-frequency core losses and conductor eddy currents due to VFD	Add an output reactor or filter between the VFD and motor to reduce additional heating.
DC motor overheats after a new dc drive replaces an M-G set	Output of the dc drive contains higher ripple voltages	Install a line choke in the dc bus to the motor to reduce ripple current.
Overheating of plant electrical distribution equipment (wires, transformers, switches, etc.)	Low power factor due to reactive currents drawn by ac motors and by harmonic currents from rectifiers and/or VFDs	If only linear loads are being served, correct with power capacitors. If nonlinear loads are being served and harmonics are present, correct the power factor with a harmonic filter or with synchronous condensers
Poor voltage regulation within the plant or building	Low power factor results in extra current being drawn	Add power-factor correction at the load with capacitors or synchronous condensers, if only linear loads are present, or with harmonic filters or with synchronous condensers if the loads are nonlinear, to reduce reactive and harmonic currents.
Low power factor and/or power-factor penalties	• Non-power factor corrected fluorescent and high-intensity discharge lighting ballasts (40%–80% PF) • Arc welders (50%–70% PF) • Solenoids (20%–50% PF) • Lifting magnets (20%–50% PF) • Small dry-type transformers (30%–95% PF) • Induction motors (55%–90% PF) • Nonlinear loads, e.g., rectifiers	Add power-factor correction at the load with capacitors or synchronous condensers if only linear loads. If loads are nonlinear, add harmonic filters to reduce reactive and harmonic currents or use synchronous condensers.
Excessive current being drawn by equipment	Improper voltage	Change the transformer tap; adjust the substation voltage by adding a voltage regulator or capacitor bank or harmonic filter or synchronous condenser. Use larger size conductors to reduce voltage drops.

Source: Trans-Coil, Inc. 1991 [B231]

5.5.2 Capacitors for motor power-factor improvement

An example of the effect of installing power-factor correction capacitors at the motor is provided in table 5-23. In this example, the maximum circuit length from the incoming service transformer, secondary unit substation, or load center switchboard to the motor terminals is 180 ft (IEEE Std 241-1990, Chapter 3, Table 30). The motors used in this example are NEMA Design B premium high-efficiency motors, rated 1800 r/min, in drip-proof enclosures (NEMA MG 1-1993 [B164]).

a) The energy savings are dependent upon the percent ampere reduction due to the additional kvar.

b) The energy savings are dependent upon the motor design, the conductor size, the conductor length.

c) The payback economics depends upon the capacitor design and features plus the labor for installation.

d) The energy savings are also dependant upon the capacitor watts loss and the reduction in heat gains to be removed by the air-conditioning system. Note that these items are not calculated for this example.

5.5.3 Capacitors for plant power-factor improvement (Square D Bulletin D-412D [B234])

An improvement power factor can provide both economic and system advantages. Direct economic advantages are attained when monetary incentives, such as a power-factor penalty, are enforced. Operational benefits, such as improved system efficiency, release of system capacity, reduction of power losses, and voltage improvement, may also be obtained.

5.5.3.1 Reduced power cost

Electric utilities are obligated to provide the necessary power to their customers. This includes both the customers active and reactive power requirements. However, the reactive power component does *not* register on the utilities revenue metering. This means the utility must spend extra money in generation and transmission equipment to provide a power component for which there is no direct return.

Consequently, many utilities now include a power-factor adjustment clause in their rate structures to compensate for providing this component. These clauses, or *penalties*, provide a strong economic incentive for improving the power factor and often account for a significant portion of the total power bill.

By locally furnishing reactive kvar, the consumer can often enjoy substantial savings by avoiding these penalties.

Example: Determine required kvar, savings, and payback period for an industrial plant subjected to the following conditions:

Demand charge: $5.00 kW billing demand

Table 5-23—Capacitors for motor power factor improvement

Motor rating		Manufacturer's data[a]						Calculated data						
hp	kW	Amperes, full-load, @ 460 V	Efficiency, full-load, nominal	Power factor, full load (cos ø)	Max kvar[b]	Installed kvar std unit	kVA input, orig, no correction	kW input	ø degrees ($\cos^{-1} ø$)	sin ø	kvar, orig, no correction	kvar, new, with correction	kVA new, with correction	Power factor, corrected
7.5	5.6	9.4	91.7	81.5	2.5	2	7.5	6.1	35.4	0.579	4.3	2.3	6.5	93.8
10	7.5	12.7	91.7	80.5	3.6	3	10.1	8.1	36.4	0.593	6.0	3.0	8.6	94.2
15	11.2	18.6	93.0	81.5	5.3	5	14.8	12.1	35.4	0.579	8.6	3.6	12.6	96.0
25	18.7	29.8	94.1	83.5	6.9	5	23.7	19.8	33.4	0.550	13.0	8.0	21.4	92.5
50	37.3	57.3	94.5	86.5	12.0	10	45.7	39.5	30.1	0.502	22.9	12.9	41.6	95.0
75	56.0	86.6	95.4	85.0	18.8	17.5	69.0	58.7	31.8	0.527	36.4	18.9	61.7	95.1
100	74.6	113.0	95.4	86.5	25.4	25	90.0	77.9	30.1	0.502	45.2	20.2	80.5	96.8
125	93.3	136.0	95.4	90.5	19.7	17.5	108.4	98.1	25.2	0.426	46.2	28.7	102.2	96.0
150	111.9	167.0	96.2	87.5	34.0	30	133.1	116.4	29.0	0.485	64.5	34.5	121.4	95.9
200	149.2	216.0	96.2	90.0	36.5	35	172.1	154.9	25.8	0.435	74.9	39.9	160.0	96.8
250	186.5	293.0	96.2	83.0	81.2	75	233.4	193.7	33.9	0.558	130.2	55.2	201.4	96.2
300	223.8	348.0	96.2	84.0	84.1	75	277.3	232.9	32.9	0.543	150.6	75.6	244.9	95.1
400	298.4	428	95.8	91.0	66.0	60	341.0	310.3	24.5	0.415	141.4	81.4	320.8	96.7
450	335.7	447	95.8	92.0	58.4	50	380.0	349.6	23.1	0.392	149.1	99.1	363.4	96.2
500	373.0	542	95.8	90.0	89.9	75	431.8	388.6	25.8	0.435	187.8	112.8	404.6	96.0
600	447.6	643	95.8	91.0	96.7	75	512.3	466.2	24.5	0.415	212.6	137.6	486.1	95.9

[a]Horizontal performance data for "Premium Efficiency Energy $aver® Motors" from GE Motors as of 6/93 for Type KX, NEMA Design B, normal starting torque, continuous 40 °C ambient, 460 V, 3-phase, nominal speed of 1800 r/min, drip-proof, frames 182–449 to 300 hp and frames 509–5011 for 400–600 hp (GEP-500J [B105]). Always consult the motor manufacturer for the latest product design and performance data.

[b]Recommended maximum capacitor rating when capacitor and motor are switched as a unit.

CAUTION: *In no case should power factor improvement capacitors be applied in ratings exceeding the maximum safe value specified by the motor manufacturer.*

Table 5-23—Capacitors for motor power factor improvement (*Continued*)

Motor rating hp	Line current, amperes with PF correction	Percent ampere reduction in line current	125% orig. full-load amperes[c]	Wire size, AWG or kcmil	Conductor resistance, Ω/1000 ft[d]	Power loss, I^2R, per phase (W)			Demand savings per circuit (W)	Annual energy savings (kWh)	Annual cost savings[e] ($)
						No PF correction	With PF correction	Difference			
7.5	8.2	12.8	11.8	12	1.6200	25.8	19.6	6.2	18.6	162.9	16
10	10.8	15.0	15.9	12	1.6200	47.0	34.0	13.0	39.0	341.6	34
15	15.8	15.1	23.3	10	1.0180	63.4	45.7	17.7	53.1	465.2	47
25	26.9	9.7	37.3	8	0.6404	102.4	83.4	19.0	57.0	499.3	50
50	52.2	18.7	71.6	4	0.2590	153.1	127.0	26.1	78.3	685.9	69
75	77.4	10.6	108.3	2	0.1640	221.4	176.8	44.6	133.8	1172.1	117
100	101.0	10.6	141.3	1/0	0.1040	239.0	191.0	48.0	144.0	1261.4	126
125	128.3	5.7	170.0	2/0	0.0835	278.0	247.4	30.6	91.8	804.2	80
150	152.4	8.7	208.0	4/0	0.0534	268.1	223.2	44.9	134.7	1180.0	118
200	200.8	7.0	270.0	300	0.0385	323.2	279.4	43.8	131.4	1151.1	115
250	252.8	13.7	366.3	500	0.0244	377.0	280.7	96.3	288.9	2530.8	253
300	307.4	11.7	435.0	2@4/0	0.0534/2	582.0	454.1	127.9	383.7	3361.2	336
400	402.6	5.9	535.0	2@300	0.0385/2	634.7	561.6	73.1	219.3	1921.1	192
450	456.1	4.4	596.3	2@350	0.0333/2	681.9	623.5	58.4	175.2	1534.8	153
500	507.8	6.3	677.3	2@400	0.0297/2	785.0	689.0	96.0	288.0	2523	252
600	610.1	5.1	803.8	2@600	0.0209/2	778.0	700.0	78.0	234.0	2050	205

[c]Wire Sizing Rules, NEC [B173], Section 4309-22 (a).
[d]Source: IEEE Std 241-1990, Table 65.
[e]Savings based on $0.10/kWh.

Billing demand: actual demand · 0.95/actual power factor

Monthly power bills should be first tabulated showing the monthly demand power factor, billing demand, and demand charge for the actual and desired power factor. A twelve-month period is normally evaluated to account for both winter and summer demand peaks.

Demand charges incurred loaned on the actual power factor				
Month	Actual demand	Power factor (kW)	Billing demand (kW)	Demand charge ($)
Jan.	227	0.6350	340	1 698.00
Feb.	257	0.6508	375	1 876.00
Mar.	226	0.6628	324	1 620.00
Apr.	240	0.6780	336	1 681.00
May	219	0.6733	309	1 545.00
June	219	0.6729	309	1 546.00
July	235	0.6339	352	1 761.00
Aug.	251	0.5837	409	2 043.00
Sept.	239	0.5485	414	2 070.00
Oct.	252	0.5502	435	2 176.00
Nov.	238	0.5716	396	1 978.00
Dec.	249	0.5758	411	2 054.00
			Annual demand charge	$22 048.00

The tabulation is again performed using a desired power factor of 0.95 as a minimum power factor required to eliminate the penalty.

Demand charges incurred based on the desired power factor				
Month	Actual demand	Power factor (kW)	Billing demand (kW)	Demand charge ($)
Jan.	227	0.95	227	1 135.00
Feb.	257	0.95	257	1 285.00
Mar.	226	0.95	226	1 130.00
Apr.	240	0.95	240	1 200.00
May	219	0.95	219	1 095.00
June	219	0.95	219	1 095.00
July	235	0.95	235	1 175.00
Aug.	251	0.95	251	1 255.00
Sept.	239	0.95	239	1 195.00
Oct.	252	0.95	252	1 260.00
Nov.	238	0.95	238	1 190.00
Dec.	249	0.95	249	1 245.00
			Annual demand charge	$14 260.00

Thus, by improving the monthly power factor to 95%, a savings of $7788 will occur. An additional reduction in the power bill will also result since I^2R is also reduced.

The required capacitive kvar to improve the power factor to 95% is determined for each month by using monthly demand power factor and table 5-26. The results of the calculations are summarized as follows:

Month	Actual demand	Power factor actual (kW)	Billing demand desired (kW)	Required kvar
Jan.	227	0.6350	0.95	201.5
Feb.	257	0.6508	0.95	215.4
Mar.	226	0.6628	0.95	181.0
Apr.	240	0.6780	0.95	181.3
May	219	0.6733	0.95	168.5
June	219	0.6729	0.95	168.8
July	235	0.6339	0.95	209.5
Aug.	251	0.5837	0.95	266.7
Sept.	239	0.5485	0.95	285.8
Oct.	252	0.5502	0.95	299.6
Nov.	238	0.5716	0.95	263.4
Dec.	249	0.5758	0.95	271.7

From the above tabulation, at least 299.6 kvar (300 kvar being the nearest available capacitor rating) is required to achieve a 95% power factor for the entire year.

The cost for providing this equipment will depend on the method of correction selected, equipment, and labor cost. For this example, a value of $50.00/kvar is selected.

The time required for the power-factor improvement project to pay for itself is as follows:

$$\text{payback period} = \frac{\$ \text{ cost}}{\$ \text{ savings/year}}$$

$$\text{payback period} = \frac{(300 \cdot \$50)}{(\$22\ 048 - \$14\ 260)}$$

$$= 1.93 \text{ years}$$

5.5.3.2 Release of system capacity

When power-factor capacitors are located at the terminals of an inductive load, they will deliver all or most of the reactive power required by the load. This means a reduction in system current will occur, permitting additional load to be connected to the system without increasing the size of transformers, switchboards, and other distribution equipment.

Often, this release in system capacity is reason enough to warrant an improvement in power factor, especially when conductors or panels are overheating or where overcurrent devices frequently open.

The percent released capacity resulting from an improvement in power factor is as follows:

$$\% \text{ released system capacity} = 100 \cdot \left(1 - \frac{\text{pf}_\text{o}}{\text{pf}}\right)$$

where

pf$_\text{o}$ is the original power factor
pf   is the final power factor after correction

Example: Determine the system capacity released by improving the power factor from 0.6 to 0.90.

$$\% \text{ released system capacity} = 100 \cdot \left(1 - \frac{0.6}{0.9}\right)$$

$$= 33.3\%$$

This means that, after adding capacitors that correct the system to a 0.90 power factor, the kVA load or line current is reduced by 33.3% of what it was before power-factor correction. Putting this another way, with the required capacitors on line, an additional 33.3% kVA load or line current can be added to this system without exceeding the amount utilized before power-factor correction.

5.5.3.3 Reduced power losses, when capacitors are located at the load

Another benefit resulting from a power-factor improvement project is the reduction of power system losses. This is especially true for older power systems where the kilowatt (I^2R) losses can account for as much as 2–5% of the total load.

Since power losses are proportional to the current squared, and the current is proportional to the power factor, an improvement in power factor will cause a reduction in system losses and reduce power bills.

This reduction can be approximated as

$$\% \text{ loss reduction} = 100 \cdot 1 - \frac{\text{original power factor}^2}{\text{desired power factor}^2}$$

5.5.3.4 Voltage improvement

When capacitors are added to the power system, the voltage level will increase. The percent voltage rise associated with an improvement in power factor can be approximated as follows:

$$\% \text{ voltage rise} \approx \frac{\text{capacitor kvar} \cdot \text{transformer } \%IZ}{\text{transformer kVA}}$$

Under normal operating conditions, the percent voltage rise will only amount to a few percent. Therefore, voltage improvement should not be regarded as a primary consideration for a power-factor improvement project.

Table 5-24—Suggested maximum capacitor ratings—used for high-efficiency motors and older design (pre-T-frame) motors[a]

Induction motor horse-power rating	Number of poles and nominal motor speed in r/min											
	2 3600 r/min		4 1800 r/min		6 1200 r/min		8 900 r/min		10 720 r/min		12 600 r/min	
	Capacitor (kvar)	Current reduction (%)	Capacitor (kvar)	Current reduction (%)	Capacitor (kvar)	Current reduction (%)	Capacitor (kvar)	Current reduction (%)	Capacitor (kvar)	Current reduction (%)	Capacitor (kvar)	Current reduction (%)
3	1.5	14	1.5	15	1.5	20	2	27	2.5	35	3	41
5	2	12	2	13	2	17	3	25	4	32	4	37
7-1/2	2.5	11	2.5	12	3	15	4	22	5	30	6	34
10	3	10	3	11	3	14	5	21	6	27	7.5	31
15	4	9	4	10	5	13	6	18	8	23	9	27
20	5	9	5	10	6	12	7.5	16	9	21	12.5	25
25	6	9	6	10	7.5	11	9	15	10	20	15	23
30	7	8	7	9	9	11	10	14	12.5	18	17.5	22
40	9	8	9	9	10	10	12.5	13	15	16	20	20
50	12.5	8	10	9	12.5	10	15	12	20	15	25	19
60	15	8	15	8	15	10	17.5	11	22.5	15	27.5	19
75	17.5	8	17.5	8	17.5	10	20	10	25	14	35	18
100	22.5	8	20	8	25	9	27.5	10	35	13	40	17
125	27.5	8	25	8	30	9	30	10	40	13	50	16

Source: Commonwealth Sprague Capacitors PF-2000B [B74].
[a]For use with 3-phase, 60 Hz, NEMA Design B motors (NEMA MG 1-1993 [B164]) to raise full-load power factor to approximately 95%.

Table 5-24—Suggested maximum capacitor ratings—used for high-efficiency motors and older design (pre-T-frame) motors[a] (*Continued*)

Induction motor horse-power rating	Number of poles and nominal motor speed in r/min											
	2 3600 r/min		4 1800 r/min		6 1200 r/min		8 900 r/min		10 720 r/min		12 600 r/min	
	Capacitor (kvar)	Current reduction (%)	Capacitor (kvar)	Current reduction (%)	Capacitor (kvar)	Current reduction (%)	Capacitor (kvar)	Current reduction (%)	Capacitor (kvar)	Current reduction (%)	Capacitor (kvar)	Current reduction (%)
150	30	8	30	8	35	9	37.5	10	50	12	50	15
200	40	8	37.5	8	40	9	50	10	60	12	60	14
250	50	8	45	7	50	8	60	9	70	11	75	13
300	60	8	50	7	60	8	60	9	80	11	90	12
350	60	8	60	7	75	8	75	9	90	10	95	11
400	75	8	60	6	75	8	85	9	95	10	100	11
450	75	8	75	6	80	8	90	9	100	9	110	11
500	75	8	75	6	85	8	100	9	100	9	120	10

Source: Commonwealth Sprague Capacitors PF-2000B [B74].
[a]For use with 3-phase, 60 Hz, NEMA Design B motors (NEMA MG 1-1993 [B164]) to raise full-load power factor to approximately 95%.

Table 5-25—Suggested maximum capacitor ratings— T-frame NEMA Design B motors[a]

Induction motor horse-power rating	Number of poles and nominal motor speed in r/min											
	2 3600 r/min		4 1800 r/min		6 1200 r/min		8 900 r/min		10 720 r/min		12 600 r/min	
	Capacitor (kvar)	Current reduction (%)	Capacitor (kvar)	Current reduction (%)	Capacitor (kvar)	Current reduction (%)	Capacitor (kvar)	Current reduction (%)	Capacitor (kvar)	Current reduction (%)	Capacitor (kvar)	Current reduction (%)
2	1	14	1	24	1.5	30	2	42	2	40	3	50
3	1.5	14	1.5	23	2	28	3	38	3	40	4	49
5	2	14	2.5	22	3	26	4	31	4	40	5	49
7-1/2	2.5	14	3	20	4	21	5	28	5	38	6	45

Source: Commonwealth Sprague Capacitors PF-2000B [B74].
[a]For use with 3-phase, 60 Hz, NEMA Design B motors (NEMA MG 1-1993 [B164]) to raise full-load power factor to approximately 95%.

Table 5-25—Suggested maximum capacitor ratings—
T-frame NEMA Design B motors[a] (Continued)

Induction motor horse-power rating	Number of poles and nominal motor speed in r/min											
	2 3600 r/min		4 1800 r/min		6 1200 r/min		8 900 r/min		10 720 r/min		12 600 r/min	
	Capacitor (kvar)	Current reduction (%)	Capacitor (kvar)	Current reduction (%)	Capacitor (kvar)	Current reduction (%)	Capacitor (kvar)	Current reduction (%)	Capacitor (kvar)	Current reduction (%)	Capacitor (kvar)	Current reduction (%)
10	4	14	4	18	5	21	6	27	7.5	36	8	38
15	5	12	5	18	6	20	7.5	24	8	32	10	34
20	6	12	6	17	7.5	19	9	23	10	29	12.5	30
25	7.5	12	7.5	17	8	19	10	23	12.5	25	17.5	30
30	8	11	8	16	10	19	15	22	15	24	20	30
40	12.5	12	15	16	15	19	17.5	21	20	24	25	30
50	15	12	17.5	15	20	19	22.5	21	22.5	24	30	30
60	17.5	12	20	15	22.5	17	25	20	30	22	35	28
75	20	12	25	14	25	15	30	17	35	21	40	19
100	22.5	11	30	14	30	12	35	16	40	15	45	17
125	25	10	35	12	35	12	40	14	45	15	50	17
150	30	10	40	12	40	12	50	14	50	13	60	17
200	35	10	50	11	50	11	70	14	70	13	90	17
250	40	11	60	10	60	10	80	13	90	13	100	17
300	45	11	70	10	75	12	100	14	100	13	120	17
350	50	12	75	8	90	12	120	13	120	13	135	15
400	75	10	80	8	100	12	130	13	140	13	150	15
450	80	8	90	8	120	10	140	12	160	14	160	15
500	100	8	120	9	150	12	160	12	180	13	180	15

Source: Commonwealth Sprague Capacitors PF-2000B [B74].
[a]For use with 3-phase, 60 Hz, NEMA Design B motors (NEMA MG 1-1993 [B164]) to raise full-load power factor to approximately 95%.

5.5.4 Synchronous condensers for power-factor improvement

When a low-power-factor problem has been identified in the facility, the traditional method has been to install capacitor banks and controllers. Although this has worked in the past, it is becoming more difficult to apply capacitors in an environment high with harmonic content caused by variable-speed drives and other nonlinear loads.

Table 5-26—Multipliers to determine capacitor kilovars required for power-factor correction

Original power factor	Corrected power factor																				
	0.80	0.81	0.82	0.83	0.84	0.85	0.86	0.87	0.88	0.89	0.90	0.91	0.92	0.93	0.94	0.95	0.96	0.97	0.98	0.99	1.0
0.50	0.982	1.008	1.034	1.060	1.086	1.112	1.139	1.165	1.192	1.220	1.248	1.276	1.306	1.337	1.369	1.403	1.440	1.481	1.529	1.589	1.732
0.51	0.937	0.962	0.989	1.015	1.041	1.067	1.094	1.120	1.147	1.175	1.203	1.231	1.261	1.292	1.324	1.358	1.395	1.436	1.484	1.544	1.687
0.52	0.893	0.919	0.945	0.971	0.997	1.023	1.050	1.076	1.103	1.131	1.159	1.187	1.217	1.248	1.280	1.314	1.351	1.392	1.440	1.500	1.643
0.53	0.850	0.876	0.902	0.928	0.954	0.980	1.007	1.033	1.060	1.088	1.116	1.144	1.174	1.205	1.237	1.271	1.308	1.349	1.397	1.457	1.600
0.54	0.809	0.835	0.861	0.887	0.913	0.939	0.966	0.992	1.019	1.047	1.075	1.103	1.133	1.164	1.196	1.230	1.267	1.308	1.356	1.416	1.559
0.55	0.769	0.795	0.821	0.847	0.873	0.899	0.926	0.952	0.979	1.007	1.035	1.063	1.093	1.124	1.156	1.190	1.227	1.268	1.316	1.376	1.519
0.56	0.730	0.756	0.782	0.808	0.834	0.860	0.887	0.913	0.940	0.968	0.996	1.024	1.054	1.085	1.117	1.151	1.188	1.229	1.277	1.337	1.480
0.57	0.692	0.718	0.744	0.770	0.796	0.822	0.849	0.875	0.902	0.930	0.958	0.986	1.016	1.047	1.079	1.113	1.150	1.191	1.239	1.299	1.442
0.58	0.655	0.681	0.707	0.733	0.759	0.785	0.812	0.838	0.865	0.893	0.921	0.949	0.979	1.010	1.042	1.076	1.113	1.154	1.202	1.262	1.405
0.59	0.619	0.645	0.671	0.697	0.723	0.749	0.776	0.802	0.829	0.857	0.885	0.913	0.943	0.974	1.006	1.040	1.077	1.118	1.166	1.226	1.369
0.60	0.583	0.609	0.635	0.661	0.687	0.713	0.740	0.766	0.793	0.821	0.849	0.877	0.907	0.938	0.970	1.004	1.041	1.082	1.130	1.190	1.333
0.61	0.549	0.575	0.601	0.627	0.653	0.679	0.706	0.732	0.759	0.787	0.815	0.843	0.873	0.904	0.936	0.970	1.007	1.048	1.096	1.156	1.299
0.62	0.516	0.542	0.568	0.594	0.620	0.646	0.673	0.699	0.726	0.754	0.782	0.810	0.840	0.871	0.903	0.937	0.974	1.015	1.063	1.123	1.266
0.63	0.483	0.509	0.535	0.561	0.587	0.613	0.640	0.666	0.693	0.721	0.749	0.777	0.807	0.838	0.870	0.904	0.941	0.982	1.030	1.090	1.233
0.64	0.451	0.474	0.503	0.529	0.555	0.581	0.608	0.634	0.661	0.689	0.717	0.745	0.775	0.806	0.838	0.872	0.909	0.950	0.998	1.068	1.201
0.65	0.419	0.445	0.471	0.497	0.523	0.549	0.576	0.602	0.629	0.657	0.685	0.713	0.743	0.774	0.806	0.840	0.877	0.918	0.966	1.026	1.169
0.66	0.388	0.414	0.440	0.466	0.492	0.518	0.545	0.571	0.598	0.626	0.654	0.682	0.712	0.743	0.775	0.809	0.846	0.887	0.935	0.995	1.138
0.67	0.358	0.384	0.410	0.430	0.462	0.488	0.515	0.541	0.568	0.596	0.624	0.652	0.682	0.713	0.745	0.779	0.816	0.857	0.905	0.965	1.108
0.68	0.328	0.354	0.380	0.406	0.432	0.458	0.485	0.511	0.538	0.566	0.594	0.622	0.652	0.683	0.715	0.749	0.786	0.827	0.875	0.935	1.078
0.69	0.299	0.325	0.351	0.377	0.403	0.429	0.456	0.482	0.509	0.537	0.565	0.593	0.623	0.654	0.686	0.720	0.757	0.798	0.846	0.906	1.049
0.70	0.270	0.296	0.322	0.348	0.374	0.400	0.427	0.453	0.480	0.508	0.536	0.564	0.594	0.625	0.657	0.691	0.728	0.769	0.817	0.877	1.020
0.71	0.242	0.268	0.294	0.320	0.346	0.372	0.399	0.425	0.452	0.480	0.508	0.536	0.566	0.597	0.629	0.663	0.700	0.741	0.789	0.849	0.992
0.72	0.214	0.240	0.266	0.292	0.318	0.344	0.371	0.397	0.424	0.452	0.480	0.508	0.538	0.569	0.601	0.635	0.672	0.713	0.761	0.821	0.964
0.73	0.186	0.212	0.238	0.264	0.290	0.316	0.343	0.369	0.396	0.424	0.452	0.480	0.510	0.541	0.573	0.607	0.644	0.685	0.733	0.793	0.936
0.74	0.159	0.185	0.211	0.237	0.263	0.289	0.316	0.342	0.369	0.397	0.425	0.453	0.483	0.514	0.546	0.580	0.617	0.658	0.706	0.766	0.909
0.75	0.132	0.158	0.184	0.210	0.236	0.262	0.289	0.315	0.342	0.370	0.398	0.426	0.456	0.487	0.519	0.553	0.590	0.631	0.679	0.739	0.882

Source: Commonwealth Sprague Capacitors PF-2000B [B74].
NOTE—To use table 5-26: With known kW demand and initial power factor (PF), enter table in original PF column and read to corrected PF value. This provides a multiplier which, used with kW demand, produces the kvar capacitor required. Example: 410 kW at 73% PF. To find kvar to obtain 95% PF, enter at 0.73 and go to 95% column. Read 0.607. Multiply kW 410 by 0.607 = 249 kvar. Use 250 kvar standard rating.

Table 5-26—Multipliers to determine capacitor kilovars required for power-factor correction (*Continued*)

Original power factor	Corrected power factor																				
	0.80	0.81	0.82	0.83	0.84	0.85	0.86	0.87	0.88	0.89	0.90	0.91	0.92	0.93	0.94	0.95	0.96	0.97	0.98	0.99	1.0
0.76	0.105	0.131	0.157	0.183	0.209	0.235	0.262	0.288	0.315	0.343	0.371	0.399	0.429	0.460	0.492	0.526	0.563	0.604	0.652	0.712	0.855
0.77	0.079	0.105	0.131	0.157	0.183	0.209	0.236	0.262	0.289	0.317	0.345	0.373	0.403	0.434	0.466	0.500	0.537	0.578	0.626	0.685	0.829
0.78	0.052	0.078	0.104	0.130	0.156	0.182	0.209	0.235	0.262	0.290	0.318	0.346	0.376	0.407	0.439	0.473	0.510	0.551	0.599	0.659	0.802
0.79	0.026	0.052	0.078	0.104	0.130	0.156	0.183	0.209	0.236	0.264	0.292	0.320	0.350	0.381	0.413	0.447	0.484	0.525	0.573	0.633	0.776
0.80	0.000	0.026	0.052	0.078	0.104	0.130	0.157	0.183	0.210	0.238	0.266	0.294	0.324	0.355	0.387	0.421	0.458	0.499	0.547	0.609	0.750
0.81		0.000	0.026	0.052	0.078	0.104	0.131	0.157	0.184	0.212	0.240	0.268	0.298	0.329	0.361	0.395	0.432	0.473	0.521	0.581	0.724
0.82			0.000	0.026	0.052	0.078	0.105	0.131	0.158	0.186	0.214	0.242	0.272	0.303	0.335	0.369	0.406	0.447	0.495	0.555	0.698
0.83				0.000	0.026	0.052	0.079	0.105	0.132	0.160	0.188	0.216	0.246	0.277	0.309	0.343	0.380	0.421	0.469	0.529	0.672
0.84					0.000	0.026	0.053	0.079	0.106	0.134	0.162	0.190	0.220	0.251	0.283	0.317	0.354	0.395	0.443	0.503	0.646
0.85						0.000	0.027	0.053	0.080	0.108	0.136	0.164	0.194	0.225	0.257	0.291	0.328	0.369	0.417	0.477	0.620
0.86							0.000	0.026	0.053	0.081	0.109	0.137	0.167	0.198	0.230	0.264	0.301	0.342	0.390	0.450	0.593
0.87								0.000	0.027	0.055	0.083	0.111	0.141	0.172	0.204	0.238	0.275	0.316	0.364	0.424	0.567
0.88									0.000	0.028	0.056	0.084	0.114	0.145	0.177	0.211	0.248	0.289	0.337	0.397	0.540
0.89										0.000	0.028	0.056	0.086	0.117	0.149	0.183	0.220	0.261	0.309	0.369	0.512
0.90											0.000	0.028	0.058	0.089	0.121	0.155	0.192	0.233	0.281	0.341	0.484
0.91												0.000	0.030	0.061	0.093	0.127	0.164	0.205	0.253	0.313	0.456
0.92													0.000	0.031	0.063	0.097	0.134	0.175	0.223	0.283	0.426
0.93														0.000	0.032	0.066	0.103	0.144	0.192	0.252	0.395
0.94															0.000	0.034	0.071	0.112	0.160	0.220	0.363
0.95																0.000	0.037	0.079	0.126	0.186	0.329
0.96																	0.000	0.041	0.089	0.149	0.292
0.97																		0.000	0.048	0.108	0.251
0.98																			0.000	0.060	0.203
0.99																				0.000	0.143
																					0.000

Source: Commonwealth Sprague Capacitors PF-2000B [B74].

NOTE—To use table 5-26: With known kW demand and initial power factor (PF), enter table in original PF column and read to corrected PF value. This provides a multiplier which, used with kW demand, produces the kvar capacitor required. Example: 410 kW at 73% PF. To find kvar to obtain 95% PF, enter at 73% and go to 95% column. Read 0.607. Multiply kW 410 by 0.607 = 249 kvar. Use 250 kvar standard rating.

Another approach is to apply a rotary brushless synchronous machine referred to as a condenser. The condenser, in conjunction with a power-factor controller, applies kvars to the system by overexciting the field of the synchronous machine to provide reactive power to the source, thus compensating for the poor power-factor load.

Using a rotary condenser for power-factor correction provides advantages. This type of power-factor correction is very smooth due to the synchronous condensers power-factor controller automatically tracking the system power factor and adjusting the synchronous condensers field as correction is needed. The synchronous condenser method of power-factor correction eliminates the problem of system over-voltage and voltage instability.

The condenser is a low-impedance source and appears inductive to the variable-speed drives. It is also tolerant of harmonic current and voltage; consequently, resonance with the system impedance is avoided.

Purchasing and installing a synchronous power-factor condenser is about the same price, and in some cases, less expensive than capacitor banks.

If system voltage and current harmonics are a problem in addition to the power-factor issue, then a solution would be the use of a synchronous motor-generator set with controls that will totally isolate critical loads from these problems.

Table 5-27—Matrix of available synchronous power-factor condenser equipment for power-factor correcting

System voltage	Correction kvar																				
	100	150	200	250	300	400	500	600	750	1000	1250	1500	1750	2000	2500	3000	3500	4000	5000	6000	7000
480	X	X	X	X	X	X	X	X	X	X	X										
2 400						X	X	X	X	X	X	X	X	X	X	X					
4 160										X	X	X	X	X	X	X	X	X	X	X	X
6 900											X	X	X	X	X	X	X	X	X	X	X
13 800														X	X	X	X	X	X	X	X

Source: Bulletin Kato SC 8-92 [B139]
NOTE—To reach higher total kvars, units may be paralleled.

5.5.5 Synchronous motors for power-factor improvement (Kato SC 8-92 [B139]; SM 2-93 [B138]; SM 1-94 [B137])

The key advantage of using synchronous motor is its ability to operate at unity or leading power factor with high efficiency. The synchronous motor is configured with a brushless excitation system that can be either manually controlled by an external dc source or automatically controlled by an external power-factor controller, which constantly monitors the motors power factor and adjusts the excitation to maintain the selected power factor. This constant monitoring and control of the motor's excitation provides smooth power-factor control over the motor's entire load range from zero to full load. Operation is free from any potentially damaging transients or nuisance trips of a variable-speed drive systems as is possible when applying capacitor banks for plant power-factor improvement.

Typical applications for use of a synchronous motor are

— Centrifugal compressors
— Conveyors
— Fans
— Pumps
— Refiners

A matrix of available synchronous motors is presented in table 5-28.

Table 5-28—Matrix of available synchronous motors

Speed (r/min)	hp @ 0.80 power factor, lead																			
	250	300	350	400	450	500	600	700	800	900	1000	1250	1500	1750	2000	3000	3500	5000	6000	8000
1800	X	X	X	X	X	X	X	X	X	X	X	X	X	X	X	X	X	X	X	X
1200	X	X	X	X	X	X	X	X	X	X	X	X	X	X	X	X	X	X	X	
900	X	X	X	X	X	X	X	X	X	X	X	X	X	X	X	X	X	X		
720	X	X	X	X	X	X	X	X	X	X	X	X	X	X	X	X	X			
600	X	X	X	X	X	X	X	X	X	X	X	X	X	X	X					
514	X	X	X	X	X	X	X	X												

Source: Bulletin Kato SM 1-94 [B137]
NOTE—Typical voltages: 460, 4000, 13 200

The following is an example of a synchronous motor application:

A manufacturing plant has a substation rated at 1500 kVA. The present load has been averaging 750 kVA of which a major portion of this load is induction motors. The plants electrical billing indicates that the highest 30 min kVA demand has been 825 kVA with 700 kW. This corresponds to a 0.85 power-factor load. With this load, the plant is absorbing 435 kvar from the utility.

The plant is expanding and plans on adding a 200 hp air compressor and 100 kVA of additional load. Now is the time to add a power-factor-correction device to increase the power factor and decrease the kVA demand from the utility. By raising the power factor, the plant will experience a dollar savings from the utility. Note that the amount of savings depends on the billing structure.

If the compressor is fitted with a synchronous motor instead of the normal induction type and slightly oversized so as to export kvars to the plant, this dollar savings will become a reality. The synchronous motor would be designed to export up to 340 kvars while running the compressor. Instant power-factor correction *without* the use of capacitors! The synchronous motor is provided with an automatic synchronizing device located on the motor shaft and an external solid-state field regulator, which provides smooth control of the motor field to maintain the plant power factor at no less than 0.98 at all times from low plant load to full expected plant load.

5.6 References

This chapter shall be used in conjunction with the following publications:

ANSI/ASHRAE/IESNA 90.1-1989, Energy Efficient Design of New Buildings Except Low-Rise Residential Buildings.[4]

IEEE Std 141-1993, IEEE Recommended Practice for Electric Power Distribution for Industrial Plants (IEEE Red Book) (ANSI).[5]

IEEE Std 241-1990, IEEE Recommended Practice for Electric Power Systems in Commercial Buildings (IEEE Gray Book) (ANSI).

IEEE Std 399-1990, IEEE Recommended Practice for Industrial and Commercial Power Systems Analysis (IEEE Brown Book) (ANSI).

National Electrical Code® Handbook, 6th edition.[6]

5.7 Bibliography

Additional information may be found in the following sources:

Acme Electric Company, Transformer Division, 4815 West 5th St., Lumbarton, NC 28358

[B1]   OPTI-MISER Sample Work Sheet.

Allen Bradley Co., Standard Drives Business, 6400 West Enterprise Dr., Mequon, WI 43092

[B2]   "Energy Savings with Adjustable Frequency Drives," publication DGI-2.1, Dec. 1984.

[B3]   Computer program: "AC Drives Energy Comparison Program Energy 4."

---

[4]This publication is available from the Customer Service Dept., American Society of Heating, Refrigerating and Air-Conditioning Engineers, 1791 Tullie Circle, NE, Atlanta, GA 30329.

[5]IEEE publications are available from the Institute of Electrical and Electronics Engineers, 445 Hoes Lane, P.O. Box 1331, Piscataway, NJ 08855-1331.

[6]This publication is available from Publications Sales, National Fire Protection Association, 1 Batterymarch Park, P.O. Box 9101, Quincy, MA 02269-9101.

American Consulting Engineers Council (ACEC), 1015 15th St., N.W., Suite 802, Washington, DC 20005

[B4]   "Guidelines for the Design and Purchase of Energy Management and Control Systems for New and Retrofit Applications."

American Council for an Energy-Efficient Economy (ACEEE), 1001 Connecticut Ave., N.W., Suite 801, Washington, DC 20036, and 2410 Shattuck Ave., Suite 202, Berkeley, CA 94704

[B5]   "Energy Efficient Motor Systems."

[B6]   "State of the Art of Energy Efficiency: Future Directions."

[B7]   "Financing Energy Conservation."

[B8]   "Energy Efficiency in Buildings."

[B9]   "Efficiency Standards for Lamps, Motors, and Lighting Fixtures," Item A901.

[B10]   "Conformance with ASHRAE/IES, Standard 90.1P in New Commercial Buildings," Item A892.

[B11]   "Commercial Building Equipment Efficiency: A State of the Art Review," Item A882.

[B12]   "Guide to Energy Efficient Office Equipment."

American Society of Heating, Refrigerating and Air-Conditioning Engineers (ASHRAE), 1791 Tullie Circle, NE, Atlanta, GA 30329

[B13]   ASHRAE Handbook, 1988: Equipment.

[B14]   ASHRAE Handbook, 1990: Refrigeration Systems and Applications.

[B15]   ASHRAE Handbook, 1991: HVAC Applications.

[B16]   ASHRAE Handbook, 1992: HVAC Systems and Equipment.

[B17]   ASHRAE Handbook, 1993: Fundamentals.

[B18]   ANSI/ASHRAE/IESNA 100.3-1985, Energy Conservation in Existing Buildings—Commercial.

[B19]   ANSI/ASHRAE/IESNA 100.4-1984, Energy Conservation in Existing Facilities—Industrial.

[B20]   "Design Guide for Cool Thermal Storage," Code 90369.

[B21]   "Energy Conservation with Chilled-Water Storage," by D. Fiorino, published in *ASHRAE Journal*, vol. 35, no. 5, May 1993.

[B22]   "Field Inspection of Building Components—A Tool for Cost-Effective Measures in Retrofitting Buildings" by M. D. Lyberg and S. A. Ljungberg, published in *ASHRAE Transactions* 1992, vol. 98, pt. 1.

[B23]   "Monitored Commercial Building Energy Dated: Reporting the Results" by D. E. Claridge, J. S. Habert, R. J. Sparks, R. E. Lopez, and K. Kissock, published in *ASHRAE Transactions* 1992, vol. 98, pt. 1.

[B24]   "Monitoring Approaches For Energy Conservation Impact Evaluation," by D. R. Landsberg and J. A. Amalfi, published in *ASHRAE Transactions* 1992, vol. 98, pt. 1.

[B25]   "Off-Peak Air-Conditioning: A Major Energy Saver," by Calvin D. Mac-Cracken, published in *ASHRAE Journal*, vol. 33, no. 12, Dec. 1991.

American National Standards Institute (ANSI), 11 West 42nd St., New York, NY 10036.

[B26]   ANSI C84.1-1989, American National Standard Voltage Ratings for Electric Power Systems and Equipment (60 Hz).

Australian National Energy Management Program, Commonwealth Department of Primary Industries and Energy, Energy Division, Edmund Barton Building, Canberra, Australia 2601

Advisory Booklets:

[B27]   No. 1, The energy management program

[B28]   No. 2, Involving employees in energy management programs

[B29]   No. 3, The energy audit

[B30]   No. 4, Energy conservation: checklists

[B31]   No. 5, Fuel substitution: considerations

[B32]   No. 6, Lighting management

[B33]   No. 7, Electric motors and machines

[B34]   No. 8, Saving diesel in transport

[B35]   No. 9, Factory heating and ventilation

[B36]   No. 10, Compressed air management

[B37]   No. 11, Heat recovery

[B38]   No. 12, Boilers

[B39]   No. 13, Oil, coal and gas-fired furnaces

[B40]   No. 14, Utilization of steam

[B41]   No. 15, LPG as a road transport fuel

[B42]   No. 16, Saving energy in commercial buildings

[B43]   No. 17, Financial evaluation of energy management programs

[B44]   No. 18, Saving energy in agriculture

Building Officials & Code Administrators International, Inc. (BOCA®), 4051 W. Flossmoor Road, Country Club Hills, IL 60478-5795.

[B45]   BOCA® National Energy Conservation Code, 1993.

214

California Energy Commission, State of California, Publication Office, P.O. Box 944295, MS-13, Sacramento, CA 94244-2950

[B46]   P103-82-001, Energy Efficiency in California's Commercial, Industrial and Agricultural Sectors: Progress and Prospects.

[B47]   P300-84-018, Energy Savings Potential in California's Existing Office and Retail Buildings.

[B48]   P400-80-066, Guide to Energy Budgets, Division 2, For First Generation Non-residential Building Standards.

[B49]   P400-80-069, Guide to HVAC Systems, Division 5, For First Generation Non-residential Building Standards.

[B50]   P400-80-070, Guide to HVAC Equipment, Division 6, For First Generation Nonresidential Building Standards.

[B51]   P400-80-071, Guide to Service Water Heaters, Division 7, For First Generation Nonresidential Building Standards.

[B52]   P400-83-001, Analysis Methodology for Determining Energy Savings Due to Daylighting.

[B53]   P400-86-007, Tracking Utility Costs by Microcomputer.

[B54]   P400-86-008, How to Organize and Communicate Your Energy Data.

[B55]   P400-88-005, Energy Efficiency Manual Designing for Compliance. 2nd Generation Nonresidential Standards. Applies to all Office, Retail, and Wholesale Buildings.

[B56]   P400-89-003, Simplified Compliance Approach for Office Buildings, 2nd edition.

[B57]   P500-82-054, Cogeneration Handbook.

[B58]   P500-83-010, Commercial Solar Greenhouse Heating System.

[B59]   P500-84-011, Industrial Demonstration of Photovoltaic and Hot Water Cogeneration in San Diego.

Canadian Electrical Association (CEA), 1 Westmount Square, Suite 1600, Montreal, Quebec, Canada H3Z 2P9

[B60]   CEA77-80, "Combined Generation of Heat and Electricity."

[B61]   CEA016U244, "Potential Savings Available by Substitution of Standard Efficiency Motors with Energy Efficient Motors for Canada.

[B62]   CEA131U293, "Investigating Adjustable Speed Drives in Commercial and Industrial Applications."

[B63]   CEA232U709, "Optimizing Building Design Parameters for Energy Efficiency."

[B64]   CEA433U492, "Efficiency Improvements in Small Horsepower Single Phase Electric Motors.

[B65]   CEA434U, "Energy Efficient Equipment Register Phase I."

[B66]   CEA502U499, "Adjustable Speed Drives: Industrial Demonstration."

[B67]   CEA912U, "State of the Art Study on Variable Speed Drives for Pumps and Fans."

[B68]   CEA891U710, "Minimum Efficiency Levels for Electric Motors: AC Polyphase 1–200 HP Motors.

[B69]   CEA8923U736, "Survey of Commercial and Industrial Adjustable Speed Drive Owner Experience."

[B70]   CEA9038U828, "The Evaluation of Compact Fluorescent Lamps for Energy Conservation."

[B71]   CEA9101U829, "Energy Consumption and Desktop Computers: A Study of Current Practices and Potential Energy Savings."

The Chartered Institution of Building Services Engineers (CIBSE), Delta House, 222 Belham High Road, London SW12 9BS United Kingdom

[B72]   Building Energy Code:

— Part 1, Guidance Towards Energy Conserving Design of Buildings and Services.

— Part 2, Calculation of Energy Demands and Targets for the Design of New Buildings and Services.

— Part 3, Guidance Towards Energy Conserving Operation of Buildings and Services.

— Part 4, Measurement of Energy Consumption and Comparison With Targets for Existing Buildings and Services.

[B73]   "Guide, Volume B, Installation and Equipment Data," 5th edition.

Commonwealth Sprague Capacitor, Inc., Power Capacitor Department, North Adams, MA 01247

[B74]   "Power Factor Correction: A Guide for the Plant Engineer," PF-2000B, 1987.

Consolidated Edison Co. of NY, Inc. (Con Ed), 666 Fifth Ave., Suite 280, New York, NY 10102-0001. Con Edison's Enlightened Energy Rebate Program:

[B75]   High-Efficiency Cool Storage Systems.

[B76]   High-Efficiency Electric Air Conditioning.

[B77]   High-Efficiency Gas Air Conditioning.

[B78]   High-Efficiency Lighting.

[B79]   High-Efficiency Motors.

[B80]   High-Efficiency Steam Air Conditioning.

Copper Development Association, Inc., 260 Madison Avenue, New York, NY 10016

[B81]   Computer program: "Copper Busbar Design Guide."

Eaton Corporation, Industrial Drives Division, 3122 14th Ave., Kenosha, WI 53141-1412

[B82]   "V-Belt Drive Efficiency" Technical Bulletin No. 700-4006.

Edison Electric Institute, Inc. (EEI), 701 Pennsylvania Ave., N.W., Washington, DC 20004-2696

[B83]   "Demand-Side Management: Commercial Markets and Programs."

[B84]   "Demand-Side Management: Evaluation of Alternatives."

[B85]   "Demand-Side Management: Industrial Markets and Programs."

[B86]   "Demand-Side Management: Overview of Key Issues."

[B87]   "Demand-Side Management: Technology Alternatives and Marketing Methods."

[B88]   "Electric Applications Handbook."

[B89]   "Electric Materials Handling Vehicles."

[B90]   "Guidelines for Compliance with ASHRAE and DOE New Building Standards."

[B91]   "Handbook For Electricity Metering," 9th edition.

[B92]   "Marketing the Industrial Heat Pump."

[B93]   "A Method for Economic Evaluation of Distribution Transformers," April 1981, Report by Task Force on Distribution Transformer Evaluation of the EEI T&D Comm.

Electric Power Research Institute (EPRI), 3412 Hillview Ave., Palo Alto, CA 94304

[B94]   "Commercial Cool Storage: Reducing Cooling Costs With Off-Peak Electricity," EU 3024.

[B95]   "Drivers of Electricity Growth and the Role of Utility Demand-Side Management," EPRI TR-102639, 1993.

[B96]   "Electrotechnology Reference Guide," Rev. 2, EPRI TR-101021, Aug. 1992.

[B97]   "Research and Development Plan for Advanced Motors and Drives," EPRI RP-2918-15, 1993.

[B98]   "Power Quality Considerations for Adjustable Speed Drives," Part 4, CU.1008D.4.91, CU.3036R.11.91.

General Electric Motors and Industrial Systems, Fort Wayne, IN 46801. (Publications are available from AKI, 400 Washington Blvd., Fort Wayne, IN 46802.

[B99]   "AC Drives Program User's Guide," GEE-1010.

[B100] "Annual Estimated Cost Savings," GET-6711C.

[B101] "Buyers Guide to A$D Adjustable Speed AC Induction Motors," GEP-9012A.

[B102] "Buyers Guide to Medium AC Induction Motors," GEP-772D.

[B103] "Energy $aver® Program User's Guide," GEE-1012.

[B104] "GE Motors: AC Motor Selection and Application Guide," GET-6812B (20M 12/93).

[B105] "GE Motors Stock Motor Catalog," GEP-500J.

[B106] "How to Maximize the Return on Energy Efficient Motors," GEA-10951C.

[B107] "How to Specify and Evaluate Energy-Efficient Motors," GEA-10951.

[B108] "Impact of Rewinding on Motor Efficiency," GET-8014.

[B109] "Integral Horsepower Motors," GEP-1087H.

[B110] "Premium Efficient Motors Slash Electric Bill," GER-3729.

[B111] "Understanding Premium-Efficiency Motor Economics," GEK-100906.

[B112] "Vertical Induction Motor Selection and Application Guide," GEP-1839C.

Honeywell, Inc., Minneapolis, MN

[B113] "Energy Conservation with Comfort, The Honeywell Energy Conserver's Manual and Workbook," 2nd edition.

International Electrotechnical Commission (IEC), Case Postale 131, 3, rue de Varembé, CH-1211, Genève 20, Switzerland/Suisse

[B114] IEC Pub 34-12 (1980), Rotating Electric Machines—Part 12: Starting performance of single-speed three-phase cage induction motors for voltages up to and including 660 V.

Illuminating Engineering Society of North America (IESNA), 120 Wall Street, Fl 17, New York, NY 10005-4001

[B115] Lighting Handbook: Reference and Application, ISBN 0-87995-102-8.

[B116] IES LEM-1-82, Lighting Power Limit Determination, ISBN 0-87995-012-9.

[B117] IES LEM-2-84, Lighting Energy Limit Determination, ISBN 0-87995-017-X.

[B118] IES LEM-3-87, Design Considerations for Effective Building Lighting Energy Utilization, ISBN 0-87995-025-0.

[B119] IES LEM-4-84, Energy Analysis of Building Lighting Design & Installation, ISBN 0-87995-018-8.

Ingersoll-Rand Company, 255 E. Washington Ave., Washington, NJ 07882, 1969

[B120] Gibbs, Charles W., ed., *Compressed Gas and Air Handbook*, 3rd edition, 1969.

Institute of Electrical and Electronic Engineers (IEEE), 445 Hoes Lane, P.O. Box 1331, Piscataway, NJ 08855-1331.

[B121]   IEEE Std 18-1992, IEEE Standard for Shunt Power Capacitors (ANSI).

[B122]   IEEE Std 112-1991, IEEE Standard Test Procedure for Polyphase Induction Motors and Generators (ANSI).

[B123]   IEEE Std 113-1985, IEEE Guide on Test Procedures for DC Machines (ANSI).[7]

[B124]   IEEE Std 114-1982, IEEE Standard Test Procedure for Single-Phase Induction Motors.[8]

[B125]   IEEE Std 115-1995, IEEE Guide: Test Procedures for Synchronous Machines.

[B126]   IEEE Std 268-1992, American National Standard for Metric Practice (ANSI).

[B127]   IEEE Std 519-1992, IEEE Recommended Practices and Requirements for Harmonic Control in Electric Power Systems (ANSI).

[B128]   IEEE Std 1046-1991, IEEE Application Guide for Distributed Digital Control and Monitoring for Power Plants (ANSI).

[B129]   IEEE Std 1068-1990, IEEE Recommended Practice for Repair and Rewinding of Motors for the Petroleum and Chemical Industry (ANSI).

[B130]   IEEE Std C57.21-1990 (Reaff 1995), IEEE Standard Requirements, Terminology, and Test Code for Shunt Reactors Rated Over 500 kVA (ANSI).

[B131]   IEEE Std C57.94-1982 (Reaff 1987), IEEE Recommended Practice for Installation, Application, Operation, and Maintenance of Dry-Type General Purpose Distribution and Power Transformers (ANSI).

[B132]   IEEE Std C57.12.00-1993, IEEE Standard General Requirements for Liquid-Immersed Distribution, Power, and Regulating Transformers (ANSI).

[B133]   IEEE Std C57.12.01-1989, IEEE Standard General Requirements for Dry-Type Distribution and Power Transformers Including Those With Solid Cast and/or Resin-Encapsulated Windings.

[B134]   IEEE Std C57.120-1991, IEEE Standard Loss Evaluation Guide for Power Transformers and Reactors (ANSI).

[B135]   "Adjustable Speed AC Drive Systems" edited by K. Bose, ISBN 0-37942-145-2.

[B136]   "Load Management" edited by Talukdar, and W. Gellings, ISBN 0-87942-214-9.

Kato Engineering Company, P.O. Box 8440, Mankato, MN 56002-8447

[B137]   AC Synchronous Motor Bulletin, Kato SM 1-94.

[B138]   AC Synchronous Motor Bulletin, Kato SM 2-93.

[B139]   Synchronous Generator Bulletin, Kato SC 8-92.

---

[7]This standard has been withdrawn; however, copies can be obtained from Global Engineering, 15 Inverness Way East, Englewood, CO 80112-5704, tel. (303) 792-2181.
[8]This standard has been withdrawn; see footnote 7 for further information.

National Electrical Contractors Association (NECA), 3 Bethesda Metro Center, Suite 1100, Bethesda, MD 20814. Electrical Design Library (EDL) publications:

Title	Index/Date
[B140] The All-Electric Building Option	302561 14K/6/88
[B141] Basics of Building Automation	12/73
[B142] Building Energy Audits	302527 12M/9/79
[B143] Cogeneration: Technical Issues and Teamwork	302562 14K/9/88
[B144] Cogeneration Utility Systems	302532 14M/12/80
[B145] Electric Heat in Perspective	10/74
[B146] Electric Load Management	302511
[B147] Electric Load Management Update	302557 14K/6/87
[B148] Energy Efficient Motors & Controls	9/77
[B149] Energy Monitoring and Control Systems	302569 14K/6/90
[B150] Heat Pumps Revisited	6/76
[B151] Heat Pumps: The Cost-Effective Option	302566 14K/9/89
[B152] Microprocessor Based Energy Controls	302528 12M/12/79
[B153] Minimizing Life-Cycle Utility Costs	302545 12M/6/84
[B154] Packaged Cogeneration Systems (PCS)	302548 14K/3/85
[B155] Photovoltaic (PV) Systems	302534 14M/6/81
[B156] Power Distribution Systems	1977
[B157] Practical and Profitable Building Conservation	302571 14K/12/90
[B158] Saving with Adjustable Speed Drives	302565 14K/6/89
[B159] Specifying Computerized Control Systems	6/75
[B160] Variable Speed Motor Controls	302537 6/82
[B161] Zonal Electric Heating	302547 12M/12/84

National Electrical Manufacturers Association (NEMA), 1300 17th St., Suite 1847, Rosslyn, VA 20009

[B162] NEMA CP 1-1992, Shunt Capacitors.

[B163] NEMA ICS 3.1-1990, Safety Standards for Construction and Guide for Selection Installation, and Operation of Adjustable-Speed Drive Systems.

[B164] NEMA MG 1-1993, Motors and Generators.

[B165] NEMA MG 2-1989, Safety Standard for Construction and Guide for Selection, Installation and Use of Electric Motors and Generators.

[B166] NEMA MG 10-1994, Energy Management Guide for Selection and Use of Fixed-Frequency Medium AC Squirrel-Cage Polyphase Induction Motors.

[B167] NEMA MG 11-1992, Energy Management Guide for Selection and Use of Single-Phase Motors.

[B168] NEMA ST 20-1992, Dry-Type Transformers for General Applications.

[B169] NEMA TP 1-1996, Guide for Determining Energy Efficiency for Distribution Transformers.

[B170] NEMA cat. no. 0235-041812, Statistical Methods for Motor Efficiency Data.

[B171] NEMA cat. no. 10408, "Total Energy Management," A Practical Handbook in Energy Conservation and Management," 2nd edition.

[B172] "New York Life's Energy Automation Pays Off," 10/87.

National Fire Protection Association (NFPA), 1 Batterymarch Park, P.O. Box 9101, Quincy, MA 02269-9101

[B173] NFPA 70-1996, National Electrical Code® (NEC®).

[B174] 1995 NEC® Committee Report on Proposals.

National Insulation Contractors Association (NICA), 99 Canal Center Plaza, Suite 222, Alexandria, VA 22314

[B175] The NICA Insulation Manual: "ii File," 93/94 (Insulation Industry File).

National Lighting Bureau (NLB), 1300 17th St., Suite 1847, Rosslyn, VA 20009

[B176] Guide to Energy Efficient Lighting System.

[B177] Guide to Lighting Controls.

[B178] Performing a Lighting System Audit (revised edition).

Niagara Mohawk Power Corp., Syracuse, NY. Niagara Mohawk Reducing Plan for Business:

[B179] Drives Program.

[B180] Lighting Existing Structures.

[B181] Lighting New Construction.

[B182] Commercial and Industrial Energy Analysis Program.

[B183] Commercial and Industrial High-Efficiency Lighting Program.

[B184] Commercial and Industrial High-Efficiency Motors and Adjustable Speed.

[B185] Commercial and Industrial Space-Conditioning Incentive Program.

Public Service Electric and Gas Company (PSE&G): 80 Park Plaza, P.O. Box 570, Newark, NJ 07101

[B186] "The Difference is Night and Day" (The New PSE&G Cool Storage Program).

Reliance Electric Company, 24703 Euclid Avenue, Cleveland, OH 44117

[B187] "Motor Application," Bulletin B-2615.

[B188] "Power Line Considerations for Variable Frequency Drives," Technical Paper D-7122.

[B189] Robechek, J. D., "Adjustable Speed Drives as Applied to Centrifugal Pumps," Technical Paper D-7100-1, Reliance Electric Co., Cleveland, OH, Oct. 1981.

[B190] Robechek, J. D., "Fan Control for the Glass Industry Using Static Induction Motor Drives," Technical Paper D-7102, Reliance Electric Co., Cleveland, OH, Oct. 1981.

Sheet Metal and Air Conditioning Contractors' National Association (SMACNA), 4201 Lafayette Center Dr., P.O. Box 221230, Chantilly, VA 22022-1230.

[B191] "Energy Conservation Guidelines," 1091.

[B192] "Energy Recovery Equipment and Systems," 1104.

[B193] "Retrofit of Building Energy Systems and Processes," 1312.

State Electricity Commission (SEC) of Victoria, Energy Business Centre, 349-351 North Road, Caulfield South, Victoria 3162, Australia

[B194] "Cogeneration," C1.

[B195] "Compressed Air Energy Cost Savings Recommendations."

[B196] "Compressed Air Savings Manual," B3, 1991.

[B197] "Efficient Electric Hot Water," E4.

[B198] "Electric Infra-Red Heating," 74 502-1450.

[B199] "Electric Steam Raising," 74-502-1440.

[B200] "Electricity for Materials Handling," July 1977.

[B201] "How To Save On Electricity Costs In Small Business."

[B202] "Ice Storage For Industry," B2.

[B203] "Induction Heating and Melting for Industry."

[B204] "Induction Metal Joining, Productivity and Energy Efficiency," B6.

[B205] "Industrial Thermal Insulation."

[B206] "Infra-Red Heating at Work in Industry," UDC 621.365.46.

[B207] "Manage Your Load & Reduce Electricity Costs," B5.

[B208] "Quality of Supply: Problems, Causes & Solutions," April 1991.

[B209] "Radio Frequency Dielectric Heating," Dec. 1985.

[B210] "Radio Frequency Heating of Dielectric Materials," 74-502-1140.

[B211] "Resistance Heating at Work in Industry," UDC 621.365.3.

[B212] "Solar Boosted Electric Hot Water," 7610-252-9252, April 1991.

[B213]   "Tungsten Halogen Heating," B4.

[B214]   "Ultra-Violet Heating at Work in Industry."

[B215]   "Waste Heat Recovery," Mar. 1986.

The Electrification Council (TEC), 701 Pennsylvania Ave., N.W., Washington, DC 20004-2696

[B216]   Cogeneration Marketing Issues & Options.

[B217]   The Commercial & Industrial Electric Heat Pump Option Program.

[B218]   Electric Process Heating For Industry Program.
  — Book #1: Fundamentals," TEC-1300.
  — Book #2: Metal Heating," TEC-1301.
  — Book #3: Heating of Liquids," TEC-1302.
  — Book #4: Heating of Non-Metals," TEC-1303.
  — Book #5: Melting Metals and Non-Metals," TEC-1304.

[B219]   Energy Efficient Motor (EEM) Reference Guide.

[B220]   Energy Management Action (EMA) Program Course.

[B221]   Motors & Motor Controls Course, 3rd edition.

Underwriters Laboratories Inc. (UL), 333 Pfingsten Road, Northbrook, IL 60062

[B222]   UL 810-1995, Capacitors.

[B223]   UL 916-1994, Energy Management Equipment.

[B224]   UL 1004-1994, Electric Motors.

[B225]   UL 1561-1994, Dry-Type General Purpose and Power Transformers (Revised 1995).

[B226]   UL 1562-1994, Transformers, Distribution, Dry-Type Over 600 Volts.

U.S. Government Publications, Superintendent of Documents, Government Printing Office Washington, DC 20025

[B227]   U.S. Department of Agriculture, Rural Electrification Administration: "Evaluation of Large Power Transformer Losses," Bulletin No. 65-2.

[B228]   U.S. Department of Transportation, Federal Aviation Administration: "Energy Conservation for Airport Buildings," Advisory Circular Number 150/5360-11, May 31, 1984.

[B229]   U.S. Department of Energy, Bonneville Power Administration, Electric Sales Clearinghouse, Washington State Energy Office, 809 Legion Way, S.E., P.O. Box 43165, Olympia, WA 98504-3165:
  — "Adjustable Speed Drive Applications Guidebook," by EBASCO: E. A. Mueller, Report Number DOE/BP-34906-1/Jan. 1990.

— Technology Update: Reducing Power Factor Cost, 1991; Buying an Energy Efficient Motor, 1991; Two-Speed Motor Installation, 1991; Optimizing Your Motor Drive System, 1991; Determining Electric Motor Load Factor, 1992; "Motor Master" Software and Reference Guide, 1992; Energy Efficient Electric Motor Selection Handbook, 1992; Electric Motor Rewind Guidebook, 1993; Simplified High-Efficiency Motor Testing Guidebook, 1993.

[B230] U.S. General Services Administration, Public Buildings Service, Washington, DC: "Quality Standards for Design and Construction," PBS P 3430.1, Oct. 20, 1983.

Miscellaneous publications:

[B231] "A Guide to Power Quality," Trans-Coil, Inc., Milwaukee, WI, 1991.

[B232] Ambrose, E. R., *Heat Pumps and Electric Heating*. John Wiley & Sons, Inc., 1965.

[B233] Andreas, John C., *Energy-Efficient Electric Motors* (Selection and Application), 2nd edition, Marcel Dekker, Inc., 1992.

[B234] Bulletin D-412D, "Power Factor Correction Capacitor—Applications," Product Data, Square D Company, 1991.

[B235] Bulletin B-6150, "Mark V Solid-State Elevator System," Westinghouse Elevator Company, 1977.

[B236] Campbell, Sylvester J., *Solid-State AC Motor Controls* (Selection and Application). Marcel Dekker, Inc., 1987.

[B237] Chalmers, B. J., *Electric Motor Handbook*. Butterworths (Publishers), Boston, 1988.

[B238] Cochran, Paul L., *Polyphase Induction Motors* (Analysis, Design, and Application). Marcel Dekker, Inc., 1989.

[B239] "Cold Air, Sleeping Giant," *Heating/Piping/Air Conditioning*, Penton Publishing Co., Cleveland, OH, Apr. 1994.

[B240] "Electrical Maintenance Hints," The Westinghouse Electric Corp., Trafford, PA, 1984.

[B241] Fink, D. G., and Beaty, H. W., editors, *Standard Handbook for Electrical Engineers*, 13th edition. McGraw-Hill Book Co., 1993.

[B242] Franklin, A. C., and Franklin, D. P., *The J&P Transformer Book*, 11th edition, Oxford, England: Butterworth Heinemann, 1983.

[B243] Keeler, R. M., *Thermal Storage System Design*. McGraw-Hill, Inc. 1993.

[B244] Kloeffler, R. G., Kerchner, R. M., Brenneman, J. L., *Direct-Current Machinery*, revised edition, The Macmillan Company, 1948.

[B245] Leonhard, W., *Control of Electrical Drives*. Springer-Verlag, 1985.

[B246] Lloyd, Tom C., *Electric Motors and Their Application*. John Wiley & Sons, 1969.

[B247] Longland, T., Hunt, T. W. and Bracknell, A., *Power Capacitor Handbook*. Butterworths (Publishers), Boston, 1984.

[B248] Nailen, R. L., *Managing Motors*. Chicago, Barks Publications, Inc., 1991.

[B249] Smeaton, R. W., editor, *Motor Application & Maintenance Handbook*, 2nd edition. McGraw-Hill, Inc., 1987.

[B250] Smith, Craig, B., Editor, "Efficient Electricity Use," Pergamon Press, Inc., 1976.

[B251] Pillai, S. K., *A First Course On Electrical Drives*, Wiley Eastern Limited, 1982 (R1983).

[B252] Publication ADG-044A, "Adjustable Speed AC and DC Drives," 2nd edition, Emerson Electric Co., Industrial Controls Division, p. B5-1.

Annex 5A

(normative)

Manufacturing end-use applications of electricity by category and sector, 1990 (billion kWh)

SIC	Industry	Total electricity consumption	End-use application			
			Motor drive[a]	Electrolytics[b]	Process heating[c]	Lighting and other[d]
	Process industries					
28	Chemicals and allied products	170.8	130.6	33.3	0.8	6.1
26	Paper and allied products	110.9	103.4	0.0	1.5	6.0
20	Food and kindred products	57.3	49.3	0.0	0.7	7.3
29	Petroleum and coal products	44.8	42.5	0.0	0.7	1.6
22	Textile mill products	28.4	23.4	0.0	0.7	4.3
21	Tobacco products	2.1	1.7	0.0	0.0	0.4
	Total process industries	414.4	350.8	33.3	4.4	25.8
	Materials production					
33	Primary metals production	152.0	60.8	63.2	23.1	4.9
32	Stone, clay, and glass products	32.3	23.3	0.0	6.9	2.1
	Total materials production	184.3	84.1	63.2	30.0	7.0
	Materials fabrication: Metals					
37	Transportation equipment	36.8	16.2	0.2	15.8	4.6
35	Machinery except electrical	35.6	18.3	0.2	12.8	4.3
36	Electrical equipment	32.8	20.8	0.0	4.8	7.2
34	Fabricated metal products	32.3	11.9	3.9	14.0	2.5
38	Instruments, related products	14.7	11.0	0.0	1.6	2.1
39	Misc. manufacturing industries	5.1	3.8	0.0	0.4	0.9
	Total metals fabrication	157.4	82.1	4.3	49.4	21.5
	Materials fabrication: Non-metals					
30	Rubber and miscellaneous plastics	36.0	25.7	0.0	4.4	5.9
24	Lumber and wood products	20.9	15.9	0.0	0.3	4.7
27	Printing and publishing	18.3	12.7	0.0	1.3	4.3
23	Apparent, textile products	5.6	4.4	0.0	0.3	0.9
25	Furniture and fixtures	6.1	4.7	0.0	0.3	1.1
31	Leather and leather products	0.9	0.5	0.0	0.0	0.4
	Total non-metals fabrication	87.8	63.9	0.0	6.6	17.3
	Total materials fabrication	245.2	146.0	4.3	56.0	38.8
	Total manufacturing	844.0	580.9	100.8	90.4	71.6

Sources: EPRI TR-101021, 1992 [B96]. (EPRI's sources are as follows: U.S. Dept. of Commerce, *1989 Annual Survey of Manufacturers;* U.S. Bureau of the Census; Federal Reserve Board, *Industrial Production and Capacity Utilization,* June 1991; Resource Dynamics Corporation estimates; and EPRI's Industrial Market Information System.)

NOTE—Totals may not equal sum of components due to independent rounding.

[a]Pumps, conveyors, fans, compressors, plus for crushing, grinding, stamping, trimming, mixing, cutting, and milling operations.

[b]Electrochemical processes for separating, reducing, and refining metals and chemicals.

[c]Cooking, softening, melting, distilling, annealing, and fusing materials and products.

[d]Lighting, space heating, plant indication and control, power systems for communication and instrumentation, emergency and computer power supply systems.

Annex 5B

(informative)

Steam generator system description

5B.1 Transformer steam generator

These units are available in sizes from 10–60 kW. Water is heated by passing through tubular secondary windings of a transformer. The hot water is then fed through a short-circuiting tube of high-resistance metal to produce steam. The steam is then passed into a steam chest that also contains hot water under pressure from the steam.

5B.2 Resistance element boiler

These units consist of a drum maintained to a preset level with water. Electric elements of heated resistance wire are withdrawable for occasional cleaning. A steam pressure switch controls the elements and hence maintains steam production. These boilers are rated from about 3–2500 kW and are suitable for small- to medium-sized requirements and for operation under very high pressure if required.

The resistance element boiler is the only one that can operate on direct current.

5B.3 Electrode boilers (low-voltage)

These units use the principle that a current will pass between electrodes immersed in water and the heat produced will change some of the water to steam.

Load control may be varied from 10–100% by raising or lowering the level of water in the boiler, alternatively the electrodes or an insulating sleeve around them can be raised or lowered. Electrode boilers for low-voltage operation are rated from 40–2400 kW.

A feature of these boilers is that in the event of loss of water the unit fails safe as no current can pass.

5B.4 Electrode boilers (medium-voltage)

Electrode boilers for medium voltage are rated from 1000–20 000 kW.

5B.5 Jet boilers

These units are a type of electrode boiler that has been specially developed for medium-voltage use. Instead of the electrodes being immersed in water, they are connected electrically by jets of water produced by a circulation pump.

Control is achieved by moving a shield over the jets to divert water away from the electrodes or by controlling the rate of flow from the circulation pump. Outputs from no-load to full-load are possible. Sizes range from 1200–50 000 kW.

Annex 5C

(informative)

Compressed air systems worksheet (SEC of Victoria [B196])

5C.1 Compressed air energy audit calculation sheet

Compressor specification (from nameplate or manual)

1.	Motor rating	_____ kW*
2.	Rated free air delivery	_____ ft^3/m (L/s)
3.	Rated operating pressure	_____ lbf/in^2 (kPa)

Compressor power requirements (refer to procedures in text)

1.	No load	_____ kW
2.	Full load	_____ kW
3.	Average load	_____ kW

Compressor operating hours

1.   Total number of operating hours per month   _____ h/month

Compressor energy consumption

1.   Energy use per month =
     Average load × operating hours per month   _____ kWh/month
                                                          (*a*)

Compressed air leakage (refer to text)

1.   Measured air leakage   _____ ft^3/m (L/s)

2.   Percent air leakage = $\dfrac{\text{measured air leakage}}{\text{average air delivery}} \times 100\%$ _____ %

3.   Power use attributable to air leaks =
     Measured air leakage (kW) – measured no load (kW)   _____ kW

4.   Air leakage energy consumption =
     kW air leakage × number of operating hours per month   _____ kWh

NOTE—Compressor input power is compressor output or motor nameplate rating divided by motor efficiency.

*Convert hp to kW by multiplying hp by 0.746 kW/hp

Energy savings

Pressure reduction savings (refer to text)

1. Interpolate using "14.5 lbf/in^2 (100 kPa) pressure reduction
    produces 8% energy saving"   _____ %

2. Potential energy saving (kWh) = energy use per month
    × percent saving through pressure reduction kWh   _____ kWh

    (*b*)

Air intake losses (refer to text)

1. Interpolate using "0.145 lbf/in^2 (1 kPa) reduction of intake
    pressure loss produces 0.5% energy saving"   _____ %

2. Potential energy saving (kWh) = energy use per month
    × percent saving through intake pressure reduction   _____ kWh

    (*c*)

3. Interpolate using "5.4 °F (3 °C) of air intake temperature
    reduction produces 1% energy saving"   _____ %

4. Potential energy saving (kWh) = energy use per month
    × percent saving through air intake temperature reduction   _____ kWh

    (*d*)

Pipe resistance losses (refer to text)

1. Interpolate using "For a 50% excess of maximum
    air velocity, energy use is increased about 2%"

    Maximum velocity for distribution minimum should not
    exceed 20 ft/s (6 m/s)

    Maximum velocity for branch line should not exceed
    32 ft/s (10 m/s)   _____ %

2. Potential energy saving (kWh) = energy use per month
    × percent saving through pipe resistance reduction   _____ kWh

    (*e*)

Subtotal energy savings
Sum the energy savings calculated above: *b* + *c* + *d* + *e* =   _____ kWh

(*f*)

Adjusted energy savings subtotal

> Since the potential energy savings outlined above are interrelated
> so that a savings in one area will cause a reduced energy saving
> in another area, the above energy savings subtotal should be
> about 10% lower: $0.9 \times f =$ _____ kWh
>
>   (g)

Waste heat recovery (refer to text)

> Assume that approximately 80% of compressed air plant
> energy consumption can be recovered as heat energy:
>
> Potential energy savings (kWh) = $0.8 \times$ energy use per month
> – adjusted energy saving subtotal = $0.8\,(a - g) =$ _____ kWh
>
>   (h)

Other energy saving opportunities (refer to text)

1. Switching and control system improvements   _____ kWh
    (i)

2. Dry air   _____ kWh
    (j)

3. Misuse of air reduction   _____ kWh
    (k)

4. Maintenance improvements   _____ kWh
    (l)

Total potential energy savings per month: $g + h + i + j + k + l =$ _____ kWh

5C.2 Compressed air system components

a) *Intercooler.* An intercooler is fitted between the stages of a multistage compressor. Its purpose is to cool the air between stages and also to condense surplus water vapor before it reaches the next compression stage.

b) *Aftercooler.* When the air leaves the compressor it is very hot and, if passed immediately to the receiver and pipework, it will cool and the water vapor within it will condense. Cooling the air immediately after it leaves the compressor condenses most of the water before it reaches the receiver. Aftercoolers are air- or water-cooled and may form part of the compressor package or be a separate free-standing item.

c) *Receiver.* From the aftercooler, air is piped to the receiver that acts to store compressed air so as to buffer the compressor from the fluctuating factory demands.

d) *Filter and dryer.* Compressed air, required to be of high quality (and extra dry, such as for instrument use), is passed through either

    1) A regenerative dryer;
    2) A refrigerated dryer;
    3) A desiccant dryer; or
    4) A combination refrigerated/desiccant dryer

before being distributed by the pipework system.

e) *Separator.* Following the air receiver, a separator is sometimes fitted to separate and discharge moisture that has condensed from the compressed air after it has further cooled in the receiver.

f) *Pipework.* The pipework delivers the quantity of compressed air to each point of use. It must also adequately drain away condensation that may form due to further cooling of the air within it. Pressure regulation are installed at each point of use or at the start of a branch where all items of equipment require a low pressure.

g) *Equipment.* Finally, at the point of use, the cool, dry, compressed air is used to power the equipment.

Annex 5D

(informative)

Refrigeration system description

5D.1 Direct process cooling

Many processes, including water chilling are cooled directly within the refrigeration machine, and the process load stream (e.g., chilled water) is pumped to the evaporator of the machine for heat removal. This refrigeration cycle is commonly used when all process loads require about the same temperature level. A standard two-stage compressor can produce temperatures to +19 °F (–7 °C); three stages down to –9 °F (–23 °C); and four stages down to –51 °F (–46 °C). Between stages, economizers may be added to improve efficiency.

5D.2 Indirect process cooling

For some processes it is more practical to pump a cooling medium to the process. This is done by cooling a brine in the refrigeration machine and circulating the fluid to various heat exchangers. Where many processes are dispersed, a central cooling plant can be used that may have lower energy consumption than several smaller systems.

5D.3 Multiple remote process cooling

Complex industrial processes require cooling at many temperatures and capacities, separately sited. A central refrigeration facility of complex design can serve these process loads.

5D.4 Direct liquid refrigeration

Direct liquid refrigeration is used extensively in ice rinks. (Note that "brine" systems are also used.) The refrigerant is supplied direct to the rink floor tubing in liquid form. There is no chilled water or brine pump in the direct liquid refrigeration system. Discharge pressure from the compressor is used to pressurize the tank that supplies cold refrigerant under lower pressure to the ice-rink tubing.

5D.5 Direct expansion (DX) liquid coolers

In DX systems the refrigerant evaporates inside tubes of a direct-expansion cooler. DX units are used with positive-displacement compressors.

5D.5.1 Small DX application

Typically used for data-processing installations. The DX systems include

a) A glycol-water mixture circulating from the condenser through a closed-circuit loop that goes to a "dry cooler."

b) Water that is pumped from the condenser to an outside cooling tower for evaporative cooling.

c) High-pressure refrigerant that is piped from the compressor to outside remote air-cooled condensers and returned to the high-pressure side of the expansion valve.

5D.5.2 Large DX application

DX air-conditioning with helical screw compressors are used for large commercial building applications. The screw compressors operate with stability down to 10% of full load. With DX, the refrigerant is piped directly to air-handling units equipped with DX cooling coils. The chilled water piping loop is eliminated. The refrigerant is piped from the screw compressor to an evaporative condenser. A conventional water-cooled condenser and cooling tower are not used in this cycle.

Annex 5E

(informative)

Alternating-current single-phase small (fractional-horsepower) motors rated 1/20–1 hp, 250 V or less

Application	Motor type	Horse-power	Speed (r/min)	Starting torque	Efficiency
Fans					
Direct drive	Permanent split capacitor	1/20–1	1625, 1075, 825	Low	High
	Shaded pole	1/20–1/4	1550, 1050, 800	Low	Low
	Split phase	1/20–1/2	1725, 1140, 850	Low	Medium
Belted	Split phase	1/20–1/2	1725, 1140. 850	Medium	Medium
	Capacitor start—induction run	1/8–3/4	1725, 1140, 850	Medium	Medium
	Capacitor start—capacitor run	1/8–3/4	1725, 1140, 850	Medium	High
Pumps					
Centrifugal	Split phase	1/8–1/2	3450	Low	Medium
	Capacitor start—induction run	1/8–1	3450	Medium	Medium
	Capacitor start—capacitor run	1/8–1	3450	Medium	High
Positive displacement	Capacitor start—induction run	1/8–1	3450, 1725	High	Medium
	Capacitor start—capacitor run	1/8–1	3450, 1725	High	High
Compressors					
Air	Split phase	1/8–1/2	3450, 1725	Low or medium	Medium
	Capacitor start—induction run	1/8–1	3450, 1725	High	Medium
	Capacitor start—capacitor run	1/8–1	3450, 1725	High	High
Refrigeration	Split phase	1/8–1/2	3450, 1725	Low or medium	Medium
	Permanent split capacitor	1/8–1	3250, 1625	Low	High
	Capacitor start—induction run	1/8–1	3450, 1725	High	Medium
	Capacitor start—capacitor run	1/8–1	3450, 1725	High	High
Industrial	Capacitor start—induction run	1/8–1	3450, 1725, 1140, 850	High	Medium
	Capacitor start—capacitor run	1/8–1	3450, 1725, 1140, 850	High	High
Farm	Capacitor start—induction run	1/8–3/4	1725	High	Medium
	Capacitor start—capacitor run	1/8–3/4	1725	High	High

Application	Motor type	Horse-power	Speed (r/min)	Starting torque	Efficiency
Major appliances	Split phase	1/6–1/2	1725, 1140	Medium	Medium
	Capacitor start—induction run	1/6–3/4	1725, 1140	High	Medium
	Capacitor start—capacitor run	1/6–3/4	1725, 1140	High	High
Commercial appliances	Capacitor start—induction run	1/3–3/4	1725	High	Medium
	Capacitor start—capacitor run	1/3–3/4	1725	High	High
Business equipment	Permanent split capacitor	1/20–1/4	3450, 1725	Low	High
	Capacitor start—induction run	1/8–1	3450, 1725	High	Medium
	Capacitor start—capacitor run	1/8–1	3450, 1725	High	High

Source: NEMA MG 11-1992 [B167]

Annex 5F

(informative)

Typical characteristics and applications of fixed-frequency medium ac polyphase squirrel-cage induction motors

Classification	Locked rotor torque (percent rated load torque)	Breakdown torque (percent rated load torque)	Locked rotor current (percent rated load current)	Slip (%)	Typical applications	Relative efficiency
Design B Normal locked rotor torque and normal locked rotor current	70–275[b]	175–300[b]	600–700	0.5–5	Fans, blowers, centrifugal pumps and compressors, motor-generator sets, etc., where starting torque requirements are relatively low	Medium or high
Design C High locked rotor torque and normal locked rotor current	200–250[b]	190–225[b]	600–700	1–5	Conveyors, crushers, stirring machines, agitators, reciprocating pumps and compressors, etc., where starting under load is required	Medium
Design D High locked rotor torque and high slip	275	275	600–700	5–8	High-peak loads with or without flywheels, such as punch presses, shears, elevators, extractors, winches, hoists, oil-well pumping and wire-drawing machines	Medium
Design E[a] IEC 34-12 Design N locked rotor torques and current	75–190[b]	160–200[b]	800–1000	0.5–3	Fans, blowers, centrifugal pumps and compressors, motor-generator sets, etc., where starting torque requirements are relatively low	High

Source: NEMA MG 10-1994 [B166]
NOTE—Design A motor performance characteristics are similar to those for Design B motors except that the locked-rotor starting current is higher than the values shown in this table.
[a]NEMA Design E motors are similar to IEC Design N as specified in IEC 34-12 (1980) [B114].
[b]Higher values are for motors having lower horsepower ratings.

Annex 5G

(informative)

Example of a 300 hp induction motor adjustable frequency drive: Efficiency vs. frequency

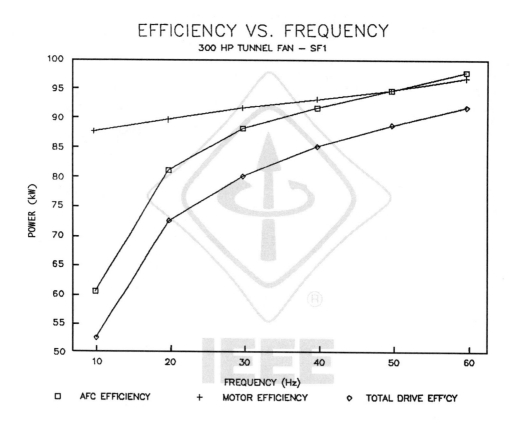

Source: Courtesy of Reliance Electric Co.

Annex 5H

(informative)

Example of a 300 hp induction motor adjustable frequency drive: Kilowatts vs. frequency

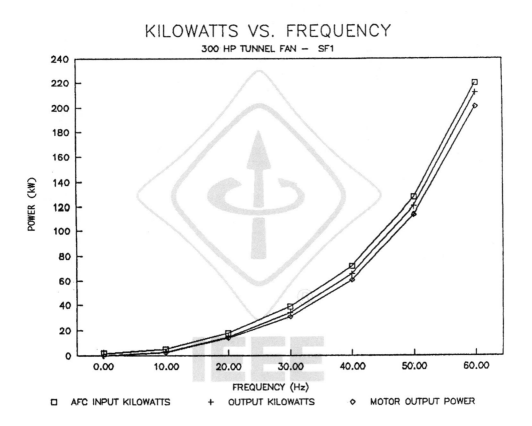

Source: Courtesy of Reliance Electric Co.

Annex 5I

(informative)

Typical range of efficiencies for dry-type transformers: 25–100% load

Rating (kVA)	100% load		75% load		50% load		25% load	
	Copper windings, 80 °C rise	Aluminum windings, 150 °C rise	Copper windings, 80 °C rise	Aluminum windings, 150 °C rise	Copper windings, 80 °C rise	Aluminum windings, 150 °C rise	Copper windings, 80 °C rise	Aluminum windings, 150 °C rise
500	98.65	97.67	98.72	98.01	98.63	98.20	97.91	97.84
1000	98.86	98.01	98.92	98.29	98.85	98.43	98.27	98.05
1500	99.07	98.28	99.11	98.50	99.03	98.62	98.50	98.26
2000	99.21	98.55	99.26	98.73	99.22	98.81	98.83	98.46
2500	99.22	98.63	99.25	98.79	99.19	98.84	98.75	98.48
3000	99.20	98.72	99.24	98.86	99.18	98.94	98.75	98.64

Source: Courtesy of ABB Power T&D Company, Inc., Bland, VA
NOTE—This table considered only 60 Hz units with a high-voltage BIL equal to or greater than 95 kV.

Chapter 6
Metering for energy management

6.1 Background

For the first three-quarters of this century, the monitoring of electrical power and energy was dominated by conventional electromechanical voltmeters, ammeters, and watthour meters. Only in the last decades have solid-state microprocessor-based digital devices become available for application in the commercial and industrial marketplace.

These new devices perform the tasks of up to 24 conventional indicating meters for about the price of three. Communication via means such as an RS-485 data link to a PC allows monitoring of up to 70 values including times and dates, min./max. history, temperature indications, and energy management alarms.

Complex waveform analysis can also be carried out for harmonic problems typically associated with adjustable speed drives that have been installed on fans and pumps for energy management savings. Since metering systems are absolutely essential to a successful energy management process, consideration should be given to applying the latest in metering technology.

It should be noted that meters alone do not save money, since they can be costly to install and maintain. However, through proper monitoring, recording, and analysis, the use of meters can lead to corrective actions that produce the desired result of reducing energy per unit of production or per service performed.

Experience has shown that a 1–2% reduction in consumption can be achieved after meters are installed just by letting the users know that they are being monitored. Up to a 5% total reduction can occur when the users then become proactive in better managing the use of their energy. Ultimately, up to a 10% total reduction can be achieved when metering is tied directly to the process through a Programmable Logic Controller (PLC) or Distributed Control System (DCS), in a closed loop automated process control arrangement.

6.2 Relationships between parameters in an electric power system survey

If only limited instrumentation is available, unknown parameters of the power system can be calculated on the basis of known quantities. In order to be valid, though, the calculations must be based on simultaneous readings of the known quantities. For example, power factor can be calculated if power (in kilowatts) and apparent power (in kilovoltamperes) are known, provided that the wattmeter and kilovoltampere meter readings were taken simultaneously, or reasonably close together under assured steady-state conditions. (See figure 6-1.)

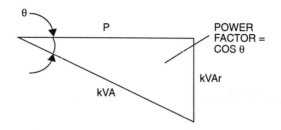

Figure 6-1—Relationship between electrical parameters

6.3 Units of measure

It is not unusual to find that even today there exists a basic misunderstanding of the difference between units of Power (a rate of consumption), and units of Energy (the total quantity consumed). The engineering data presented in table 6-1 is an attempt to clarify this issue.

Note that the units of power, multiplied by a specific period of time such as minutes or hours, yield units of energy or work. In other words, a unit of power acting for a specific period of time yields a specific quantity of energy or useful work.

Table 6-1—Engineering data

Units of power (rate)	Typical use	Units of energy or work (quantity)
kW	Electricity	kWh
SCFM	Air	SCF
GPM	Water	gal
ton	Refrigeration	ton-h
lb/h	Steam	lb
hp	Work	hp-h

6.4 Typical cost factors

A typical cost factor or "rule of thumb" is useful to have handy for estimating the cost of operating a specific piece of equipment for a year, and in comparing the costs of supplying heat energy from a variety of sources. The formulas for developing several cost factors are presented in a) through d) in the following listing, along with examples. Note that the term MMBtu means one million British Thermal Units. The term M stands for 1000 in the language of the financial world, thus MM stands for one million.

a) $/MMBtu fuel

$Gallon fuel oil   or $MSCF natural gas

Gallon MMBtu   MSCF   MMBtu

Example:

$$\frac{\$1.00 \text{ Gallon}}{\text{Gallon } 0.148 \text{ MMBtu}} = \$6.76/\text{MMBtu}$$

b) $/MMBtu electricity

$$\frac{\$ \quad kWh}{kWh \quad MMBtu}$$

Example:

$$\frac{\$0.06 \quad kWh}{kWh \quad 0.003413 \text{ MMBtu}} = \$17.58/\text{MMBtu}$$

c) $/1kW-y

$$\frac{h \quad \$}{y \quad kWh}$$

Example:

$$\frac{8760 \text{ h} \quad \$0.06}{y \quad kWh} = \$526/\text{kW-y}$$

d) $/1hp-y

$$\frac{kW \quad h \quad \$}{hp \quad y \quad EFF \quad kWh}$$

Example:

$$\frac{0.746 \text{ kW} \quad 8760 \text{ h} \quad \$0.06}{hp \quad y \quad 0.89 \quad kWh} = \$441/\text{hp-y}$$

Comparing the cost between heating with fuel at $1.00 per gallon and electricity at $0.06 per kilowatthour, electricity is almost three times more expensive. Note, however, that the efficiency when burning the fuel will be somewhat less than 100% and must be taken into consideration. See a) and b) above.

The term $/1 kW-y is useful when determining the cost of operating a one-kilowatt load for a period of one year (in this case 8760 h, or any other value that represents the time of operation). See c) above.

The term $/1 hp-y is useful when determining the cost of operating a one-horsepower motor for a period of one year (in this case 8760 h). Note the efficiency of the motor must be estimated and used in the formula. The value of 0.89 is typical for normal efficiency motors 50 hp and greater, running at an average of 75% load or greater over the time period specified. See d) above.

A good rule of thumb for installation of a permanent kilowatthour meter is based on the cost of electricity, the size and efficiency of the load, the hours of operation, and the cost of the meter. For example, at $0.06/kWh and 8760 h of operation per year with an efficiency of 89%, a 200 hp motor will consume $200 \times \$441.00$ or \$88 200.00 worth of electricity per year.

Experience has shown that a 1% drop in energy consumption can be expected just from the fact that the load is being metered. In other words, someone really cares how much energy is being used. A 1% reduction will save \$882.00 per year. Assuming the installed meter cost is \$1000.00 installed, the simple payback is less than 14 months, well under the typical two-year-or-less guideline widely followed in industry today.

6.5 Six reasons to meter

6.5.1 Charge out energy to individual departments

This is the most basic reason to meter. Each month the total energy bill is proportioned to the various departments. This data is used to compare costs against the department's budget and thus develop a variance-dollar value. Also included is the use of meters for revenue billing when energy is sold to a third party such as in a cogeneration arrangement.

6.5.2 Accountability for energy used

Trending of energy consumption per unit of production or per service performed is the basis for initial analysis and resulting corrective actions.

6.5.3 Efficiency of utility equipment and systems

The experience industry has gained from a formal utility test process provides the following guideline values:

a) Centrifugal air compressors at 125 psig:
    3.2 to 4.0 kWh/1000 standard cubic feet (sf^3)
b) Centrifugal chiller drives producing 45 °F water:
    0.6 to 1.0 kW/ton refrigeration or
    0.6 to 1.0 kWh/ton-h refrigeration
c) Refrigeration delivered to the conditioned space:
    1.0 to 2.0 kW/ton refrigeration or
    1.0 to 2.0 kWh/ton-h refrigeration
d) Steam boilers at 250 psig:
    Number 6 fuel oil = 7.1 to 9.0 gal/1000 lb steam
    Natural gas = 1000 to 1300 sf^3/1000 lb steam
    Pulverized coal = 100 to 120 lb/1000 lb steam
    (Fuel consumption values will be about 25% higher at 600 psig.)

6.5.4 Provide information for audits of energy projects

With funding becoming increasingly difficult to obtain, audits of cost reduction energy management projects have been required more frequently over the past five years.

6.5.5 Maintenance work: Identify location of performance problems, and provide feedback to managers

The collection of energy-consumption data in support of maintenance work is seen as a viable tool, greatly aiding in the identification of equipment performance problems. As a side issue, performance problems associated with the people operating the equipment are also identified, allowing managers to take any necessary corrective action.

6.5.6 Identify potential future additional energy savings

With the long-term goal of management being one of continual improvement, metering and trending systems provide data on which to base resource allocation decisions.

6.6 The importance of audits

Audits based on a well-designed metering system will frequently yield surprising results and may often identify considerable savings. The results obtained from the audit will normally be in direct proportion to the effort expended. For example, in kilowatt demand control, many equipment suppliers gloss over the importance of a survey of plant loads to identify those that are suitable candidates for load shedding. In some cases, the vendors may acknowledge the need for such a survey but underestimate the time and manpower required to do a thorough job.

It is thus apparent that a formal, standardized survey method needs to be used if truly useful information is to be obtained. The survey sheet shown in table 6-2 is typical and is used to tabulate all loads 5 hp (or 5 kVA) and larger. One sheet should be used for each common bus switching point or motor control center (MCC).

This survey form involves using up-to-date electrical drawings where available, supported by field checks to verify exact conditions whenever questions exist. Approximately one man-hour, on the average, is required to complete each sheet, with some follow-up time to resolve occasional questions that arise when reviewing the survey results. Actual running kilovoltampere and kilowatt data are obtained using a variety of electrical meters described later in this chapter.

Typical survey results can be presented in many forms; two examples are shown in table 6-3 and figure 6-2. Table 6-3 lists electrical system usage by function for a typical industrial plant and points out the major users. In this example, approximately 34% of the total load is associated with the production and distribution of chilled water for air conditioning.

A second approach is illustrated in figure 6-2 using block diagrams to detail the efficiency of the plant electrical system. Data for this diagram is collected using portable electrical meters, combined with the data form shown in table 6-2.

Table 6-2—Typical form for recording electrical load

Description and equipment number	Ownership	Connected kVA	Running kVA	Running kW	Cubicle	Operating cost/day	Estimate of time On (min)	Estimate of time Off (min)	Conditions under which device can be down	Remarks (Utilization, spares, etc.)
Supply fan K-7801.05		100			1D		10		For kWd control, or as part of cycling for kWd reduction	Consider as one load to be shed
Return air fan K-7801.06a		25			2C			50		
Return air fan K-7801.06b		25			2D					
Elevator		75			2E		0		None	Not available
Air handling unit for S-3	S3	7.50			3D		55	5	kWd control only	Affects yarn quality
Service panel K-15		30			31L		0		None	Not available
Lighting panel K-78		30			31R		0		None	Review need for photo cells on some lighting circuits
Unit substation KL-8			MCC K6-8-1							
	Date (DD/MM/YR)_____						Prepared by WLS		L-5-4230-05 Ref Drawing Number_____ Page 1 of 1 Rev 0	

NOTE—Surveying plant loads is essential to understanding where energy is used. This typical form is for recording electrical load data. Similar forms can be used for each motor control center in the plant and for all loads of 5 hp or 5 kVA or more. On an annual basis to keep the data current, approximately 1 man-hour per form is needed.

Table 6-3—Typical electrical system usage by function for a typical industrial textile plant

	%[a] of total load		% of total load
HVAC fans[b]	13	Chip drying	3
Chillers	12	Staple drawing	3
Compressed air	11	Polymerization	2
Texturizing	8	Filament draw twist	1
Extruding and metering	6	Staple tow drying	1
Quench air	6	Filament spinning	1
Cooling water pumps	5	Waste treatment	1
Staple spinning	5	Nitrogen, inert gas	1
Lighting	5	Cooling tower fans	1
Filament spin draw	4	Beaming	1
Chilled water pumps	3	Miscellaneous	7
		Total	100.0

[a]Percent values are rounded to the nearest whole number.
[b]Heating, ventilation, and air conditioning.

Note in figure 6-2 that the overall plant efficiency is only approximately 81%, which means that 19% of all electrical energy is wasted as heat. This waste heat may require additional air conditioning to remove the heat from the production areas. From this typical diagram, it appears that effort should quickly be directed toward improving the 53.3% efficiency of the adjustable speed drive systems, which use 7% of all the plant's electrical energy.

With regard to audits of the electrical power system, metering results can be used to compile an energy profile that can

a)   Aid in establishing and refining energy use by product line, department, or area.

b)   Establish and improve energy-use accountability.

c)   Allow measurement of cost reductions.

d)   Help determine equipment capabilities and load factors for future modifications and plant expansion.

e)   Provide data for analyzing results that vary from established standards.

In addition, an energy profile likely will reveal areas where conservation projects are most beneficial.

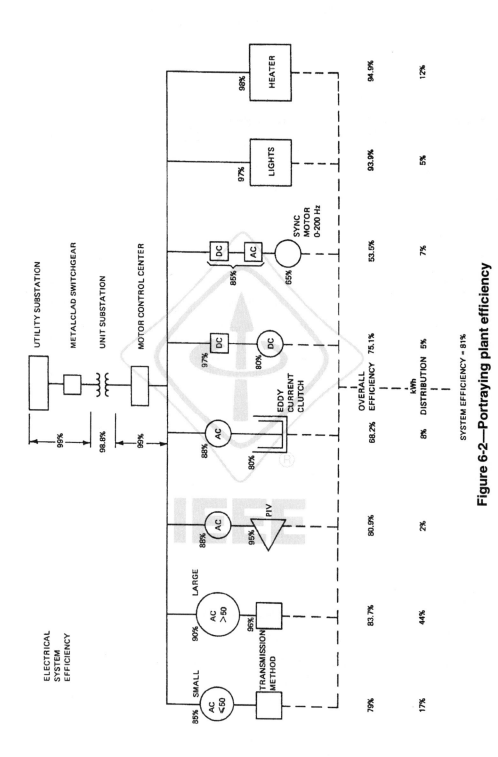

Figure 6-2—Portraying plant efficiency

6.7 Utility meters

Before taking steps to reduce a plant's electric bill, it is important to become familiar with the instruments utilities use to meter electric power consumption. The basic unit is the watthour meter, and the electromechanical style industrial meters are very similar to those used in residential service. They are more complex due to the polyphase power source and use of instrument transformers to obtain voltage and current for the meter.

Watthour meters measure electrical energy through the interaction of magnetic fluxes generated by the voltage and current coils acting to produce eddy currents in the rotating aluminum disk. Eddy current flow generates magnetic lines of force that interact with the flux in the air gap to produce turning torque on the disk. The meter is thus a carefully calibrated induction motor, the speed of which depends on the energy being measured. Each revolution of the meter disk represents a fixed value of watthours. The register counts these revolutions through a gear train and displays this count as watthours. Jewel or magnetic bearings are used to support the disk and the gear assembly is designed to impose minimum load on the disk.

The upper part of figure 6-3 is a simplified diagram showing how the kilowatt demand value is obtained from the utility kilowatthour meter. Each time the meter disk makes one complete revolution under the influence of the voltage and current coils, the photocell energizes the relay and transfers contacts A and B, providing kilowatthour output pulse. At the end of the demand interval, usually 15, 30, or 60 min, a clock pulse is given for 2–6 s, signifying that a new interval has begun.

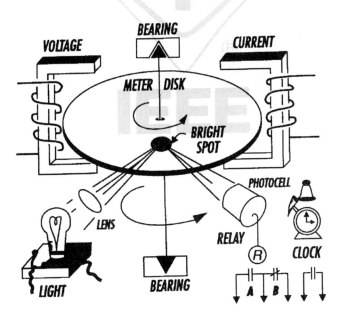

Figure 6-3—Simplified watthour meter

The A, B, and clock contacts are used by the demand monitoring or control equipment to develop the curve shown in the lower part of figure 6-3. The kilowatthour pulses are merely added to each other over the demand interval. A line connecting the tops of the columns of pulses describes the accumulation of kilowatthour pulses during the interval. The actual kilowatthour per pulse and the number of total pulses recorded during the interval will depend on the plant load and the Potential Transformer (PT) and Current Transformer (CT) ratios for the specific kilowatthour meter.

It is important to note at this point that the slope of the line in figure 6-4, mathematically speaking, is defined as rise over run, which is kilowatts divided by time (one-half hour in this case), which is equal to kilowatthour demand. That is to say, the total kilowatthour of energy consumed, divided by the time over which it was consumed, yields average kilowatt demand (kWd) for power over that time interval. This means that by detecting the slope of the line early in the interval, corrective action can be taken to reduce the slope by shutting off loads, lowering the average kWd to a more acceptable value. It follows that a flat line of zero slope would indicate no further energy consumption, a zero demand for power. Taken to the extreme, a line with a negative slope would indicate negative energy consumption, with a reversal of power flow; i.e., on-site generation of power back into the utility company's transmission system. This is unlikely to be realized since most utility meters are ratcheted to prevent reverse registration. Meters would be added to separately register kilowatthour being exported to the utility system.

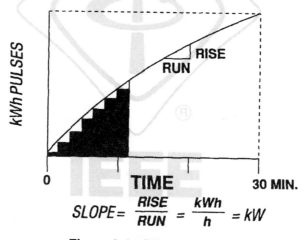

Figure 6-4—Demand curve

The exact length of the demand interval will vary by utility, one of the most common being 30 minutes. This interval length is most frequently associated with the time for power company generators, transformers, and transmission lines to build up sufficient heat due to overload conditions to do permanent damage to the equipment.

However, as kWd peaks increase, and as a few utility customers attempt "peak-splitting," the demand intervals are being reduced to 15 or even five minutes. In a few tariffs, a "sliding interval" is used, where there is no identified beginning and end to the interval. The kWd peak is then the highest average kWd for any successive 30 minutes during the power company billing period. There are some extreme situations where the utility will refuse to supply the customer with kilowatthour pulse information. In that case, PTs and CTs and appropriate kilowatthour meter or other transducer must be installed in order to obtain kilowatt demand information. If the utility meter is accessible, an optical transducer can be placed over the face of the meter and the transducer can "read" the utility meter. The transducer output can then be sent to the EMS system. The EMS can then be programmed to properly interpret the meter reading.

Utilities would like to supply power on a smooth continuous basis, with no peaks and valleys that require large swings in the amount of generating equipment being called into service. For that reason, utility bills reflect the effect of such swings in demand using a concept of load factor. Load factor is defined as the ratio of average demand to peak demand and is illustrated by the truck traveling along the mountain road in figure 6-5. With large peaks and deep valleys the truck, representing the utility, has to use a lot of gas to reach the top of the mountain peaks, and has to brake hard to slow down in the deep valleys. This arrangement yields a relatively low load factor of 0.5.

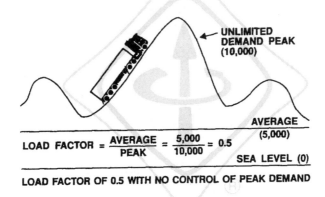

Figure 6-5—Load factor without peak demand control

The load factor can be improved by limiting the peak values and filling in the valleys as shown in figure 6-6. Now a small car, representing a smaller kilowatt load on the utility, can be used to travel over the smaller mountain peaks with less braking required in the shallow valleys. This improved load factor is now 0.63. The ideal load factor is 1.0, meaning that the demand is absolutely constant with no peaks and valleys. This, of course, is not actually possible. About the best that can be expected will be approximately 0.98 for a monthly period, and 0.85 for an annual value.

The face of a conventional electromechanical kilowatthour meter is shown in figure 6-7. The diagram illustrates a PT and a CT connected to one of the three input phases to the meter. The CTR on the meter face stands for "CT ratio" and the 800 to 5 indicates that when 800 A flow through the phase conductor, 5 A flow through the CT secondary into the meter. The PTR on the meter face stands for "PT ratio" and the 60/1 indicates that with 120 V on the secondary serving the meter potential coils, the primary is connected to a 7200 V supply (120 V × 60 = 7200 V).

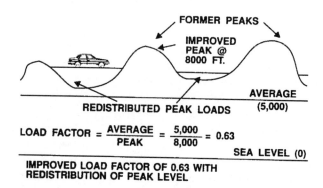

Figure 6-6—Load factor with redistribution of peak load

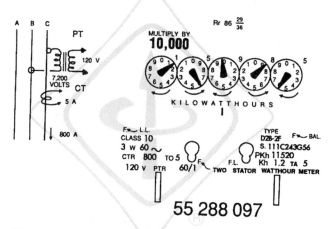

Figure 6-7—Electromechanical kilowatthour meter

Reading the dial to obtain the kilowatthour value is sometimes a great source of confusion, because the dials turn in opposite directions. The rightmost dial turns one complete revolution, which advances the left adjacent dial one number. In this example, the rightmost dial is read as 5 (and remains a 5 until the hand actually touches the mark for the 6). The second dial from the right is read as an 8, and will become a 9 when the rightmost dial hand points straight up to 0. The third dial from the right is read as a 9 (not a 0 as you might expect), and will not become a 0 until the hand on the second dial from the right points straight up to 0.

Using the same reasoning, the second dial from the left is read as a 5, and the leftmost dial is read as a 1. It becomes apparent that the meter dials must be read from right to left to determine the correct values from the dial hand position. The meter readings would thus be recorded as 15985. That value would be used as the current reading, from which a previous reading would be subtracted. That difference is then multiplied by 10 000 to obtain the kilowatthour that passed through the primary feeders during the specified time interval, such as a week or month.

Six sample meter faces are presented in figure 6-8 with their meter readings.

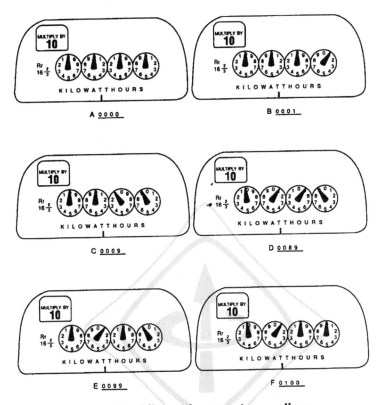

Figure 6-8—kilowatthour meter readings

6.8 Timing of meter disc for kilowatt measurement

The disc of a watthour meter makes from one-half to one rotation per watthour depending on the meter constant Kh. The number of watthours for any period is then the product of the number of disc rotations times the meter constant. The speed rotation of the disc, therefore, indicates the usage rate or the watts being used. The kilowatt is a more reasonable quantity for consumption rate and the following formula can be used when timing a meter disc:

$$\text{kilowatts} = \left(\frac{(PTR) \cdot (CTR) \cdot (Kh \text{ Wh/rev}) \cdot [3600 \text{ (s/h)}]}{1000 \text{ W/kW}}\right) \bullet \left(\frac{rev}{s}\right)$$

$$\text{kilowatts} = (3.6) \times (PMC) \times \frac{(rev)}{s}$$

where

　　PTR is potential transformer ratio

CTR is current transformer ratio

Kh is the meter constant found on the face of the meter (typically 1.2 to 1.8)

rev is the number of disc revolutions during observation period

s is the period of observation in seconds

PMC[1] is primary meter constant = $Kh \cdot (\text{PTR} \cdot \text{CTR})$

Note that timing the disc to determine passing kilowatts and kilowatthours is becoming more difficult with replacement of the electromechanical watthour meters with new digital technology. Some new digital meters offer a flashing bar to simulate the time between rotations. Other digital meters read and display kilowatt values directly.

6.9 Demand meters

6.9.1 Introduction

Gas and water are sold by the cubic foot and electricity by the kilowatthour. The cost of producing and delivering these commodities is not strictly proportional to volume in one case or quantity of electricity in the other. Because electricity cannot be stored, the cost of serving a customer who runs, for example, a 100 hp motor for one hour a day is much greater than the cost of serving one who operates a 10 hp motor ten hours a day, though both use the same number of total kilowatthours. The one-hundred horsepower motor requires a larger generator at the power house, heavier cables to carry the electrical energy perhaps many miles from the power house to the motor, and larger transformers, switches, fuses, and protective apparatus at several points along the line. All of this necessitates an expense that must be recovered either in a higher rate to all users, or in a proper charge for service plus a lower rate for kilowatt-hours.

For the purposes of billing for this "standby expense" the term *demand* has been adopted. The demand meter is installed along with the conventional watthour meter either as a separate device or as an attachment for the watthour meter. The customer then pays an amount determined by the maximum demand for the previous month in addition to the kilowatt-hours, but the rate for the latter is generally made low so that, if the load has been reasonably steady, the total bill is less than it would be otherwise. Thus, there is a powerful incentive for a larger use of electricity at a steady and continuous rate with benefit to all concerned.

The demand charge is based on the highest load for a short time period, say 15 or 30 minutes, which occurred during the month, but may not necessarily be fixed by the actual number of kilowatt-hours used during that period.

---

[1]This value is often stamped on the meter nameplate and identified by the symbol *PKh*.

6.9.2 Demand measurement in general

The demand is the load that is drawn from the source of supply at the receiving terminals averaged over a suitable and specified interval of time, such as 15 or 30 minutes. Demand is expressed in kilowatts, kilovoltamperes, amperes, or other suitable units.

It will thus be noted that as demand is measured in watts, etc., and not in watthours, etc., it is displayed by an indicating and not by an integrating device.

The demand is usually measured in kilowatts; but where power factor is an important item, the reactive kilovoltamperes and sometimes kilovoltamperes demands are measured. It is very seldom that a customer is billed on instantaneous demand, unless the short high peaks may be of such a nature as to interfere with the stability of the system.

6.9.3 The demand meter as a standard

It will be understood from all the above that the general conception of demand is not a distinct quantity. However, it can be expressed in fundamental units for the purpose of billing. It almost seems unnecessary to measure the kilowatt-hours and demand with high accuracy, when the best rate schedule possible can reflect cost conditions only approximately. However, accurate meters have come to be expected in order to establish a customer's confidence in the utility. Buying and selling being based upon measurement and price, it is necessary to have a standard of measurement agreed upon by both parties to the transaction. In order to be acceptable as a standard, a meter must be such as to facilitate calibration in terms of standard quantities. The important point is whether the unit used is readily reproduced for calibration and whether it is acceptable to both the utility and the customer.

The time intervals over which the power is averaged are determined in the following manner. One o'clock to 1:15 is one interval, 1:15 to 1:30 is another interval, 1:30 to 1:45 is still another interval, and so on; the entire day and month are divided up into 15-minute blocks. The block that shows highest or maximum demand is the one that is generally used for billing purposes.

There are a number of ways in which the average power may be determined; it being very unlikely that during any 15-minute period of time the power is absolutely steady. More likely, it is high one minute and low the next. The type of meter that is generally used is the so-called integrating demand meter, which gives the demand in kilowatts directly. If the kilowatthours that are consumed in a 15-minute period are known, it is a comparatively simple matter to determine the average power, for example, if from 1:00 to 1:15 the kilowatthours consumed were 62, the power would be 62 divided by 15/60, or 248 kW.

If from 1:15 to 1:30 the kilowatthours consumed were 84, the average kilowatts would be 336. It would involve considerable labor to read all the watthour meters exactly every 15 minutes, but the demand meter does this automatically.

Indicating demand meters generally consist of a standard watthour meter, the usual integrating register being replaced by a combination watthour and demand register.

They are herein termed "indicating" because they use pointers and scales as in indicating instruments to distinguish them from recording and printometer meters, and also because demand is expressed by definition in average watts or power over a time interval and not watthours or energy.

All principal models of watthour meters, both single or polyphase, can be equipped with demand registers that fit interchangeably with the usual mountings of the ordinary integrating dial.

These demand registers usually retain the integrating register mechanism and dial, and, in addition, there is a separate gear train also driven by the meter element, which pushes the demand pointer forward over a graduated dial, marked in kilowatts (not kilowatthours). The registers for alternating currents also contain a miniature timing motor of the synchronous type, operated from the same supply circuit, whose function is to measure the required time intervals over which the demand is averaged.

6.9.4 The use of instrument transformers with watthour and demand meters

Instrument transformers are used for two reasons—first, to protect operators from contact with high-voltage circuits, and second, to permit the use of meters with a reasonable amount of insulation and reasonable current-carrying capacity. It would be difficult and expensive to make a watthour meter with a 6600 V potential coil or a 1000 A current coil. The secondaries of potential transformers are wound for 120 V, and the secondaries of the current transformers for 5 A, thus making possible the use of 120 V, 5 A meters. The transformer insulates the meter from the line and, even if the current in a 2300 V circuit was only 5 A, it would not be safe to connect the current coils of the watthour meters directly into the line. Instead, a 5-to-5 current transformer would be used.

Instrument transformer ratios are often stated. A 2400–120 V potential transformer has a 20:1 ratio. A 75–5 A current transformer has a 15:1 ratio.

In making installations, due consideration must be given to the marked ratios of the potential and current transformers that are to be used with the watthour meter, for these determine the watthour constant, the register ratio, and the dial multiplier.

In the majority of cases, such things as phase-angle errors and ratio errors can be ignored.

The secondary voltage that is impressed upon the potential coil of the meter is in phase with the primary voltage, and the current that passes through the current coil is in phase with the primary current.

All instrument transformers should be grounded on the secondary side. This prevents a static voltage above ground on the secondary due to the relations of transformer capacitance. It is also a precaution against danger from high voltage resulting from puncture of insulation by lightning or other abnormal stresses. Any point of the secondary may be grounded in polyphase groups, but the preference is a neutral point or common wire between two transformers.

6.9.5 Potential transformers

The potential transformer is in principle an ordinary transformer, especially designed for close regulation so that the secondary voltage under any condition will be as nearly as possible a fixed percentage of the primary voltage.

The secondary voltage cannot generally be exactly proportional to the primary voltage or exactly opposite to the primary voltage in phase, on account of the losses in the transformer and the magnetic leakage.

There are two classes of errors inherent in voltage transformers: ratio error and phase-angle error. The part of these errors due to the exciting current is constant for any particular voltage. It can be reduced to a minimum by choosing the best quality of iron and working as a low magnetic density. The part of the errors due to the load current varies directly with load and can be minimized by making the resistance and reactance of the winding very low. In any transformer, the ratio error can be neutralized for one particular condition of load by compensating the transformer by an amount just sufficient to make up the voltage drop at the specified load.

The effect of the phase displacement of the secondary voltage need not be considered when using voltmeters, frequency meters, etc. With wattmeters and watthour meters whose indications depend not only on the voltage, but also on its phase relation to the line current, the phase-angle error has some effect.

These ratio errors and phase-angle errors are small and it is very seldom that they are taken into consideration when potential transformers are used with watthour meters.

The secondary terminals of the voltage transformers should never be short-circuited. If they should become short-circuited, a heavy current will flow that will likely damage the winding.

6.9.6 Current transformers

The current transformer is connected directly in series with the line and the current coils of the meters are connected in its secondary. When current is flowing in the primary winding, the secondary winding must be kept closed, by short-circuiting the secondary winding or else closed through the current coils of the meters. If the secondary should be opened, there is a possibility of dangerously high voltages being set up in the secondary winding, and there is also a possibility of the calibration of the transformer being changed by oversaturating the iron. It is important to remember that for a given primary current a definite secondary current is going to flow in the secondary winding, depending upon the ratio of the transformer. If the primary current increases, the secondary current also increases in proportion. Only a very small component of the primary current is required to magnetize the iron of the transformer, and any other flux that would be set up by the primary current is neutralized by the secondary current. It is obvious that if the secondary was open, no secondary current could flow and consequently, no secondary flux could neutralize that set up by the primary current; consequently, the iron would be magnetized to a very high density, and the voltage generated in the secondary winding would become enormous. In this respect, the current transformer differs from the potential transformer, where the voltage across the primary is practically constant, and when the secondary current is reduced to zero the primary current also is reduced and the only current left is the small magnetizing current. In the case of a current transformer the primary current will continue to be forced through even though its secondary is opened. The opening of the secondary does not in most cases introduce enough impedance in the line to appreciably reduce the primary current.

A current transformer is going to do everything possible to maintain its flow of secondary current, and for any given primary current the secondary current will remain practically constant, even though the impedance of the secondary load will be varied over comparatively wide limits.

Current transformers also have ratio and phase-angle errors, but in most meter installations these are ignored because they are too small to take into consideration.

6.10 Paralleling of current transformers

For the purpose of totalization, the secondaries of current transformers are sometimes paralleled, and the total secondary current is sent through the coil of a meter. When current transformers are paralleled, it is essential that they be of the same ratio.

When the transformers are paralleled, the circuits must be of the same frequency and from the same source of supply. The sum of the secondary currents that flow to the meter coil will be a vector sum. The method is quite accurate and simple. In case there is no load on one of the transformers very little current will leak through this winding because of its very high impedance as compared to the meter coil itself.

There may be some questions as to why these currents add in the meter coil and do not pass back through the other transformers, especially if the transformers are unequally loaded, but probably the best explanation is that there is no other place for these secondary currents to flow. As stated before, a current transformer is going to do everything possible to keep its secondary currents flowing, and if the impedance of its burden increases, the transformer automatically raises its voltage until the current does flow.

If several current transformers are connected in parallel on the secondary side, the meter coil must, of course, have current capacity to carry the resultant current, or else the current transformers must be designed so that the resultant current will be within the meter rating, usually 5 A or less.

6.11 Instrument transformer burdens

The burden on an instrument transformer is simply the load on its secondary. It is usually expressed in voltamperes at rated voltage or current, as the case may be and at a specified power factor.

The phase angle and ratio of a voltage transformer are affected by its secondary burden. This burden consists of the potential circuits of the watthour meter, wattmeter, voltmeter, etc. Variations of the line voltage also affect the errors, but since the voltage is practically constant, the small variations need not be considered.

In the current transformer, the phase-angle and ratio errors depend upon the secondary burden and also upon the primary current. The primary current may vary over a comparatively wide range because of the changes of the load, and the errors are different for each value of current, consequently the watthour meter can be corrected for only one value of current. The secondary burden consists of the current coils of wattmeter, watthour meter, ammeter, etc.

The power factor of the line in no way affects the ratios and phase angles of the instrument transformers, and should not be confused with the power factor of the secondary burdens.

Typical performance curves of the various types of instrument transformers can usually be obtained from the manufacturers. If the burdens are known, the errors can be determined from the curves.

Detailed information on instrument transformers can be found in IEEE Std C57.13-1993 [B9][2].

6.12 Multitasking solid-state meters

For many decades, the only type of meter was an electromechanical device that measured power consumption. These meters perform their task satisfactorily, operate reliably, and have remained virtually unchanged for many years. Even today, electromechanical meters are the dominant measurement device on power systems, and utilities continue to use and purchase them. However, there is a growing niche of applications where a more advanced device has advantages. This niche has fostered the development of the solid-state meter.

6.13 Metering location vs. requirements

An electrical power distribution system diagram for a typical building or industrial plant is shown in figure 6-9. This type of diagram is known as a "single line diagram" (SLD) because one single line is used to represent the three phase conductors, neutral conductor and/or grounding conductor. Various symbols are used to represent transformers, reactors, resistors, fuses and circuit breakers, motors and etc. The standard symbols are shown in IEEE Std 315-1975 and 315A-1986 [B8].

The incoming supply power from the utility is at the left, feeding the primary substation transformer T1. If the primary substation T1 is owned by the utility, the primary feed may be one of several high voltage levels such as 35 kV, 69 kV, 100 kV, etc. The secondary of T1 is commonly one of several medium voltages such as 6.9 kV, 12.47 kV, or 15 kV. This is usually the feed voltage to the various secondary substations in the facility. Individual large (200 hp and larger) synchronous or induction motors may be fed by a unit substation such as T2, at one of several medium voltages such as 2.3 kV or 4.16 kV.

---

[2]The numbers in brackets correspond to those of the bibliography in 6.19.

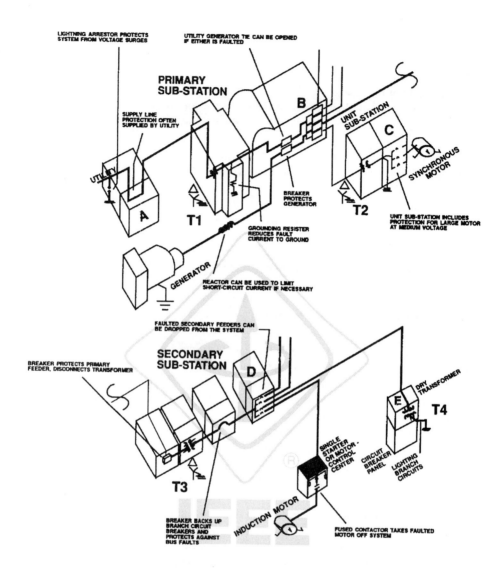

Figure 6-9—Electrical power distribution system

Further along in the system, the remaining motor loads are typically fed by one or more secondary substations such as T3. The most common secondary voltage is 480 V with a grounded wye transformer connection. Lighting circuits are usually powered from a lighting panel supplied by a small (10–45 kVA) dry-type transformer T4, with a primary rated at 480 V and a secondary of 208/120 V with a grounded wye transformer connection. Occasionally, lighting circuits will be operated at 277 V phase to neutral, supplied through a three-phase 480 V circuit breaker on the secondary side of transformer T3.

Permanent metering of voltage, current, kilowatts, and kilowatthours can be installed at locations A through E. Potential and current transformers are always used at locations A, B, and C, and may also be used at D and E. Portable survey meters must be connected to the secondary terminals of these instrument transformers at Locations A, B, and C, and may be directly connected at D and E. The clamp-on CT supplied with an analyzer that has an insulation rating of 600 V can safely be attached around energized conductors at locations D and E.

Note that when portable survey metering is connected to the secondary terminals of potential and current transformers at locations A, B, and C, the appropriate multiplier must be included to reflect the ratio factors of the transformer(s).

The following list of checkpoints should be used to make an effective and safe energy survey:

a) Make detailed plans prior to the actual survey.

b) Use prepared forms so records can be easily compared on a year-to-year basis.

c) On constant load where efficiency can be determined in a short time, consider using analog type indicating devices.

d) Where loads fluctuate or have varying duty cycles, use an automatic logging system or chart-type recording devices.

e) Check new equipment when installed to determine if it meets specifications and to provide data for reference in future surveys.

f) Be sure to make the survey when equipment is in use at its normal production load.

g) Make sure equipment has reached normal operating conditions: correct temperature, speed under normal load, etc.

h) Have test equipment neatly arranged on a portable cart and be sure to include extra test leads, equipment fuses, etc.

i) Make certain that personnel doing the testing are qualified to work on energized electrical equipment even if tests are done while equipment is de-energized and have the necessary safety equipment such as safety glasses, rubber gloves, and hard hats.

j) Refer to the procedures in NFPA 70E-1995 [B13] for comprehensive information on standard safety requirements.

6.14 Metering techniques and practical examples

6.14.1 Motor loads

It is not unusual to find motors to be 40–60% of a building's total electrical load, and upwards of 70–90% of the electrical load in an industrial plant. Because motors are such a large user of electricity, it is useful to review the operation of the squirrel cage induction motor and outline a simple method to determine output horsepower while running.

The induction motor is made up of two basic parts (see figure 6-10). On the outside is a stationery coil of wire wound around a core of laminated steel known as the stator. The synchronous speed of an induction motor is the speed at which stator flux rotates around the stator when connected to three phase ac power. It is determined by the frequency (in hertz) of stator current and the number of poles to which the motor stator is connected (see table 6-4). At 50 and 60 Hz, the following synchronous speeds are obtained:

$$\text{Synchronous speed} = \frac{120 \times \text{frequency}}{\text{number of poles}} \text{ expressed in speed in rpm}$$

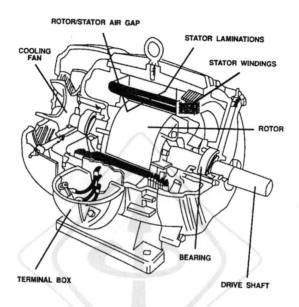

Figure 6-10—Three-phase ac induction motor

Table 6-4—Typical motor speeds

Number of poles	Synchronous speeds (rpm)	
	60 Hz	50 Hz
2	3600	3000
4	1800	1500
6	1200	1000
8	900	750
10	720	600
12	600	500

The magnetic field produced in the stator winding is continuously pulsating across the air gap and into the second basic part of the motor, the rotor. This rotor is made up of steel laminations and usually cast aluminum bars with end rings. As the magnetic flux cuts across the rotor bars, it induces a voltage between them in much the same way as a voltage is induced in the secondary winding of a transformer.

Because the rotor bars are part of a closed circuit through the end rings, a heavy current is induced in them. This rotor current in turn produces its own magnetic field, interacting with the rotating magnetic field of the stator, causing the rotor to be pulled around. The rotor never quite catches up with the rotating magnetic field of the stator, and in fact, if it did the transformer action across the air gap would stop and no energy would be transferred to the motor shaft.

6.14.2 Slip

The phenomenon of slip is very important to the operation of induction motors. At no-load the rotor almost manages to keep up to the synchronous speed of the rotating magnetic field in the stator. The only energy transferred across the air gap is that required to overcome drag from friction in the bearings and air resistance. As the load increases, the speed of the rotor tends to fall more and more behind the magnetic field, causing more lines of magnetic flux to be cut and more energy to be transferred across the air gap and to the shaft of the rotor. The actual speed of the rotor is always just the correct rate to transfer the amount of energy needed to drive the load at that moment in time.

At full load, the speed of the rotor has slowed to match that on the motor nameplate. The difference in speed between the synchronous rotating magnetic field in the stator and the actual speed of the rotor is known as slip. A typical low-slip three-phase four-pole motor (synchronous speed of 1800 rpm), would be rated at 1750 rpm at full load. The National Electrical Manufacturers Association (NEMA) standard for low-slip motors (torque design A and B) requires that slip does not exceed 5% at full load.

Motor manufacturers can change the speed-torque curve of their motors. Four different speed-torque curve characteristics have been partially standardized by NEMA as Design A, B, C, or D motors (see table 6-5 and figure 6-11). These characteristics were assigned when fixed 60 Hz frequency was dominant in industry. Table 6-5 shows the speed-torque curves for these motors at 60 Hz.

Table 6-5—NEMA motor design characteristics

NEMA design	Starting current	Starting torque	Rated load slip
A	Normal	Normal	Low[a]
B	Normal	Normal	Low
C	Normal	High	Low
D	Low	Very High	High[b]

[a]5% or less.
[b]Available 5–8% slip or 8–13% slip.

The Design B motor is used most often in industrial applications. Figure 6-11 shows its speed-torque curve with important points identified.

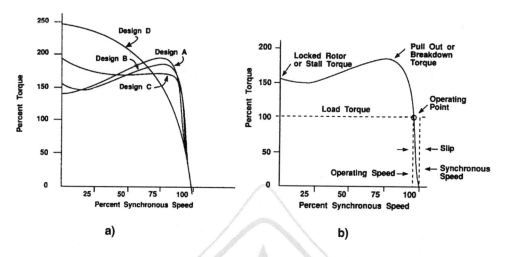

Figure 6-11—Typical speed-torque curves for NEMA design motors

6.15 Motor power

Motors are rated in horsepower. This is a measure of how much mechanical work the motor can do in a specified amount of time, and was developed when James Watt was trying to persuade industrialists to replace animal power by his steam engine. After many tests, he concluded that an average horse, when working a treadmill in a mine, could do about two million ft-lb of work in one hour. He rounded this off to 33 000 ft-lb/min (or 550 ft-lb/s) and used this as the definition of one horsepower.

One horsepower is also about 746 W, so one convenient rule for making quick estimates for the equivalent electric power is three-quarters of a kilowatt is equal to one horsepower. (The input kilowatts to the motor would be somewhat larger because of losses.) A related rule of thumb is that one horsepower output uses one kilovoltampere at the input. These are only approximations to use when more precise information is not readily available.

Some formulas for determining horsepower include the following:

$$\text{horsepower} = \frac{(\text{Speed in rpm}) \times (\text{Torque in lb-ft})}{5252}$$

$$\text{horsepower} = \frac{(\text{Work done in one hour } [\text{in ft-lb/h}])}{1\ 980\ 000}$$

$$\text{horsepower} = \frac{(\text{Number of kilowatts})}{(0.746)}$$

6.16 Motor surveys

When induction motors are operating in the 70–80% efficiency range, the reduction in power achieved for each percentage point improvement is approximately 1 kW per 100 hp of actual running load.

Thus if a 100 hp motor runs continuously for one year, the savings from improving its efficiency one percentage point would be

$$8760\frac{h}{y} \times \frac{1\ kW}{100\ hp} \times \frac{\$0.06}{kWh} \times 100\ hp \times 0.01\ pu\ =\ \$526/y$$

If a suitable smaller hp motor is already available, the one-time change-out costs are estimated at $500 to $600, material and labor. To identify the actual motor candidates, a formal motor survey is required.

The detection and change out of large underloaded induction motors to smaller and/or higher efficiency induction motors will contribute greatly to improved system efficiency and power factor. Such underloaded motors can be detected by the use of devices such as a stroboscopic digital readout tachometer.

Note that a digital-style tachometer is required to enable reading the speed to the nearest single digit. Thus the commonly used stroboscope tachometer, with the speed determined by reading the markings on a 4 in plastic top dial, is not suitable for detecting motor slip r/min.

The actual running speed of an induction motor depends on the motor design and the load on the output shaft. The speed at full load is indicated on the motor nameplate. Calculated slip from nameplate data is usually accurate to within only ±1% due to variations in voltage levels and balance at the motor's terminals. Full-load slip r/min is the difference between the synchronous r/min and the full-load r/min:

slip r/min = synchronous r/min – running r/min

For a four-pole motor with a full-load speed of 1700, it would be

1800 – 1700 = 100 r/min slip.

Measurement of slip to determine motor load is a useful technique when dealing with the typical energy audit involving a number of motors because the variations in (1) motor nameplate data, (2) line voltage at the motor, and (3) temperature of the motor stator may tend to average out on a statistical basis. The slip technique is not nearly as accurate or useful when the load of a single motor is being evaluated.

The accuracy of nameplate data is addressed in NEMA MG 1-1993 [B11] in Section 12.46, and the text has not been changed since the early sixties. The standard states, "The variation from the nameplate or published data speed of alternating current, single and polyphase, integral horsepower motors shall not exceed 20 percent of the difference between synchronous speed and rated speed when measured at rated voltage, frequency and load with an ambient temperature of 25 degrees C." In practice, most manufacturers can hold a 5% tolerance of the rated nameplate values; however, there are exceptions.

Most motor manufacturers, when specifying the full load rpm, take their readings at 25 °C or on a cold motor. A common error in reading the full load rpm with a precision tachometer with the motor hot is that the reading gives a misleading overload appearance. The standard temperature rise of induction motors can be measured by either the thermometer method (T) or the resistance method (R). The (T) method is usually used for nameplate marking. The difference between the (T) and (R) methods is that the (R) method uses a resistance type detector embedded in the windings for determining the motor temperature rise. In the case of (R) data, the allowable temperature rise is about 10 °C higher than that of the (T) method for the same point in timer measurement.

The increase percent slip per degree Celsius rise is the range of 0.342 to 0.380 for motors 10 hp and above. For example, a 10 hp three-phase 1725 rpm motor is running at 2 °C above an ambient of 25 °C. The true full load slip is thus calculated as follows:

(0.342% per degree Celsius rise) (20 °C) + 6.84% increase
(1.0684) (1800 − 1725) + (1.0684) (75) = 80.13 rpm full load slip
The corrected full load speed is thus (1800 − 80.13) = 1719.87 rpm

The variation in slip due to changes in voltage is proportional to the inverse square of the voltage ratio. Thus when the operating voltage is above 460 V, the full-load slip range is reduced. Likewise, when the operating voltage is below 460 V, the full-load slip range is increased. For example, a motor with a nameplate voltage of 460 V running at a 5% reduction in line voltage of 437 V would have an increase of the full-load slip to 110.8% of nameplate. An increase of 5% to 483 V would decrease the slip range to 90.7% of nameplate rating. It is always desirable to obtain voltage and current readings at each motor starter or motor disconnect when conducting a motor survey, thus allowing for the voltage correction to be made as necessary.

A check can also be made of motor loading by taking the ratio of measured running amperes compared to full-load nameplate amperes. This ratio is valid if the measured amperes are above 55–65% of full-load rating, which is the range where amperes are linear with load.

Another magnitude check on motor load is to measure true rms motor input power using a clamp-on kilowattmeter. Comparing the measured kilowatts with a calculated kilowatt based on an *assumed efficiency value* usually available from the motor manufacturer upon request will yield yet another check on the actual motor loading. For example,

$$\text{Motor load} = \frac{\text{measured kilowatt}}{(\text{Nameplate horsepower}) \, (0.746) \, (0.88)}$$

The most accurate determination of motor load is to have actual dynamometer test data from a certified lab of a certified motor rewind shop. This data is then kept on file to allow for further comparisons. This technique is expensive and most likely to be used only for large and/ or critical motors in a facility.

In summary, the slip measurement technique is useful for energy audits to identify potential savings from changing out underloaded motors to smaller, high-efficiency motors running at a higher percent of full load. When dealing with specific motors, considerations should be given to additional measurements of voltage, amperage, and temperature and the use of the appropriate correction factors detailed above.

As detailed in figure 6-12, measurement of motor output is based on the principle that slip r/min is linear from 10% load to 110% load. For example, the four-pole motor discussed earlier was found to be running at 1760 r/min. If the nameplate output is 10 hp, what was the output at that time?

Full-load slip = 1800 − 1700 = 100 r/min = 100% load.

Running slip = 1800 − 1760 = 40 r/min.

$$\text{Running load} = \frac{\text{running slip}}{\text{full-load slip}} = \frac{40}{100} = 0.40.$$

Output load = 10 hp × 0.40 = 4 hp.

Note that below about 50% of full-load amperes, the motor current is not a valid measurement of motor load. Due to the decrease in motor power factor, more and more of the apparent current is reactive amperes as the motor load decreases.

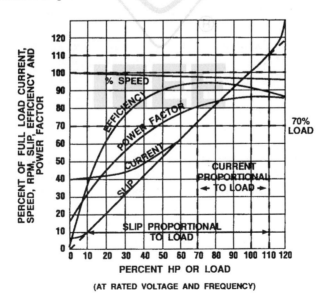

Figure 6-12—Typical NEMA Design B induction motor curve

The greatest benefits from using this technique with optical and mechanical digital tachometers include the following:

a) Ability to locate undcrloaded induction motors quickly

b) Ability to watch loading of equipment versus other conditions such as throughput, filter conditions, temperature, and pressure

c) Ability to determine approximate motor efficiency at any load when used in conjunction with conventional kilowatt monitoring on the motor input leads

d) Ability to assist in sizing future motor requirements based on actual load data

e) Ability to assist maintenance mechanics in periodically checking motor loading, which would help detect worn bearings, clogged filters, etc.

To determine motor efficiency, an additional piece of equipment is required to provide data on kilowatt power input to the motor.

6.17 Performing a motor survey

Note from figure 6-12 that both the power factor and efficiency decrease as the motor becomes unloaded. Thus a motor load survey is one of the prime areas of consideration in an initial energy audit of your facility. To aid in that survey, a form such as shown in figure 6-13 is recommended to assure that all of the pertinent information is obtained.

Three examples are presented in figures 6-14, 6-15, and 6-16 to illustrate the use of the motor data test sheet form. The top half is for recording motor nameplate date, including lines a) through f). The lines involving a computed answer include the formula for each individual calculation. For example, line f) is obtained by subtracting line d) from line e).

The measured values recorded on lines g) through j) can be obtained using an energy analyzer. The operating speed entered on line k) is obtained by a digital tachometer as described earlier. Line l) is an estimate of operating hours per year, and is usually less than 8760 h due to machine or equipment downtime for maintenance. Line m) is for listing the cost of electricity per kilowatthour and weight, include demand charges if known, and will vary depending on location of the facility.

The calculated values for lines n) through x) are based on data from the previous lines and are obtained by using the math formulas shown on each line. Based on the hp output on line q), a calculation is then made on line y) to determine the energy savings obtained by changing to a high-efficiency motor of the same or smaller horsepower size. The formula uses the value of $100/s$, where s is the percent efficiency of the existing normal motor under the present load conditions, using the value calculated on line s). The EFF value as detailed in footnote 1 of figure 6-13 is obtained from motor manufacturer data sheets.

Replacement motor cost is the cost of the motor (list price times the appropriate multiplier, usually 0.70 to 0.95), plus an estimated labor and material change-out cost. This cost is usually in the range of $500 to $600, depending on the degree of access difficulty and if modifi-

Induction Motor Test Data Sheet

Company _____ Plant_____ Date_____

Building _____ Dept. _____ Application_____

Motor Shop Number _____

Make _____

Model/Type _____

Serial number _____

Service factor_____

Enclosure type_____

a) Full load HP_____

b) Volts _____

c) Amperes _____

d) Full load speed _____

e) Sync. speed_____

   2-pole = 3600, 4-pole = 1800, 6-pole = 1200

f) Full load slip (e–d)_____

Phase and Hz _____

Frame size _____

Insulation class _____

Efficiency rating _____

NEMA torque type _____

Temperature rise _____

Calculated Values

n) Running slip (e–k) _____

p) Percent load (n/f)(100%) _____

q) HP output (a)(p)/(100%) _____

r) kW output (q)(0.746) _____

s) Eff. percent (r/j)(100%) _____

t) kVA input (g)(h)(1.732)/(1000) _____

u) Power factor (j)/(t)(100%) _____

v) kW losses (j–r)_____

w) $/year operation (j)(l)(m) _____

x) $/year losses (v)(l)(m) _____

Measured Values

g) Average volts_____

h) Average amperes _____

j) Average kW _____

k) Operating speed, rpm _____

l) Full load operating hrs._____

m) Avg. electricity price ($kWh) _____

y) [1]Annual energy savings due to changeout with a _____hp high-efficiency motor (r)(l)(m)(100/s – 100/EFF) _____

z) [2]Replacement motor cost _____

_____

Simple payback, years (z)/(y) _____

[1]EFF is the efficiency (%) of a replacement premium motor at the appropriate load factor.

[2]Cost is the total cost of purchasing and installing an optimally sized, high-efficiency motor.

Figure 6-13—Induction motor test data sheet

cations to the mounting plate, pulleys, and belts as well as new motor heaters (motor overload relays) are required to accommodate changes in the motor frame size.

If the motor is being downsized, Table 430-152 of the National Electrical Code® (NEC®) (NFPA 70-1996 [B12]) must be followed to ensure that the branch circuit protection is not set too high for the new smaller motor.

The term *simple payback* means just that, and is not structured to take into account the time value of money and other economic variables. Motors with less than a one-year simple payback are generally excellent candidates for changing out. A payback of over two years generally requires special consideration.

Example one in figure 6-14 illustrates a 75 hp motor running a fan load of 36 hp. The calculated simple payback associated with changing to a high-efficiency 40 hp motor with an assumed nominal efficiency of 88% is 0.527 years.

Example two in figure 6-15 illustrates the new 40 hp high-efficiency motor running at a slightly lower speed, at an output of 35 hp and an efficiency of 90%. In example two, the calculation for line y) becomes the difference between line x) on example one and line x) on example two ($5420 − $1160 = $4260). The simple payback of 0.47 years is better than the expected value of 0.527 calculated in example one because the fan is running at a slightly slower speed and the power required from the motor decreases as a factor somewhere between the square and the cube function of the speed ratio.

Example three in figure 6-16 illustrates another case study involving a 100 hp motor driving cooling water pump. The nameplate data and measured values are as shown. The reader is invited to calculate values for lines n) through x). Then calculate the annual savings on line y) for a replacement 100 hp high-efficiency motor with an assumed efficiency of 90%.

With an estimated replacement cost of $4000 for the motor and $600 for installation, calculate the simple payback. Is this motor a good candidate for change out?

The accuracy of determining motor output based on slip is dependent on the accuracy of the motor nameplate data and the actual voltage serving the motor. As mentioned earlier, nameplate accuracy is required to be ±10%. This means that the data calculated in figure 6-13 are at best ±10%, which is usually more than adequate for the broad survey work designed to identify motors that are 25, 50, 75, or 100% loaded. Closer examination would require verification of the individual motor actual full-load rpm by use of a dynamometer.

The second concern is that of voltage serving the motor. Table 6-6 shows how voltage and frequency variations affect induction motors. Note that the percent slip is an inverse squared function of voltage. That is, with undervoltage the percentage of full-load slip increases, and with overvoltage the percentage of full-load slip decreases. As long as the voltage is within ±5% (506 to 437 on a 460 V base), this effect on slip is less than ±10% and usually can be ignored. The percentage of full-load slip value should be adjusted when the voltage variation is greater than ±5%.

Induction Motor Test Data Sheet

Company _____ABC_____ Plant __XYZ__ Date __MM-DD-YR @HHMIN__

Building __EXTRUSION__ Dept. __BLOCK I__ Application __FAN AT2-1__

Motor Shop Number ___1943___   Phase and Hz ___3/60___

Make __LOUIS ALLIS__   Frame size ___505___

Model/Type _____   Insulation class _____

Serial number _1186886_   Efficiency rating ___B___

Service factor ___1.15___   NEMA torque type __40 °C__

Enclosure type _OPEN DRIP PROOF [ODP]_ Temperature rise _____

a) Full load HP ___75___

b) Volts __440/220__

c) Amperes __91/182__

d) Full load speed _1,175_

e) Sync. speed __1,200__

2-pole = 3600, 4-pole = 1800, 6-pole = 1200

f) Full load slip (e–d) __25__

Measured Values

g) Average volts ___445___

h) Average amperes __75__

j) Average kW __40__

k) Operating speed, rpm _1,188_

l) Full load operating hrs. _8,000_

m) Avg. electricity price ($kWh) _0.05_

Calculated Values

n) Running slip (e–k) ___12___

p) Percent load (n/f)(100%) __48%__

q) HP output (a)(p)/(100%) ___36___

r) kW output (q)(0.746) __26.9__

s) Eff. percent (r/j)(100%) __67%__

t) kVA input (g)(h)(1.732)/(1000) _57.8_

u) Power factor (j)/(t)(100%) __0.69__

v) kW losses (j–r) __13.1__

w) $/year operation (j)(l)(m) __$16,000__

x) $/year losses (v)(l)(m) __$5,420__

y) [1]Annual energy savings due to changeout with a _40_ hp high-efficiency motor (r)(l)(m)(100/s – 100/EFF) $3,832

z) [2]Replacement motor cost $1,521 + 500 = $2,021

$$y = (26.9)(8,000)(0.05)\left(\frac{100}{67} - \frac{100}{88}\right)$$

Simple payback, years (z)/(y) __0.527__

$$= (10,760)(0.356) = \$3,832$$

[1]EFF is the efficiency (%) of a replacement premium motor at the appropriate load factor.

[2]Cost is the total cost of purchasing and installing an optimally sized, high-efficiency motor.

Figure 6-14—Example 1, motor with load

Induction Motor Test Data Sheet

Company ___ABC___ Plant ___XYZ___ Date MM-DD-YR@HHMIN

Building ___EXTRUSION___ Dept. BLOCK I Application FAN:AT2-1

Motor Shop Number ___2088___ Phase and Hz ___3/60___

Make SiEMENS—ALLIS Frame size ___324 T___

Model/Type _____ Insulation class ___B___

Serial number 51-391-771 Efficiency rating _____

Service factor ___1.15___ NEMA torque type ___B___

Enclosure type Fan-Cooled [TEFC] *(Totally enclosed)* Temperature rise ___40 °C___

a) Full load HP ___40___

b) Volts ___460/230___

c) Amperes ___50/100___

d) Full load speed ___1,180___

e) Sync. speed ___1,200___
   2-pole = 3600, 4-pole = 1800, 6-pole = 1200

f) Full load slip (e–d) ___20___

Calculated Values

n) Running slip (e–k) ___17.5___

p) Percent load (n/f)(100%) ___87.5%___

q) HP output (a)(p)/(100%) ___35___

r) kW output (q)(0.746) ___26.1___

s) Eff. percent (r/j)(100%) ___90%___

t) kVA input (g)(h)(1.732)/(1000) ___33.5___

u) Power factor (j)/(t)(100%) ___86.7___

v) kW losses (j–r) ___2.9___

w) $/year operation (j)(l)(m) $11,600

x) $/year losses (v)(l)(m) $1,160

Measured Values

g) Average volts ___460___

h) Average amperes ___42___

j) Average kW ___29___

k) Operating speed, rpm 1,182.5

l) Full load operating hrs. ___8,000___

m) Avg. electricity price ($kWh) 0.05

y) [1]Annual energy savings due to changeout with a _____hp high-efficiency motor (r)(l)(m)(100/s – 100/EFF) $4 260

z) [2]Replacement motor cost 1521 + 500 = $2,021

Simple payback, years (z)/(y) ___0.47___

$$y = \$5,420 - 1,160 = \$4,260$$

[1]EFF is the efficiency (%) of a replacement premium motor at the appropriate load factor.
[2]Cost is the total cost of purchasing and installing an optimally sized, high-efficiency motor.

Figure 6-15—Example 2, actual replacement result with new high-efficiency motor

Induction Motor Test Data Sheet

Company _____ABC_____ Plant __XYZ__ Date _MM-DD-YR @HHMIN_

Building _TEXTILE_ Dept. _BEAMING_ Application _CW PUMP 8_

Motor Shop Number ___1998___

Make _____GE_____

Model/Type _5K4445C22_

Serial number _BA093031_

Service factor_____1.0_____

Enclosure type_____ODP_____

a) Full load HP_____100_____

b) Volts _____440/220_____

c) Amperes _122.5/245_

d) Full load speed __1,775__

e) Sync. speed_____1,800_____

2-pole = 3600, 4-pole = 1800, 6-pole = 1200

f) Full load slip (e–d)_____25_____

Phase and Hz _____3/60_____

Frame size _____4440_____

Insulation class _____B_____

Efficiency rating _____

NEMA torque type _____B_____

Temperature rise _____50 °C_____

Calculated Values

n) Running slip (e–k) _____

p) Percent load (n/f)(100%) _____

q) HP output (a)(p)/(100%) _____

r) kW output (q)(0.746) _____

s) Eff. percent (r/j)(100%) _____

t) kVA input (g)(h)(1.732)/(1000) _____

u) Power factor (j)/(t)(100%) _____

v) kW losses (j–r)_____

w) $/year operation (j)(l)(m) _____

x) $/year losses (v)(l)(m) _____

Measured Values

g) Average volts_____440_____

h) Average amperes _____115_____

j) Average kW _____80_____

k) Operating speed, rpm _1,778_

l) Full load operating hrs. _7000_

m) Avg. electricity price ($kWh) _0.05_

y) [1]Annual energy savings due to changeout with a _____hp high-efficiency motor (r)(l)(m)(100/s – 100/EFF) _____

z) [2]Replacement motor cost _____

Simple payback, years (z)/(y) _____

[1]EFF is the efficiency (%) of a replacement premium motor at the appropriate load factor. → 90%

[2]Cost is the total cost of purchasing and installing an optimally sized, high-efficiency motor.

Figure 6-16—Example 3, Calculated values exercise

Induction Motor Test Data Sheet

Company _____ ABC _____ Plant _ XYZ _ Date MM-DD-YR@HRMN

Building _ TEXTILE _____ Dept. BEAMING Application CW PUMPS

Motor Shop Number _ 1998 _____ Phase and Hz _ 3/60 _

Make _ GE _____ Frame size _ 444 U _

Model/Type 5K4445C22 _____ Insulation class _ B _

Serial number BA093031 _____ Efficiency rating _____

Service factor _ 1.0 _____ NEMA torque type _ B _

Enclosure type _ ODP _____ Temperature rise _ 55 °C _

a) Full load HP _ 100 _

b) Volts _ 440/220 _

c) Amperes _ 122.5/245 _

d) Full load speed _ 1,775 _

e) Sync. speed _ 1,800 _
  2-pole = 3600, 4-pole = 1800, 6-pole = 1200

f) Full load slip (e–d) _ 25 _

Measured Values

g) Average volts _ 440 _

h) Average amperes _ 115 _

j) Average kW _ 80 _

k) Operating speed, rpm _ 1,775 _

l) Full load operating hrs. _ 7,000 _

m) Avg. electricity price ($kWh) _ 0.05 _

$$y = (65.6)(7,000)(0.05)\left(\frac{100}{82} - \frac{100}{90}\right)$$

$$= (22,960)(0.1084) = \$2,489$$

Calculated Values

n) Running slip (e–k) _ 22 _

p) Percent load (n/f)(100%) _ 88% _

q) HP output (a)(p)/(100%) _ 88% _

r) kW output (q)(0.746) _ 65.6 _

s) Eff. percent (r/j)(100%) _ 82% _

t) kVA input (g)(h)(1.732)/(1000) _ 87.6 _

u) Power factor (j)/(t)(100%) _ 91% _

v) kW losses (j–r) _ 14.4 _

w) $/year operation (j)(l)(m) $28,000

x) $/year losses (v)(l)(m) $5,040

y) [1]Annual energy savings due to changeout with a _100_ hp high-efficiency motor (r)(l)(m)(100/s – 100/EFF) $2,489

z) [2]Replacement motor cost $4,000 + 600 = $4,600

Simple payback, years (z)/(y) _ 1.85 _

[1]EFF is the efficiency (%) of a replacement premium motor at the appropriate load factor.

[2]Cost is the total cost of purchasing and installing an optimally sized, high-efficiency motor.

Figure 6-17—Example 3, Calculated payback exercise

Table 6-6—The effects of voltage and frequency variation on induction motors

Variation	Starting & max. running torque	Synchronous speed	% slip	Full-load speed	Full-load efficiency	Full-load power factor	Full-load current	Starting current	Temp. rise, full load	Max. overload capacity	Magnetic noise, no load in particular
Voltage variation: 120%	Increase 44%	No change	Decrease 30%	Increase 1.5%	0–6% decrease	Decrease 5–15 points	Increase 12%	Increase 20%	Increase 5–6 °C	Increase 44%	Noticeable increase
100% voltage	Increase 21%	No change	Decrease 17%	Increase 1%	Slight decrease	Decrease 5–10 points	Increase 2–4%	Increase 10–12%	Increase 3–4 °C	Increase 21%	Increase slightly
Functions of voltage	(Voltage)	Constant	$\dfrac{1}{(\text{Voltage})^2}$	(Synchronous speed slip)	—	—	—	Voltage	—	(Voltage)	—
90% voltage	Decrease 19%	No change	Increase 23%	Decrease 1.5%	Decrease 2 points	Increase 5 points	Increase 10–11%	Decrease 10–12%	Increase 6–7 °C	Decrease 19%	Decrease slightly
Freq. variation: 105% freq.	Decrease 10%	Increase 5%	Practically no change	Increase 5%	Slight increase	Slight increase	Decrease slightly	Decrease 5–6%	Decrease slightly	Decrease slightly	Decrease slightly
Function of frequency	$\dfrac{1}{(\text{Frequency})}$	Frequency	—	(Synchronous speed slip)	—	—	—	$\dfrac{1}{(\text{Frequency})}$	—	—	—
95% frequency	Increase 11%	Decrease 5%	Practically no change	Decrease 5%	Slight decrease	Slight decrease	Increase slightly	Increase 5–6%	Increase slightly	Increase slightly	Increase slightly
1% unbalance	Slight decrease	Slight decrease	—	Slight decrease	2% decrease	5–6% decrease	1.5% increase	Slight decrease	2% increase	—	—
2% unbalance	Slight decrease	Slight decrease	—	Slight decrease	8% decrease	7% decrease	3% increase	Slight decrease	8% increase	—	—

NOTE—This table shows general effects, which will vary somewhat for specific ratings. From *Engineer's Digest*, September 1989.

6.18 Summary

Monitoring and reporting energy consumption allows for close control while minimizing expenses. It also provides historical energy consumption data to aid in projecting future loads and developing standards for the next year. Such data are essential to financial forecasts and operating budgets.

Unless energy consumption is measured, it is next to impossible to know where to direct conservation efforts. A metering system provides that vital ingredient to a successful energy management program.

6.19 Bibliography

Additional information may be found in the following sources:

[B1] Andreas, J. C., *Energy-Efficient Electric Motors, Selection and Application*, 2d ed. New York: Marcel-Dekker, 1992.

[B2] Caywood, R E., *Electric Utility Rate Economics*. New York: McGraw-Hill, 1972.

[B3] *The Dranetz Field Handbook for Electrical Energy Management*. Edison, NJ: Dranetz Technologies, Inc., 1992.

[B4] *Electrical Engineering Pocket Handbook*. St. Louis, Mo.: the Electrical Apparatus Service Association, Inc., 1988.

[B5] *Energy Management Fact Sheet High-Efficiency Motors*. Milwaukee, Wis.: Wisconsin Electric Power Company.

[B6] *Handbook for Electricity Metering*, 9th ed. Washington D.C.: Edison Electric Institute, 1992.

[B7] *IEEE Standards Collection: Electricity Metering (C12)*, 1993 Edition.

[B8] IEEE Std 315-1975 (Reaff 1993), IEEE Graphic Symbols for Electrical and Electronics Diagrams and IEEE Std 315A-1986 (Reaff 1993), Supplement to IEEE Std 315-1975 (ANSI).

[B9] IEEE Std C57.13-1993, IEEE Standard Requirements for Instrument Transformers.

[B10] Kennedy, R. A. and Rickey, D. N. "Monitoring and Control of Industrial Power Systems," *IEEE Computer Applications in Power*, vol. 2, no. 4, Oct. 1989.

[B11] NEMA MG 1-1993, Motors and Generators.

[B12] NFPA 70-1996, National Electrical Code® (NEC®).

[B13] NFPA 70E-1995, Electrical Safety Requirements for Employee Workplaces.

[B14] *Pocket Ref.*, 2d ed. Newark, Del.: Maintenance Troubleshooting, 1996.

[B15] Stebbins, W. L. "Highly Efficient Energy Metering and Trend Analysis Techniques for Maximum Control," *Proceedings of the 14th World Energy Engineering Congress*, the Association of Energy Engineers, Atlanta, Ga., Oct. 1991.

[B16] Stebbins, W. L. "Implementing an Effective Utility Testing Process: A Keystone for Successful Energy Management," *Proceedings of the 13th World Energy Engineering Congress*, the Association of Energy Engineering, Atlanta, Ga., Oct. 1990.

[B17] Stebbins, W. L. "Metering Revised—Innovative Concepts for Electrical Monitoring and Reporting Systems," *Proceedings of the 15th World Energy Engineering Congress*, the Association of Energy Engineers, Atlanta, Ga., Oct. 1992.

[B18] Stebbins, W. L. "New Concepts in Electrical Metering for Energy Management," *IEEE Transactions on Industrial Applications*, vol. IA-22, no. 2, pp. 382-388, Mar./Apr. 1986.

[B19] Stebbins, W. L. "Utility Monitoring Systems: Key to Successful Energy Management," *Proceedings of the 6th World Energy Engineering Congress*, the Association of Energy Engineers, Atlanta, Ga., Nov. 1983.

Chapter 7
Energy management for lighting systems

7.1 Introduction

While far from the greatest user of energy, lighting is significant because it enters visibly in virtually every phase of modern life. The opportunities for saving lighting energy require that attention be paid to many nonelectric parameters.

Lighting systems are installed to permit people to see. Attention should be paid not only to economics and efficiency but also to the type of work which people do and the space in which they do it. Lighting also affects other environmental and building systems, especially those that heat and cool occupied spaces within buildings. All of these factors should be integrated into the design process.

This chapter presents the state-of-the-art in optimizing the use of lighting energy and it also alerts the lighting designer and user to the problems involved in designing an energy-effective lighting system.

7.2 Definitions of basic lighting terms

7.2.1 ballast: A device used with an electric-discharge lamp to provide the necessary voltage and current for starting and operating the lamp.

7.2.2 candlepower: Luminous intensity expressed in candelas.

7.2.3 coefficient of utilization (CU): The ratio of the amount of luminous flux received on the work-plane to the amount of luminous flux emitted by the luminaire's lamps alone.

7.2.4 color rendering: A general expression for the effect of a light source on the color appearance of objects compared with color appearance under a reference light source.

7.2.5 general lighting: Lighting designed to provide a substantially uniform level of illumination throughout an area, exclusive of any provision for special local requirements. *See:* **task lighting.**

7.2.6 glare: The sensation produced by brightnesses within the visual field that are sufficiently greater than the luminance to which the eyes are adapted to cause annoyance, discomfort, or loss in visual performance and visibility.

7.2.7 illumination: A condition where light impinges upon a surface or object.

7.2.8 lamp: A generic term for a man-made source of light.

7.2.9 lamp lumen depreciation factor (LLD): The factor used in illumination calculations to quantify the output of light sources at 70% of their rated life as a percentage of their initial output.

7.2.10 louver: A series of baffles used to shield a source from view at certain angles or to absorb unwanted light. The baffles usually are arranged in a geometric pattern.

7.2.11 lumen (lm): A unit of luminous flux.

7.2.12 luminaire: A complete lighting unit consisting of a lamp or lamps together with the parts designed to distribute the light, to position and protect the lamps, and to connect the lamps to the power supply.

7.2.13 luminaire dirt depreciation factor (LDD): The factor used in illumination calculations to relate the initial illumination provided by clean, new luminaires to the reduced illumination that they will provide due to dirt collection at a particular point in time.

7.2.14 luminance: A measure of the amount of light flux (lumens) per unit of area reflected from or transmitted through a surface.

7.2.15 luminance contrast: The relationship between the luminances of an object and its immediate background.

7.2.16 luminous efficacy: A measure of lamp efficiency in terms of light output per unit of electrical input expressed in lumens per watt.

7.2.17 polarization: The process by which light waves are oriented in a specific plane.

7.2.18 reflectance (of a surface or medium): The ratio of the reflected flux to the incident flux.

7.2.19 shielding angle (of a luminaire): The angle between a horizontal line through the light center and the line of sight at which the bare source first becomes visible.

7.2.20 task lighting: Lighting designed to provide illumination on visual tasks usually at higher levels than the surrounding area.

7.3 Concept of lighting systems

In the succeeding subclauses, the major elements of the lighting system will be analyzed. For more detailed information the reader should consult the appropriate reference in the bibliography (7.14).

A lighting system is that portion of the branch-circuit wiring system that supplies the lamps or ballasts together with the associated controls such as switches and dimmers. The system also includes the light source(s), luminaire, shielding, and optical control media, the entire space to be lighted, and the nature of the illumination required. The effectiveness of the entire system is expressed in *figures of merit*. A generalized diagram of this *system* and the associated figures of merit are given in figure 7-1. Specialized lighting terms used in this chapter are briefly described in 7.2.

The human eye is not equally sensitive to all wavelengths in the visible spectrum. Colors in the green and yellow region will produce a sensation of brightness in the eye with less radiant power than will wavelengths in the blue and red region of the spectrum. Lighting units or *photometric* units take this sensitivity response into account in their definitions. Photometric quantities therefore apply to human visual response rather than the strictly radiant effects of energy.

7.4 The task and the working space

7.4.1 Task illumination

Usually a visual task is considered the *given* in a lighting system and the design process starts from there. However, optimizing energy for lighting requires that consideration be given to the possibility of changing and improving the visual task characteristics so that the lighting requirements of the task are less stringent.

Vision requires sufficient light (brightness) and size for the eye to resolve an image, sufficient time to recognize an image, and contrast to separate the brightness of an object from its background.

A century of research information has resulted in tables of recommended illumination values for different visual tasks, scaled to reflect the accuracy with which the task can be performed. Improved instrumentation and understanding of the visual performance process are resulting in better methods of determining optimum task illumination. Measures such as task difficulty and criticalness are being combined with characteristics of the human seeing mechanism, including such variables as age and eye defects.

Most lighting recommendations are given for *illumination on the task* and presume that there is little need to illuminate the entire space to the task-illumination requirements. Recent studies involving work stations rather than general lighting have shown that under certain conditions the energy requirements for lighting spaces can be substantially reduced without sacrificing task visibility.

Generally, improving the reflectance characteristics of the task and the quality or the quantity of its illumination will improve contrast. The contrast is defined as follows:

$$\text{Luminance contrast} = \frac{(\text{background luminance}) - (\text{object luminance})}{\text{background luminance}}$$

$$\text{Luminance} = \text{illuminance*} \cdot (\text{reflection factor})$$

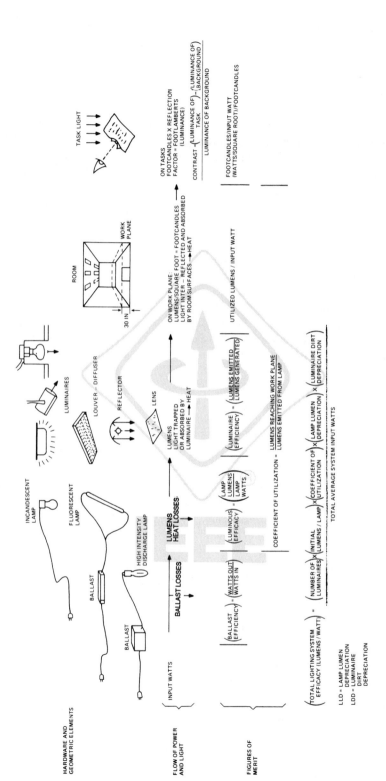

Figure 7-1—Total lighting system

*Illuminance is the preferred term for illumination.

Contrast improvement may not be simple where other than matte objects and backgrounds are involved. Even with matte surfaces, contrast may be subtly reduced by a phenomenon known as *veiling reflections,* a condition where contrast is lost because of the reflection of the light source in portions of the task. Contrast losses may be large even on ordinary tasks such as reading printed words on paper.

Design and evaluation techniques are available that can improve task contrast by reducing veiling reflections. At present, calculation techniques involve the determination of *equivalent sphere illumination* (ESI). Illumination (footcandles) is referenced to a known task illuminated by a light source consisting of a perfectly diffusing hemisphere placed over the task. ESI calculations are referenced to sphere illumination to compare whether a given lighting system is worse or better than sphere illumination in producing contrast.

7.4.2 Efficient lighting design

Since the room acts as part of the lighting system, careful attention should be paid to how light is directed from the lighting equipment to the task areas. The controlling factors in this process are the size and shape of the space, the reflectances of the surfaces, and the characteristics of the room furnishings. Light may be trapped, absorbed, reflected, or modified by any surface within the space. Substantial light may also be allowed to enter or escape by means of entrances and windows.

While the most energy-efficient room may result from the use of highly reflecting surfaces and furnishings, such a space may be psychologically and aesthetically objectionable. Trade-offs are necessary. Typical recommendations for room surface reflectances are shown in table 7-1.

Table 7-1—Recommended surface reflectances for offices

Surface	Reflectance equivalent range
Ceiling finishes	80–90%
Walls	40–60%
Furniture	25–45%
Office machines and equipment	25–45%
Floors	20–40%

It is also important to consider the luminance of the surfaces surrounding the visual task to avoid great changes in contrast, which can cause eye fatigue and discomfort. Keeping within recommended luminance ratios is particularly important where localized or task lighting is used since lighted areas are more concentrated and general room illumination may be substantially below task levels. Table 7-2 summarizes recommended ratios that may be useful in overall planning.

Table 7-2—Recommended luminance ratios

To achieve a comfortable balance in the office, it may be desirable and practical to limit luminance ratios between areas of appreciable size from normal viewpoints as follows:

1 to 1/3 between task and adjacent surroundings
1 to 1/5 between task and more remote darker surfaces
1 to 5 between task and more remote lighter surfaces

These ratios are recommended as maximums; reductions are generally beneficial.

7.4.3 Lighting design calculations

In general, the zonal cavity method of performing lighting calculations has gained rapid acceptance as the preferred way to calculate number and placement of luminaries required to satisfy a specified illuminance level requirement. However, there is still considerable merit in the point-by-point method.

7.4.3.1 Zonal cavity method

The basic lumen method is based on the definition:

$$\text{footcandle} = \frac{\text{lumens striking an area}}{\text{square feet of area}}$$

In order to take into consideration such factors as dirt on the luminaire, general depreciation in lumen output of the lamp, etc., the above formula is modified as follows:

$$\text{footcandle} = \frac{\text{lamps/luminaire} \times \text{lumens/lamp} \times \text{CU} \times \text{LLF}}{\text{area/luminaire}}$$

In the determination of CU, the cavity ratio of the room is eminently involved. Rooms are usually classified according to shape by 10 room cavity numbers. Tables prepared by luminaire manufacturers normally provide the designers with easy selection of proper CUs to be used. Light loss factor (LLF) is a factor used in calculating illuminance after a given period of time and under given conditions.

7.4.3.2 Point-by-point method

In this method, information from luminaire candlepower distribution curves is applied to the mathematical relationship. The total contribution for all luminaires to the illumination level on the task plane is summed from the direct illumination component and the reflected illumination component. Calculation details are quite time consuming. Computer programs for such computations are now available. The designer needs only to supply the details of the room, the geometry of the lighting installation, the room surface reflectances, and the candle-power data for the luminaire. A printout of footcandles at the desired points can be obtained.

7.5 Light sources

7.5.1 Light source efficacy

Table 7-3 shows the approximate allocation of energy use in various light sources. A light source can be characterized by light output, life, maintenance of light output, color, etc. A more useful measure is lamp luminous efficacy, which is the amount of visible light emitted by the lamp per watt of input power. As table 7-4 indicates, the incandescent lamp has the lowest efficacy at between 10-20 lm/W, and mercury is approximately double that at about 50 lm/W. Lamp efficacies may improve as new lamps are developed.

Table 7-3—Lamp energy data (nominal data in %) initial ratings

Lamp type	Radiated energy			Conducted and convected	Ballast
	Light	Infrared	UV		
Incandescent (100 W)	10	72	—	18	—
Fluorescent (40 W)	20	33	—	30	17
Fluorescent (40 W excluding ballast)	24	40	—	36	—
Mercury (400 W deluxe white)	15	47	2	27	9
Metal halide (400 W clear)	21	32	3	31	13
High-pressure sodium (400 W)	30	35	0	20	15
Low-pressure sodium (180 W)	29	4	0	49	18

While lamp efficacy is good indicator of the lamp's ability to convert electric energy into visible energy, it does not include the losses that take place in the luminaire and the room as the light is directed toward the task area. A true measure of system efficacy should include the utilization characteristics of the fixture, the room, and the effect of dirt accumulation. These loss factors may be light-source dependent; therefore, using the most efficient lamp may not always result in the most efficient lighting system.

7.5.2 Light source characteristics

The light source shall match the environmental constraints. For example, fluorescent lamps are generally the most efficient for an interior general lighting system where the large lighted

285

Table 7-4—Lamp lumen efficacies

Source	Wattage	Approximate lumens/watt input*
Incandescent	40 W general service	11
Incandescent	1000 W general service	22
Fluorescent	2 24 in cool white—T12	50
Fluorescent	2 48 in cool white—T12	77
Fluorescent	2 96 in (800 mA) cool white	73
High-intensity discharge (HID)	400 W phosphor-coated mercury	50
HID	1000 W phosphor-coated mercury	55
HID	400 W metal halide	75
HID	1000 W metal halide	85
HID	400 W high-pressure sodium	100

*Lamp only.

area of the lamp can be effectively utilized to distribute light. At high mounting heights or where the lighting area shall be optically controlled more tightly, smaller, higher intensity sources generally are more efficient. Table 7-5 provides general guidelines as to lamp performance and application characteristics.

7.5.3 Chromaticity and color rendering

Light sources do not emit light equally at all wavelengths in the visible spectrum, and color distortion results. Therefore, the light source color spectrum should be carefully evaluated. If possible, an actual system or a demonstration area should be used to see how the colors appear before final installation. Two measures generally indicate the color characteristics of light sources: chromaticity and color rendering.

Chromaticity is an indication of *warmth* or *coolness* of a light source and is measured in terms of a color temperature scale in kelvin (see figure 7-2). The higher the apparent color temperature of a light source, the cooler it appears to the eye. Sources with high color temperatures seem to be preferred at high illumination levels but personal preference is also involved.

Color rendering describes how well the light source makes object colors appear according to a defined standard. Often this is highly subjective and people refer to their own experience or to outdoor day-light conditions as their personal standard. Light source spectral distribution characteristics provide a clue to the color rendering ability of light sources. If the visible

Table 7-5—Lamp output characteristics

Interior general lighting (commercial and industrial) lamp selection guide

Lamp Type	Incandescent		Fluorescent				High-intensity discharge			
	Standard	Tungsten halogen	Cool white	Warm white	Deluxe cool white	Deluxe warm white	Deluxe white mercury	Warm deluxe white mercury	Metal halide	High-pressure sodium
Application	Local lighting accent, display. Low initial cost general lighting	Accent, display general lighting in lobbies, theaters, etc.	Most commercial and industrial general lighting systems	Many commercial and industrial general lighting systems	General and display lighting where good color daylight-like atmosphere important	General and display lighting where good color incandescent-like atmosphere important	Store and other commercial and industrial-general lighting	Store and other commercial-general lighting	Industrial, commercial general lighting, pools, arenas	Industrial general lighting. Some commercial applications
Lamp optical controllability potential	Excellent	Excellent	Fair	Fair	Fair	Fair	Good	Good	Excellent	Excellent
Glare control with typical luminaire	Good	Excellent	Good	Good	Good	Good	Excellent	Excellent	Good	Good
Color effects: accents	Warm colors	Warm colors	Yellow, blue, green	Yellow, green	None-excellent color balance	Yellow, orange, red-like incandescent	Red, yellow, blue, green	Red, orange, yellow, green	Yellow, green, blue	Yellow, orange
Color effects: grays	Blues	Blues	Reds	Red, blues	None	Blues	Deep reds	Deep reds, blues	Reds	Deep reds, greens, blues
Range of lamp wattages for typical applications	30–1000	100–1500	20–215	20–215	20–215	20–110	40–1000	175–1000	175–1500	35–1000
Range of initial lamp lumens for typical applications	210–23 740	3450–34 730	1300–16 000	1300–15 000	850–11 000	820–6550	1575–63 000	6500–58 000	14 000–155 000	3600–140 000
Initial lamp lumens/watt	7–24	17–24	65–85	65–85	40–60	40–60	30–65	35–60	80–100	100–140
Average rated lamp life (h)	750–2000	1500–4000	9000 (3 hours/start)—30 000 (continuous burning for F40 types)				16 000–24 000+	24 000+	7500–20 000 (10 or more hours/start)	16 000–24 000+ (10 or more hours/start)
Light output depreciation characteristics	Good	Excellent	Good	Good	Fair	Fair	Fair	Fair	Fair	Excellent

287

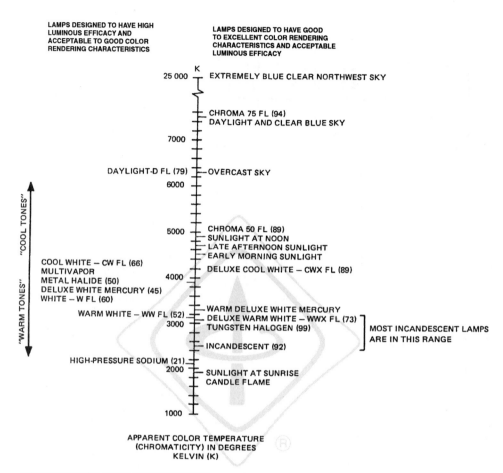

Figure 7-2—Color characteristics of light sources

spectrum is generally filled between the limits of human vision (wavelengths from 400 nm to approximately 750 nm), color rendering will be better than if the source has emissions only in a narrow band or portions of the visible spectrum. Side-by-side visual comparisons and test installations are, however, a more reliable guide to color rendering and color acceptability. The CIE color rendering index (Ra) is a frequently used measure of a light source's ability to render colors in a natural and normal way. It may also be used to compare light source colors if the sources have the same chromaticity and meet certain other criteria.

7.5.4 Details of available light sources

7.5.4.1 Incandescent

Incandescent lamps have many desirable characteristics; however, they are very inefficient converters of electricity into light. Approximately 90–95% of the energy consumed by an incandescent light source is converted not to light, but to heat. The efficacy of incandescent (general lighting types) is approximately 20 lm/W.

Recent technical advances have made possible slightly more efficient reduced wattage incandescent lamps. Improved mount structures, smoother tungsten filament wire, and the use of fill gas such as Krypton work together to provide the same lumen and life characteristics as previous lamps, but with approximately 5% less wattage.

Energy-saving potential also exists for incandescent lamps when reflector lamps can be employed. Incandescent lamps using built-in reflectors offer better utilization of the light produced compared to nonreflector types. In this family of lamps are the R lamps or indoor reflector lamps, PAR, or parabolic aluminized reflector lamps, and a newer line of indoor reflector lamps called ER or elliptical reflector types. ER lamps permit lamp wattage reductions up to 50% by improving reflector's optical efficiency.

More recent developments are the application of a tungsten-halogen cycle and the use of selective reflecting thin films. In the tungsten-halogen lamps, the gas combines with the tungsten evaporating from the filament to form a tungsten-halogen compound. When the halide contacts the hot filament, the compound dissociates and the tungsten is deposited back onto the filament. This reduces the rate of evaporation and increases lamp efficacy. The selective reflecting thin film is comprised of many layers of a particular thickness of the order of the wavelength of light. The film is transparent to visible light and highly reflective to the near and far infrared radiation (IR) greater than 7000 A. The bulb wall is shaped such that the IR is reflected back onto the filament, further heating the filament. This technique increases the lamp efficacy by a factor approaching two.

7.5.4.2 Fluorescent

Fluorescent lamps, introduced in the late 1930s, offered the user two major benefits compared to incandescent lamps: increased efficiency and significantly longer life. Typical standard fluorescent lamps today have efficacies over 80 lm/W. The characteristics of modern fluorescent lamps are as follows:

a) High lumen per watt (lm/W) efficacy (80+)
b) Long life (20 000+ h)
c) Availability in a wide variety of sizes and shapes
d) Rapid starting and relighting after momentary outage
e) Availability of high color rendering types for color-critical applications
f) Good maintenance of light output over lamp life

Since the early 1970s, there has been a line of reduced wattage replacements for standard fluorescent lamps. These lamps are now available in all popular sizes and colors for most applications. However, limitations of the reduced wattage lamps are as follows:

a)   Use only where ambient temperature does not drop below 60 °F (16 °C).
b)   Use only on high-power-factor fluorescent ballasts (CBM or other approved types).
c)   Do not use where cold air will be directed onto the lamp surface.

Wattage reductions possible with these reduced wattage lamp types are in the range of 10–20% for the popular types. Light output is also reduced by 3–18%, depending upon the specific lamp used. Typical energy savings will be approximately 5–6 W per lamp for the popular 4 ft 40 W replacement and 17.5 W per lamp for the popular 8 ft slimline 75 W replacement. Savings in energy costs normally pay back the new lamp cost in one year.

Another type of fluorescent lamp available for further wattage and light output reductions is the impedance modified fluorescent. These lamps contain capacitors that add impedance to the lamp circuit and thereby reduce the current and wattage of the lamps. These lamps operate like a *fixed-dimmer* control and allow fluorescent lighting systems to be reduced in wattage by 33%, 50%, or more. Light output is also reduced in proportion. These special lighting devices may be considered where the lower light levels can be tolerated or where energy guidelines are mandated by legislation.

Recent significant improvements in fluorescent lamps are as follows:

a)   Cathode cutout lamp
b)   32 W F32 T-8 lamp

The cathode cutout lamp has a thermal cutout switch in series with the filament. The switch opens the circuit that applies the low voltage (3.6 V) across the electrodes after the discharge has been ignited. The efficacy of the lamp is increased by 6% (81 lm/W) for the standard 40 W F40 rapid-start lamp. This lamp is not recommended for use with electronic ballasts, and it should not be operated in the dimmed mode (which substantially reduces lamp life).

The 32 W F32 T-8 lamp is rapidly replacing the standard 40 W T12 lamp. This lamp uses the rare earth 70 CRI phosphor and has a lamp efficacy of 90 lm/W at 60 Hz. When operated at higher frequency in the instant mode, its efficacy approaches 100 lm/W, but its life is rated at 15 000 h.

Growing in popularity are fixtures with three lamps in the standard 2 ft by 4 ft fixture. For maximum efficiency, the fixture uses low-loss ballasts, cone-on-cone diffusers, and special reflectors. Switching provides further potential energy reductions. A second circuit permits the inside, the two outside, or all three lamps to operate—giving fixed 33%, 55%, and 100% light output adjustment.

Application limitations of fluorescent lamps, including energy-saving types, are as follows:

a)   Limited, normally, to indoor applications due to large optical size of the lamps

b)   Temperature sensitive: high and low ambient temperatures reduce light output

c)   Highest wattage lamp: 215 W

Compact fluorescent lamps have single-ended electrical sockets such that they can be placed in a base containing a ballast with an Edison-type connector for direct replacement of incandescent lamps. The efficacy can be as high as 65 lm/W compared to 16 lm/W for a 75 W incandescent lamp. While most of the modular types are operated in the preheat mode, the electronic ballasted lamps are operated in the rapid-start mode and in principle could be dimmed.

7.5.4.3 High-intensity discharge

High-intensity discharge (HID) light sources fall into four major categories—mercury, metal halide, high-pressure sodium and low-pressure sodium. The HID group of sources offers many energy-saving opportunities due to high lumen-per-watt potential.

Mercury light sources have been available in their present form since about 1934. The efficacy and life have steadily increased to a point where they now have efficacies up to 63 lm/W and typical average life ratings exceeding 24 000 h. Improvements in phosphor technology have made mercury lamps available with color rendering properties suitable for all but the most critical applications.

Mercury lamps are available in numerous wattages ranging from 40 W to 1000 W in several phosphor colors. Typical applications include street lighting, industrial lighting, area and security lighting, and merchandising lighting, both indoors and outdoors.

The new HID lamps, however, offer significantly higher efficacy and even better color rendering properties than mercury types now available. Thus, economically, mercury has been surpassed, and few new mercury systems are being installed where cost and energy efficiency have been considered.

Metal-halide lamps were introduced in the mid-1960s and have continued to gain wide acceptance for a broad range of applications, industrial and commercial, indoor and outdoor, and have been almost universally accepted for sports lighting applications.

While operating in a similar manner to mercury lamps, metal-halide types utilize rare earth metal iodides or halides to improve both efficacy and color rendition. Metal-halide lamps offer efficacies today up to 125 lm/W of *white* light. The color rendering of these sources allows their use in virtually all applications where high light output and good color are required. The broad color spectrum of metal-halide lamps coupled with the excellent optical control potential of the sources, makes them ideal for floodlighting, area lighting, and high-quality industrial lighting.

Limitations of metal-halide lamps include relatively long restart or restrike time (8–15 min) if the lamp is turned off momentarily. Another limitation is that low-wattage lamps are only recently available. Typical wattages are 175, 250, 400, 1000, and 1500. Wattages under 100 would increase the application potential for low mounting height.

The smaller metal-halide lamp has an output of 2580 lumens at 40 W, and an efficacy of 65 lm/W. It can replace a 150 W incandescent lamp operating daily for long hours.

Typical rated lamp life is 7500–20 000 h with fair to good maintenance of light output over its life.

High-pressure sodium (HPS) lamps offer efficacies that exceed those available from most metal-halide lamps. Current available types offer efficacies up to 140 lm/W. A broad range of wattages is available with types now ranging from 35 W to 1000 W.

The main advantage of HPS compared to mercury and metal halide is shorter hot restart time. Lamps typically restart in 2–3 min after a momentary outage (see table 7-6).

Table 7-6—Lamp start times

Lamp	Initial start* (min)	Restrike* (min)
Mercury	5–7	3–6
Metal halide	3–5	10–15
High-pressure sodium	3–4	1

*Time to reach 80% lumen output.

Major disadvantages of high-pressure sodium lamps are as follows:

a) Color rendering index is low (typically 25), limiting applications for this source. This can be improved to a CRI of approximately 55–60, but at the expense of reduced life and efficacy.

b) Lamps should be removed from circuit when lamps begin to cycle ON and OFF due either to premature or normal end-of-life failure. This is a lamp/ballast compatibility problem that will eventually be resolved by the ballast manufacturers.

c) Color temperature is relatively low at 1900–2100 K

d) Stroboscopic effects are greater than other HID lamps.

Typical applications of high-pressure sodium lamps include lighting for the following:

a) Streets and highways
b) Industrial interior
c) Outdoor area (parking lots, storage, sports, and recreational)
d) Bridges and tunnels
e) Closed-circuit television

HPS is also now being used for many office and commercial interior applications, sometimes coupled with metal-halide sources. A combined HPS/metal-halide system offers high efficacy and substantially improved color rendition. The life rating of typical HPS lamps is 24 000+ h for most types.

Low-pressure sodium (LPS) is, at this time, the most efficient light source with lamp efficacies ranging up to 183 lm/W. Sizes available are 18 W, 35 W, 55 W, 90 W, 135 W, and 180 W. Major applications of LPS are security, area, and roadway lighting.

LPS is more closely related to fluorescent than HID since it is a low pressure, low-intensity discharge source and is a linear source like fluorescent. As with fluorescent, this source does not lend itself to applications where a high degree of optical control is necessary such as *long throw* floodlights.

The major disadvantage of LPS is that it is a monochromatic source, i.e., it provides light of only one color (yellow), which gives the lamp a poor color rendering index. Other disadvantages are high-ballast losses and low system efficiency. Major advantages are that this source delivers its full rated output throughout life and will restrike instantly when hot. Life ratings of LPS sources are generally 18 000 h for all but the 18 W, which is currently rated at 10 000 h.

7.5.4.4 Economic considerations

Although prices vary considerably among energy-efficient light sources such as fluorescent and HID lamps, the major factor that determines lighting costs is not lamp or fixture cost but energy cost. Today this portion of the total cost of providing light with a general lighting system is typically 80% more of the total. Before any decision is made concerning the choice of lighting system, the user should first decide what the color requirements are for the specific application. Once this is determined, the user should then complete an economic analysis for the various light sources and systems he or she wishes to consider. The economic analysis should include the initial ownership and operating costs.

Most lamp and luminaire manufacturers offer free computerized analyses for this purpose, and these sources of information should be consulted. In addition, many manufacturers offer a wide variety of literature on product applications and energy conservation for the prospective buyer. These sources will help the user make better economic judgments for specific applications.

7.6 Ballasts

7.6.1 General definitions

7.6.1.1 figures of merit: Electrical losses in ballasts may be calculated by determining the difference between the input power and the wattage delivered to the lamp.

Ballast losses = ballast input watts – lamp watts

Table 7-7—Typical HID lamp ballast input watts

Lamp type	ANSI designation	Watts	Reactor	High reactance autotransformer (LAG)	Ballast type		
					Constant wattage autotransformer (CWA)	Constant wattage regulated (CW)	High reactance regulated (regulated lag)
Mercury	H46	50	68	74	74	—	—
	H43	75	94	91–94	93–99	—	—
	H38/44	100	115–125	117–127	118–125	127	—
	H39	175	192–200	200–208	200–210	210	—
	H37	250	272–285	277–286	285–300	292–295	—
	H33	400	430–439	430–484	450–454	460–465	—
	H36	1000	1050–1070	—	1050–1082	1085–1102	—
Metal-halide	M57	175	—	—	210	—	—
	M58	250	—	—	292–300	—	—
	M59	400	—	—	455–465	—	—
	M47	1000	1050	—	1070–1100	—	—
	M48	1500	—	—	1610–1630	—	—
High-pressure sodium	S76	35	43	—	—	—	—
	S68	50	60–64	68	—	—	—
	S62	70	82	88–85	95	—	105
	S54	100	115–117	127–135	138	—	144
	S55	150 (55 V)	170	188–200	190	—	190–204
	S56	150 (100 V)	170	188	188	—	—
	S66	200	220–230	—	245–248	—	254
	S50	250	275–283	296–305	300–307	—	310–315
	S67	310	335–345	—	365	—	378–380
	S51	400	440–463	464–470	465–480	—	480–485
	S52	1000	1060–1065	—	1090–1106	—	—

Table 7-8—Typical fluorescent lamp ballast input watts

Lamp type	Nominal lamp current	Nominal lamp (W)	Standard ballasts		Energy-saving ballasts			Electronic ballasts				Circuit type
			One-lamp	Two-lamps	One-lamp	Two-lamps	One-lamp	Two-lamps	Three-lamps	Four-lamps	
F20T12	0.380	20	32	53	—	—	—	—	—	—	—	Rapid start, preheat lamp
F30T12	0.430	30	46	81	—	—	31	61	92	—	Rapid start
F30T12, ES	0.460	25	42	73	—	—	27	52	81	—	Rapid start
F32T8	0.265	32	—	—	37	71	36	58	87	112	Rapid start
F40T12	0.430	40	57	96	50	86	36	71	109	—	Rapid start
F40T12, ES	0.460	34/35	50	82	43	72	31	59	93	—	Rapid start
F48T12	0.425	40	61	102	—	—	—	—	—	—	Instant start
F96T12	0.425	75	100	173	—	158	88	140	—	—	Instant start
F96T12, ES	0.455	60	83	138	—	123	73	116	—	—	Instant start
F48T12, –800 mA	0.800	60	85	145	—	—	—	—	—	—	Rapid start
F96T12, –800 mA	0.800	110	140	257	—	237	—	—	—	—	Rapid start
F96T12, ES, 800 mA	0.840	95	125	227	—	207	—	—	—	—	Rapid start
F48 –1500 mA	1.500	115	134	242	—	—	—	—	—	—	Rapid start
F96 –1500 mA	1.500	215	230	450	—	—	—	—	—	—	Rapid start

System input (W)

Such information may be obtained from catalogs of ballast manufacturers. Note carefully the listings of the differences in ballast losses as a function of supply voltage. Ballasts designed for high-wattage lamps, which have higher starting voltages, may be more efficient on high-voltage distribution systems and vice versa. Carefully calculate the wiring and distribution losses and compare these losses against ballast losses to obtain the most efficient overall electrical system. Tables 7-7 and 7-8 provide guideline information for common types of fluorescent and high-intensity discharge lamp ballasts.

7.6.1.2 ballast factor: The ballast factor is the ratio between the light output from a lamp operating on a commercial ballast and the light output from the same lamp operating on a reference ballast. The ballast factor of a reference ballast is 100.

7.6.1.3 ballast life: Fluorescent and HID ballasts are long lived and may be expected to last 12–15 years in normal operation, provide nameplate ratings are not exceeded. However, high temperatures are detrimental to ballast life and a 10 °C (50 °F) increase in the hotspot temperature of a fluorescent lamp ballast may reduce expected life by 50%. Retrofitting old low-efficiency ballasts with new high-efficiency units may provide a double bonus: energy savings and longer ballast operating life since more efficient ballasts generally operate at lower temperatures.

Depending upon the circuit used, it may be important to ballast life to promptly replace failed lamps. This is particularly true of instant-start fluorescent systems where failed lamps can *rectify*, drawing heavy current through ballast windings, which results in ballast failures. If lamps are removed from sockets to reduce lighting energy use, ballasts will normally remain energized, except in the case of slimline or instant-start circuits where removing the lamp disconnects the ballast from the electrical system. For other fluorescent and HID systems, an energized ballast will draw some power and may negatively affect system power factor. Rapid-start fluorescent ballasts should be deenergized if lamps are removed to prevent ballast failure.

7.6.1.4 ballast efficacy factor: The ballast efficacy factor (BEF) is the relative light output divided by the power input of a fluorescent lamp ballast. Ballasts meeting the new standard will be marked with a capital "E" printed within a circle. Some ballasts are excluded from meeting these standards. Dimming ballasts, ballasts designed for use in an ambient temperature of 0 °C or less, and ballasts having a power factor of less than 90% made for residential applications are excluded. Table 7-9 shows the minimum BEF required by Public Law 100-357.

7.6.1.5 ballast interchangeability. The optimum performance of any HID lamp can be obtained only by operating it with a ballast designed for the lamp. However, in recent years a number of lamps have been developed that can be used on various types of existing ballast circuits. Special metal-halide lamps are available that are designed to be used with mercury ballasts; a special type of high-pressure sodium (HPS) lamp called *Penning Start* can also be operated on mercury ballasts.

It may be desirable to upgrade the efficiency of a lighting system by replacing a mercury lamp with either a metal-halide or an HPS type that operates at higher lumens per watt. The

Table 7-9—Ballast efficacy factors

Ballast	Nominal lamp watts	BEF
1-F40T12 120 V 277 V	40 40	1.805 1.805
2-F40T12 120 V 277 V	80 80	1.060 1.050
2-F96T12 120 V 277 V	150 150	0.570 0.570
2-F96T12H0 120 V 277 V	220 220	0.390 0.390

light generated is, therefore, available at lower cost and it may be possible to reduce the number of luminaires depending upon mounting height, spacing, and fixture considerations. Some retrofit lamp designs, however, are not as efficient as their standard counterparts. They may provide somewhat less luminous efficacy, and may have shorter life. Lamp ratings should be carefully checked and any limitations on ballast use should be noted.

7.6.2 Fluorescent ballasts

7.6.2.1 Effect of temperature and voltage variations

The lamp bulb wall temperature affects the efficiency and output of a fluorescent light more than any other environmental factor. The optimum temperature is 40 °C (104 °F). ANSI test procedures call for manufacturers to rate lamps in free air at an ambient temperature of 25 °C ± 0.5 °C (77 °F ± 0.9 °F). To determine the light output of lamps in practical applications, tests certified by the Certified Ballast Manufacturers (CBM) organization may be employed. The ballast is operated at its rated center voltage and the initial lamp light output cannot be less than 95% of the manufacturer's rated output.

Rated lumen values are based on measurements made at an ambient temperature of 25 °C (77 °F) in still air. The effect of higher or lower temperatures varies with lamp type. The relationship of lumen variations versus ambient temperature for three types of fluorescent lamps is shown in figure 7-3.

Ballasts are typically designed to operate fluorescent lamps over a range of ±10% about the rated center voltage. The power input and light output tend to decrease with decreasing input voltage.

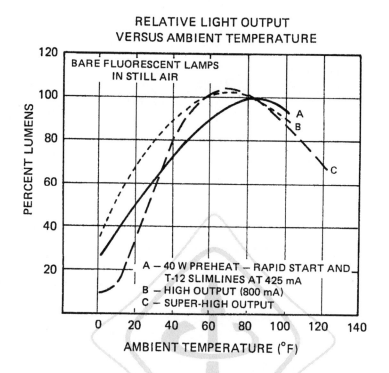

Figure 7-3—Relationship of lumens versus ambient temperature

7.6.2.2 Energy-efficient ballasts

In general, the greater the ballast size (power rating), the greater is the ballast efficiency. That is, the relative ballast losses are less for a 40 W ballast than a 20 W ballast. A two-lamp ballast is more efficient than a one-lamp ballast. The ballast efficiencies can generally be calculated from the manufacturers' catalog data. However the designer shall be certain the manufacturer specifies the test conditions at which the ballast is rated.

The internal ballast losses are determined by coil construction, the nature of the magnetic materials, and resistances of the conducting coil wire. Over the years, standard ballast designs have been the victims of cost reduction, which has tended to decrease ballast efficiencies. With increasing energy costs, ballast manufacturers have introduced energy-efficient ballasts that minimize the ballast losses.

The following details some of the new ballasts.

a) *Low-energy ballasts.* Fluorescent lamps may be operated at less than rated power and output, provided starting voltage and operating voltage requirements are met. In the case of rapid start, lamps and cathodes shall be heated regardless of the current through the lamp to provide rated lamp life. *Low energy* ballasts are low-current designs that can provide energy reductions compared to the standard units. They are useful where certain luminaire spacing to mounting height criteria shall be followed, and the desired illumination level is less than that obtained by full output operation of the lamp. Before operating fluorescent lamps at less than the rated output, it is wise to check the ambient temperatures of the air surrounding the lamps. Higher minimum ambients are required when low-energy lamps or standard lamps are used with low-energy ballasts.

b) *High/low ballasts.* This type of ballast, which is generally available only for rapid-start circuit operation, contains extra leads that can be connected or switched to provide multilevel operation of the lamp. Two-level and three-level rapid-start ballasts are available, and fixture output may be set according to the lighting requirements of the area. Operation of fluorescent lamps at less than rated output may raise minimum operating temperature requirements.

c) *Low-loss ballasts.* Ballasts may be designed to reduce internal losses by improving mechanical and electrical characteristics. More efficient magnetic circuits, closer spacing of coils, and improved insulation systems can result in loss reductions of approximately 50%, compared to conventional units. Again, the ballast manufacturers' ratings should be consulted to determine which ballast will have the least losses for the power system and lamp combination involved.

7.6.2.3 Electronic ballasts

The efficiency of fluorescent and HID lamp systems can be improved even more by utilizing electronic ballasts. Improvements occur in the following two ways:

— By way of lower internal losses within the ballasts.

— By making use of the fact that fluorescent lamp efficacy increases as a function of the frequency of the applied power.

Electronic ballasts usually operate lamps at frequencies about 20 kHz. In general, operating a fluorescent lamp with an electronic ballast will realize a 20–25% increase in system efficacy compared to a standard core-coil ballast. The electronic ballast offers improved performance that includes:

a) Reduced flicker
b) Improved voltage and thermal regulation
c) The ability to dim fluorescent lamps over a large range without affecting lamp life

Table 7-10 shows the performance of the standard and energy-efficient core-coil ballasts as well as the electronic high-frequency ballasts.

The data is for two-lamp systems. The 40 W F40T12 CW lamp is still in general use; however, the 32 W F32T8 lamp is gaining in popularity based on its high system efficiency. Table 7-10 shows that the electronic ballast can meet all necessary lamp parameters to maintain lamp life. The high harmonic data for the T8 electronic ballast was designed before

Table 7-10—Performance data of some fluorescent lamp-ballast systems

Ballast	EE mag	EE mag	Electronic ballasts		
Lamp	40W F40 T-12 CW	40W F40 T-12 CW	40W F40 T-12 CW	32W F32 T-8 41 K	32W F32 T-8 41 K
Power (W)	85	81	74	69	65
Filament voltage (V)	3.5	0	2.6	3.4	0
Lamp current crest factor	1.7	1.7	1.4	1.5	1.5
Ballast factor (%)	93	93	93	100	100
Light output (lm)	5690	5680	5670	5820	5820
Flicker (%)	30	30	1	1	1
3rd harmonic (%)	12	12	5	43	43
Ballast efficiency (%)	87	87	89	90	90
Lamp efficacy (lm/W)	77	80	86	93	100
System efficacy (lm/W)	67	70	77	84	90

harmonics became an issue. Today the technology is available to reduce the harmonic contents within a specified limit. The constraint would be cost considerations.

Electronic ballasts are available for all of the commonly used fluorescent lamps, including the 40 W F40T12, the 34 W F40T12, the 40 W F40T10, and the 32 W F32T8 (rapid and instant start), all of the F96-type lamps, and the high-output F96 lamps. There are one-, two-, three-, and four-lamp ballasts. The new generation of electronic ballasts for compact fluorescent lamps (CFLs) has power factors above 90% with harmonics below 20%.

7.6.3 High-intensity discharge (HID) ballasts

Ballast factors for HID lamps are usually close to unity, and ballast circuitry is somewhat simpler than in fluorescent ballasts, due to the widespread use of single-lamp circuits.

For mercury vapor lamps, ballasts circuits may be of the reactor, autotransformer, or regulator types. Regulator ballasts contain circuitry that is designed to operate the lamp at relatively constant wattage even though nominal input line voltage may vary.

In general, HID ballast losses range between 10% and 20% of lamp wattage. However, since each ballast circuit is somewhat different, catalog ratings should be used for more precise information.

7.6.3.1 Ballasts for high-pressure sodium (HPS) lamps

At present, four general types of ballasts are available to the HPS lamp user. Each has its advantages and disadvantages.

The lag circuit ballast is simply an inductance placed in series with the lamp. It has the poorest wattage regulation due to changes in line voltages but can be used effectively on circuits where the line voltage varies by no more than ± 5%. It has a relatively high starting current that can produce a desirably faster lamp warm-up. It is relatively inexpensive and small in size and has low power losses.

The lead circuit ballast is built with a capacitance in series with the inductance and the lamp, in combination with an autotransformer. The ballast size is kept small by keeping a portion of the primary winding common to the secondary winding. With the secondary winding providing the wattage regulation, the effectiveness of the regulation depends on the amount of coupling between the primary and secondary. A variation of line voltage by ± 10% will only result in a ± 5% variation in lamp power. The lead circuit ballast has low starting current.

The magnetic-regulator ballast is essentially a voltage-regulating isolation transformer with its primary and secondary windings mounted on the same core, and contains a third capacitive winding that adjusts the magnetic flux with change in either primary or secondary voltage. It provides the best wattage regulation with change of either input voltage or lamp voltage. However, it is the most costly and has the greatest wattage loss. Figure 7-4 shows the effect of line-voltage variation on lamp watts for various ballast types.

7.6.3.2 Electronic ballasts

The major problem with existing HPS lamp ballasts is their inability to operate the lamp at rated power. Electronically controlled ballasts have been designed and built with a solid-state control circuit and a reactor. Figure 7-5 shows the circuit of one such ballast called the *super-regulated* ballast. The use of a solid-state switching device permits the control winding to be shorted in a *phase controlled* manner, and thus provides a smooth and continuous variation in the average inductance of the ballast. The solid-state control circuit monitors lamp and line operating conditions and then establishes the proper value of ballast inductance required to operate the lamp at its rated power. If the ballast operates the lamp so that it averages reasonably close to its rated wattage or below over its life, satisfactory performance can be expected.

Because the HPS lamp requires more voltage as it ages, special attention shall be directed to the regulation of the lamp wattage, which can vary to limits that can be very damaging to the lamp. To avoid this, lamp-wattage limits have been imposed on ballasts by the lamp manufacturers. The wattage and voltage limits are defined by the generally accepted diagram called the *trapezoid*. The trapezoid shows the lamp user the range of wattage and voltage at which the lamp shall operate to give acceptable performance in life, luminous output, and stability. The trapezoid for the 400 W HPS lamp is shown in figure 7-6. For present-day ballasts operating at rated input voltage, all lamps of a particular design will operate on a smooth curve within the trapezoid called the *ballast characteristic*. A properly designed ballast will

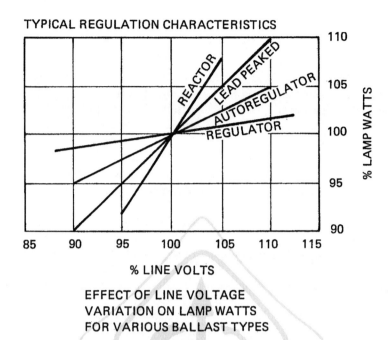

Figure 7-4—Ballast regulation characteristics

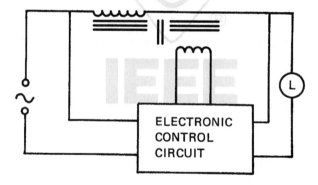

Figure 7-5—Energy-efficient electronically controlled ballast

generate a smooth wattage versus voltage curve with a *haystack* appearance so that with increasing voltage the wattage will gradually rise to a peak, then start to decrease toward the end of life. HPS lamps on electronic ballasts can be made to traverse the trapezoid by way of a straight line. If the light output of such a combination is constant, input watts can be reduced approximately 20% over the life of the lamp.

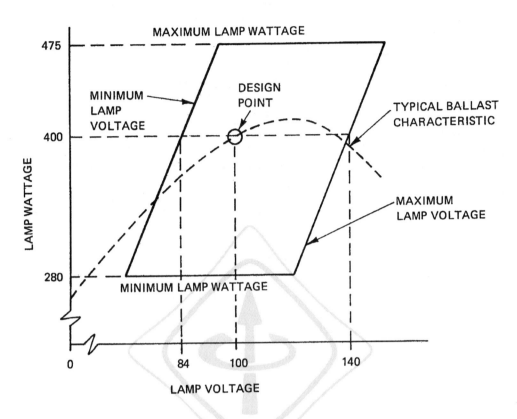

Figure 7-6—Trapezoid diagram for the 400 W HPS lamp

7.7 Luminaires

7.7.1 Efficiency criteria

The principle purpose of a lighting fixture or luminaire is to contain the light source with its necessary mounting and electrical accessories and then direct the light from the source into the area to be lighted. Luminaire efficiency is defined as the amount of light that is emitted from the luminaire divided by the amount of light generated by lamps. Efficiency, therefore, tells how well the luminaire succeeds at permitting the light generated by the lamps to escape absorption, internal reflections, and trappings by the lamps themselves and other luminaire components.

7.7.2 Glare control and utilization

The most efficient luminaire may be no luminaire at all since exposed lamps may efficiently direct their light into the space without utilizing any type of enclosure. Such lighting equipment, however, can result in excessive glare and, depending upon the room reflectances, shape of the space, and mounting height, may not put the light where it is required. There is a

303

figure of merit for general lighting systems that tells how effectively the luminaire and the room work in combination. This is the *coefficient of utilization* (CU) and is a measure of the light flux reaching the task area divided by the light flux generated by the lamps. The coefficient of utilization depends upon the intensity distribution of the luminaire, the size and shape of the space, the reflectances, and the luminaire mounting arrangement. CU data are normally provided by the luminaire's manufacturer for each type of luminaire. The CU permits rapid estimation of the illumination that can be achieved for a given system.

Where people spend long periods of time doing visual work, it is important to control the glare from the lighting system. The *visual comfort probability* or VCP is a measure of direct discomfort glare based upon the studied reactions of people to areas of brightness within their view. A VCP rating indexes the probability that a lighting system will be comfortable from the direct glare standpoint if a person stands in the worst position in the room and looks horizontally. VCPs can also be calculated for individual positions and viewing direction within a space.

Generally, light emitted from a luminaire within a cone described by an angle of 45° rotated around a vertical line from the luminaire to the floor will not contribute substantially to direct glare. Light emitted above 45° may cause glare depending upon its intensity and the luminance of the surrounding space. Figure 7-7 shows that luminaires that emit light below 45° can cause loss of contrast on the tasks, depending upon their position in the field of view. To minimize veiling reflections, light emitted from luminaires above 45° is desirable. Thus, a compromise should be made to avoid having light from the luminaire emitted between 0° and 45° to minimize veiling reflections and between 45° and 90° to minimize direct glare. The type of task or use of the space determines which zone should receive the greater emphasis.

7.7.3 Shielding media

Some unique solutions to the problem described above have been provided using new types of shielding materials that can direct light from luminaires to rather narrow zones for maximum visual effectiveness and control of glare. One example of this is a lighting material with a so-called "batwing" distribution. Figure 7-8 shows that a high percentage of light is emitted near 45° so that if work stations are placed between rows of luminaires, the major illumination on the task comes from the side. Veiling reflections will thus be minimized with a fair degree of glare control.

Other techniques can also minimize reflections. Polarizing materials can reduce the amount of light reflected from the task, thus improving task contrast. Changing the position of the task with respect to the luminaire can result in dramatic improvement, and if task locations are known, installing luminaires away from the reflected glare zone will always be beneficial.

7.7.4 Dirt effect and maintenance considerations

The efficiency of the luminaire is highly dependent upon the reflectance of its interior and exterior surfaces and the amount of dirt collected on the lamps and shielding materials. All surfaces eventually collect dust, or otherwise lose their lighting effectiveness over a period of time. In extreme cases, such losses may amount to as much as 50% over a two-year period.

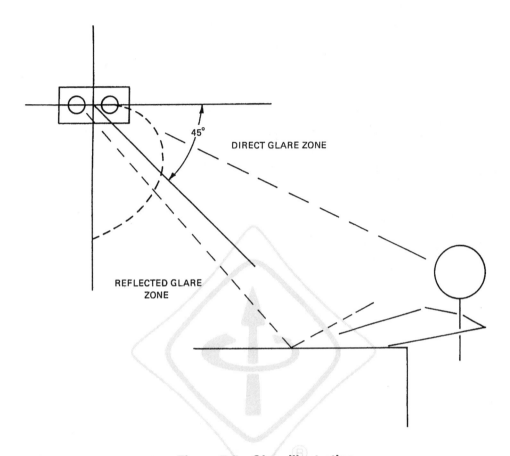

45°

DIRECT GLARE ZONE

REFLECTED GLARE
ZONE

Figure 7-7—Glare illustration

More typical, perhaps are losses of 10–25% over the same period in air-conditioned spaces with 8–12 h/day usage.

Luminaire manufacturers publish expected *luminaire dirt depreciation factors* (LDDs) to help evaluate loss of efficiency due to dirt. However, it must be recognized that lighting in an interior space depends not only on luminaire dirt depreciation but also on room surface dirt depreciation, e.g., aging and discoloration of luminaire shielding materials and permanent degradation of room furniture surfaces.

7.7.5 Thermal performance

It is well known that fluorescent lamp performance is affected by the minimum lamp wall temperature (MLWT). Light output is maximum at about 40 °C (104 °F) and decreases above and below this temperature. In nearly all fixtures, the MLWT is well above the 40 °C in which the light output is decreased by 10–20% in two-lamp and four-lamp fixtures, respec-

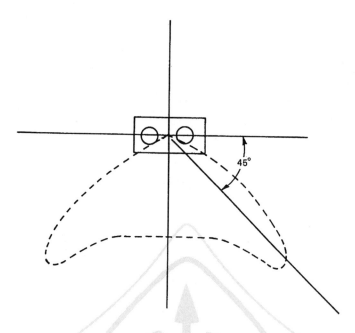

Figure 7-8—Batwing distribution

tively. Air-handling fixtures can reduce the lamp wall temperature, but they are not in wide use.

The compact fluorescent fixtures have even a greater effect on the performance of the lamps since these fixtures are more confining. Not only does the light output decrease by 20%, but the system efficacy is also decreased by 20%. Advanced fixture designs include venting and conductive methods to reduce the MLWT. This becomes important since it will increase the actual lumen output of the fixtures and expand the applications of this efficient light source.

7.7.6 Air movement

Enclosed luminaires may "breathe," pushing air out of their enclosures as lamps warm up, and drawing air in as lamps cool down. This may increase the dirt collection on the interior surface of the luminaire. In dirty areas, dirt accumulation can be substantially lessened by using luminaires that circulate room air, using heat from the lamps to provide a chimney effect. This draws dirt up and through the luminaires without depositing it on luminaire surfaces. Air-handling luminaires, i.e., those luminaires that are connected to a building heating, ventilating, and air conditioning (HVAC) system, are usually constructed so that room air is exhausted by passing it over lamp, ballast, and luminaire surfaces. Not only does this lessen dirt accumulation, but it can also make the lamps operate closer to their peak output. Ballasts are also kept cooler, which extends life.

7.7.7 New luminaires for energy-efficient light sources

a) *Fluorescent.* A new trend for lighting new buildings is the increased use of the reflectorized fixtures. This trend may be traced to an increase in the number of state and national lighting efficiency standards in recent years. However, these fixtures can create a teardrop-like distribution that may eliminate glare on a computer screen, but also reduce light to other areas.

The design of the optical reflector aids important performance criteria. At the present time, complex multiple plane designs using up to 30 reflective planes can be created using CAD/CAM technology, which yields excellent lighting distribution as well as improved fixture efficiencies.

b) *Specular retrofit reflectors.* Reflectors are available in two basic types: semi-rigid reflectors, which are secured in the fixture by mechanical means, and adhesive films, which are applied directly to the interior surfaces of the fixture. Either silver or aluminum may be used as the reflecting media. In general, reflectors increase the percentage of lamp lumens that reach the workplane. When the visual appearance of a delamped fixture is unacceptable, or when the original fixture has an unusually low efficiency, reflectors may be used as an efficiency remedy. However, the uniformity of illumination should be evaluated to ensure that adequate light is provided at the work stations.

c) *High-pressure sodium.* Proper luminaire design is the key to lighting efficiency. Newly developed luminaires use prismatic glass reflectors that are especially made for high-pressure sodium lamps. In addition to achieving maximum light utilization, they reflect the intense light source with excellent light cutoff and high-angle brightness control.

7.8 Lighting controls

7.8.1 General

The proper control of a lighting system is one of the most effective ways to save lighting energy. Control techniques may range from the simple ON/OFF switch installed on an individual luminaire to elaborate computer-controlled master switching systems designed for operating the lighting system of a large building. The installation of a proper number of switch-points is generally the key to effective control of energy by switching. Once a system is installed, additional switch-points may be difficult to add; so careful attention should be paid to this part of the lighting system during initial design.

7.8.2 Switching

It has been common practice to wire electrical power systems in such a way that the minimum amount of electric wire or circuiting is required. With a very small increase in cost, electric circuits can be wired so that full advantage of reduced lighting can be obtained.

Switching should be arranged for each area in which an individual or group might work and require higher levels of lighting than the balance of the room. Low-voltage switching techniques increase the multiple point control ability of the system; 277 V switching with barriers between different switch legs is available for use in commercial buildings.

Partial lighting could be acceptable at several periods of the day: during cleaning operations, at lunch hour, and before the staff is fully occupied in the morning. For example, office lighting may range from 50 fc to 100 fc, whereas janitors can work perfectly well and safely at 15–20 fc. The control is normally arranged so that half of the ballasts or one-third of the ballasts in a fluorescent or HID installation are turned off. In some cases, this means that half, one-third, or two-thirds of the fixtures will be extinguished. In other cases where multiple ballasts are installed in continuous rows of fixtures or paired fixtures, this simply means the fixture will be dimmed to one-half, one-third, or two-thirds light. In the latter case, the advantage of even lighting and normal ratios of minimum to average lighting are obtained and minimum shadowing occurs. This type of control is very easily accomplished by wiring branch circuits on alternate switch or circuit breaker legs or in a three-phase system by wiring one-third of the lights on each phase. Suitable arrangements that stagger the lighting can provide a current balance at the electric service. In most cases, very little additional conduit is required, and only a minimum amount of additional wire and a few additional branch breakers for each panel are needed.

It may be desirable to place switches adjacent to windows so that lights can be turned off or reduced if not needed. Significant energy savings can be made by the use of day lighting. In some instances, two zones of control, running parallel to the windows, can be justified.

It is essential to remind those responsible for controlling the light switches of the need to turn off lights. No advantage will be gained in the introduction of additional controls if they are not utilized. The use of posters, notices, meetings, and checking by supervisory staff is desirable.

7.8.3 Dimming

Various methods of dimming are available which provide a pleasing effect, utilize only the amount of light needed, and save energy. Reduced voltage is mostly used for incandescent, but more sophisticated methods are required for fluorescent and HID lighting.

It is possible to reduce the voltage of a system very simply by the use of variable tap autotransformers or by delayed-firing thyristors.

Because variable tap autotransformers have relatively low losses and may be used to control either large or small incandescent loads, such control can be used for theater or accent lighting where varying levels of illumination are desired. A less expensive way of accomplishing the same type of lighting control is to use fixed-tap autotransformers; however, the lighting levels are adjustable only in steps.

A second type of incandescent dimming uses transistors or thyristors. They reduce the voltage by reducing the time in each cycle that the voltage is applied. This method is the least expensive. One other form of voltage dimming involves the use of a saturable reactor together with an electronic control.

Dimming of the HID and fluorescent lamps can be accomplished with a phase-controlled circuit. In full-range systems using rapid-start fluorescent lamps, lamp cathodes are held at relatively constant voltage to prevent premature failure of the tube or extinguishing of the arc. Some types of electronic ballast also incorporate a dimming feature with the additional advantage that power losses through the solid-state devices are minimal.

An electronic ballast can readily dim fluorescent lamps without conditioning the input power as required to dim magnetic ballasted systems. The magnetic ballasted fluorescent systems are limited to the scheduling strategy, lumen depreciation, and simple daylighting applications. Fluorescent lamps operated on a dimming electronic ballast can perform the above more effectively and save more energy without increasing the flicker. The dimming capabilities will allow dynamic control of light levels to accommodate an electronic office that requires changes in illumination levels for hard paper tasks and VDTs.

New types of auxiliary controls were developed in 1981 that can cut the power of lamps operating on standard ballasts by 25% or more. These controls are easily installed in the luminaire or branch circuit, but a careful evaluation should be carried out beforehand to make sure that the control, ballast, and lamp work effectively without compromising system performance or component life.

7.8.4 Remote control systems

In most large buildings it is desirable to have the ability to control lighting from a central location or from central floor points. This provision enables individuals assigned the task to turn on lights in the morning and turn off lights at night, and make intermittent adjustments to the lighting as desired.

7.8.5 Automatic control systems

Automatic controls vary from a simple time clock, a photocell, or a presence detector to a sophisticated computer-controlled lighting system. A price versus benefit cost analysis will be required for each installation. The system should be programmed for normal operations and have a local manual override. Lighting should be switched in distributed groups so that areas can be lighted or darkened as conditions change.

7.9 Optimizing lighting energy

To achieve optimum use of energy in lighting systems, the most important considerations in the lighting design process are as follows:

a) Utilizing daylight
b) Understanding visual tasks
c) Considering quantity and quality of illumination

d)   Selecting proper lighting hardware (lamp, ballast, luminaire, control)

e)   Understanding maintenance characteristics

f)   Spacing criterion

The relative importance of each of these factors varies with the situation. For example, when selecting lighting fixtures for a parking lot, the visual task is all pervasive. That is, the entire parking lot requires a low, uniform wash of light while the construction characteristics dictate as few obstructions to the pavement surface as possible. On the other hand, the lighting of a conventional office requires a lighting level of sufficient magnitude to permit reading of second or third carbon copies at specific work stations while the remaining portion of the space may be lighted to a somewhat lower level.

7.9.1 Utilizing daylight

Daylight is important to the reduction of daytime consumption of energy attributable to illumination systems. If daylight is considered, the design of an electric lighting system must be modified to integrate it into the overall system.

Daylight factor is defined as the illuminance received at a point indoors, from a sky of known or assumed luminance distribution, expressed as a percentage of the horizontal illuminance outdoors from an unobstructed hemisphere of the same sky.

Daylight factors (calculated) between 2% and 5% indicate that the electric lighting should be designed to take full advantage of available daylight. This can be accomplished by using

a)   Task or task-ambient lighting (to use daylight to provide the general surround or ambient lighting)

b)   Electric lighting control systems with photosensor controls to automatically dim the light in proportion to the available daylight so as to produce a constant illuminance

For an industrial building, windows in the sidewalls admit daylight and natural ventilation, and afford occupants a view out, all of which may be desirable. Nevertheless, their uncontrolled luminance may be troublesome. There are many forms of lighting sections used by architects to admit some daylight into a factory. Each will require some control means to make the daylight useful to worker's visual tasks, thus resulting in energy savings as the ultimate goal.

7.9.2 Visual task

The designed illuminance level should be appropriate for the visual task. Some engineering judgement shall be exercised in establishing the design level, which should be adjusted to the age of the worker and the visual difficulty of the task. The maximum contrast ratio should be observed in the design.

Visual performance is the rate of information processed by the visual system as measured by the speed and accuracy with which the visual task is performed.

The visibility of a visual task is determined by the visibility of the most difficult element that must be detected or recognized in order that the task may be performed. The luminance of the task is the most important lighting factor in achieving good task visibility.

7.9.3 Illumination quantity and quality

Quantity of illumination is usually measured in lumens per square foot (footcandles) or lumens per square meter (lux). Since the eye cannot "see" footcandles, some lighting recommendations are now given in footlamberts or units of luminance. Luminance is more closely related to seeing since the eye is an excellent judge of luminance differences (contrast), a necessary element of vision.

Of equal importance is the quality of illumination. Quality factors include lighting uniformity, control of direct and reflected glare, and proper color.

a) *Direct glare.* The glare occurring because of a direct view of the luminaires. Disability glare occurs when the presence of a bright light source close to the line of sight makes the task more difficult to see.

b) *Reflected glare.* It occurs when a high-luminance luminaire is reflected from a glossy surface. Where the reflections have no useful purpose, they can be reduced by using low-luminance luminaires, or by arranging the luminaire, task, and worker so that reflections are not directed along the line of sight.

c) *Veiling reflections.* It refers to the loss of contrast on the task surface when light emitted form the source bounces off the task into the viewer's eye. A 1% loss of contrast requires an increase of 10–15% in footcandles to maintain the same visual performance.

d) *Visual comfort probability (VCP).* It describes the probable degree to which a pattern of overhead-mounted light sources will cause discomfort. A VCP rating of 80 means that 80% of the occupants will find the glare produced by the system acceptable.

Lighting quality in a truly effective lighting installation requires careful consideration of spacial relationships between task surfaces, lighting sources, viewing angles, visual tasks, room reflectances, and other factors.

A common mistake is to compromise the quality of a lighting system to maximize quantity or efficacy. This is counterproductive since glaring, uncomfortable lighting results in complaints of headaches, eye strain, and similar problems that can affect productivity.

7.9.4 Lighting hardware

Energy efficiency is only one of the factors in the lighting hardware selection process. Physical characteristics, quality of construction, ease of servicing, and lighting quality factors are most important in a given installation. However, particular attention should be given to the parts of the system that transform electric energy into visible energy (the lamp and ballast) and the luminaire, which collects and directs that visible energy into the room. Compensation for an inefficient ballast, lamp, or luminaire cannot generally be provided by manipulation of the other variables in the lighting system.

To manipulate these system variables in order to achieve optimum lighting energy effectiveness, the engineer or designer should be familiar with various lamp types, compatible energy-efficient ballasts, and up-to-date specially designed new luminaires. In the preceding subclauses, information on these subjects has been provided to assist the task of selecting an optimum lighting system.

7.9.5 Maintenance

Maintenance of lighting systems keeps the performance of the system within basic design limits, promotes safety, and when considered in the design stage, can help to reduce the electrical demand and capital cost.

Maintenance includes the replacement of failed or deteriorated lamps, ballasts, sockets, and switches; and the cleaning of luminaires and room surfaces at suitable intervals. Group relamping has visual, electrical, and financial advantages over spot lamp replacement. As a general guide, luminaires should be cleaned annually.

Ignoring maintenance will require higher initial lumens and more fixtures or higher wattage lamps, and will become expensive because of wasted energy. When costs are calculated in units of annual dollars per footcandle, an evaluation can be made to aid the selection of the system.

7.9.6 Space-mounting height ratio

The ratio of luminaire spacing to height of the luminaire above the work plane is a critical consideration in industrial lighting design. This relation is known as the spacing to mounting height ratio (S/MH). The value S/MH in lighting fixture manufacturer catalogs is the maximum spacing that will result in relatively uniform lighting at the work plane. This spacing will provide good overlap of light between adjacent fixtures, regardless of interference from workers or equipment. The wider the fixture spacing, the greater shall be the light output from each fixture. Many industrial plants are congested with columns, equipment, and piping, which cast shadows. For this reason, individual work stations shall be considered in addition to area lighting.

HID lamps are available in lumen output sizes that span a range of more than 10:1. Initial cost of a lighting system will be lower and energy costs will be less if larger lamps and fewer luminaires are used. However, caution is required because using larger lamps and spacing luminaires far apart can produce unsatisfactory light distribution. At a spacing to mounting height ratio of 1:1, for example, the overlap of light form adjacent fixtures and lamp shielding will be reasonable and reflector brightness will be within acceptable limits. When the mounting height is equal to the spacing between fixtures, the area covered by each luminaire is equal to the square of the mounting height. Multiplying the square of the mounting height by twice the desired average maintained light level yields a good approximate value for required initial lamp lumens.

Table 7-11 offers a comparison of the energy requirements of four major lighting systems. The table summarizes the results of a fixture layout study to provide an even 100 fc to illuminate a 10 000 ft^2 (926 m^2) area. The data would also hold true for similar interior lighting applications, provided there was sufficient ceiling height for proper mounting of the fixtures. Fluorescent lighting is the optimum choice in most low-ceiling areas.

Table 7-11—Energy requirements for four major lighting systems

System	Fluorescent	HPS (400 W)	Metal halide (400 W)	Mercury (400 W)	Incandescent (1000 W)
No. of fixtures	73	40	65	118	70
Power requirement (kW)	33.2	19	30	52	70
Percent of HPS power required	175	100	158	274	368

7.10 Power factor and effect of harmonics on power quality

Lighting system elements either have an inherently high power factor or are available with integral power-factor correction. Fluorescent ballasts used in commercial buildings deliver over 90% power factor. Almost all ballasts are made with capacitors that provide not only high power factor but also phase shifting in two-lamp units to reduce stroboscopic effects. High-power-factor ballasts are almost routine for fluorescent lighting, but for many of the HID luminaires they must be specified.

Most dimming systems for discharge-type lighting depend on a wave-chopping action or use only part of the cycle to reduce power to the lighting system. These units will have an inherently lower power factor in the dimming phases because of their high harmonic content. However, the effect of the low power factor upon the buildings will be reduced because of the power reduction of the system in the dimmed phases.

The major impact that electronic ballasts have on the power line is harmonic current. Harmonics appear on the power line because of the basic electronic design of the ballast. If flicker is to be reduced, electronic ballast must first produce a pure dc voltage. Pure dc is obtained by applying energy storage capacitors after the input diode rectifier. This capacitor causes current to flow in short bursts with high peak current. (The bursts are made up of a series of sine waves composed of the fundamental frequency and harmonics.) The result, if not corrected, is a low power factor (50%) and circulating current that heats up transformers and wastes energy. Current electronic ballasts address the harmonics and power factor issue by internal design (achieving THD as low as 6%, and power factor near unity).

There is a mathematical relationship between power factor and harmonics, It can be shown that when displacement power factor is ignored, harmonic currents of 20% relate to a power factor of just over 98% and harmonic distortion of 5% requires a power factor of over 99.8%. Essentially, increased harmonics contribute to lowering the power factor.

Today, the switching frequencies of the ballasts are in the range of 20–50 kHz. Fast rise times on the switching transients at these frequencies generate harmonics up into the high-frequency spectrum. This can cause interference to radio communications, sensitive medical equipment, and computers. However, ballasts chosen for use in hospitals and computer factories will cause no interference.

Harmonic distortion of power system voltages or currents is always undesirable. Harmonic components add to system losses by creating additional heating in power system equipment. High levels of harmonic voltage distortion can cause equipment misoperation or failure. Voice communication circuits may be adversely impacted. Transformer and motor derating due to harmonic current and voltage increases the capital cost of equipment for a given application.

Future utility harmonic distortion standards may provide an additional incentive for the selection of equipment with low harmonic current distortion. The revision of IEEE Std 519-1992 is expected to include proposed limits for harmonic current generation by customers, with the utility being responsible for limiting harmonic voltage distortion to specified levels.

7.11 Interaction of lighting with other building subsystems

7.11.1 General

The lighting subsystem in a building interacts primarily with both the electrical and HVAC subsystems. It has secondary effects on building acoustical controls and fire safety. Lighting reacts with the electrical system because it uses electricity as its source of energy, and it reacts with the HVAC system because of the effects of heat dissipated from the lighting subsystem. This heat generally reduces the building heating requirements.

Based on the transfer function method (TFM) of the ASHRAE publication, the instantaneous rate of heat gain from electric lighting may be calculated from

$$q_{el} = 3.41 \, W \, F_{ul} \, F_{sa}$$

where

q_{el} is the heat gain, Btu/h
W   is the total light wattage
F_{ul} is the lighting use factor
F_{sa} is the lighting special allowance factor

The total light wattage is obtained from the ratings of all lamps installed, both for general illumination and for display use.

The use factor is the ratio of the wattage in use, for the conditions under which the load estimate is being made, to the total installed wattage.

The special allowance factor is for fluorescent luminaires or for luminaires that are ventilated or installed so that only part of their heat goes to the conditioned space. For fluorescent luminaires, the special allowance factor accounts primarily for ballast losses. ASHRAE recommends a "value of 1.20 for general applications." Industrial luminaires have special allowance factors to be dealt with individually.

7.11.2 HVAC subsystem interaction

Lighting for buildings is derived either from natural sources or from artificial, electrically operated devices. Depending on building types and location, the balance between daylight and artificial light during daylight hours varies considerably. At night, artificial lighting is required in all applications.

Sunlight provides 36 lumens of light for each Btu per hour per square foot of solar radiation. By way of comparison, table 7-12 shows this relationship for more typical interior lighting sources.

Table 7-12—Comparative output of light sources

Source	Lumens/(Btu/h) per square foot (including ballast)
Incandescent lamp	6
Mercury lamp	15
Fluorescent lamp	20
Metal-halide lamp	30
High-pressure sodium (HPS) lamp	35
Daylight	36

Thus, daylight provides light with less total associated heat than do most of the other commonly used interior lighting sources. One problem with daylight is that without proper control, more light can be introduced than may be required, and, during the cooling seasons at least, more heat is introduced than is needed or would be present with more limited artificial sources. This is particularly true in the direct beam of sunlit interiors.

There are very few studies detailing the effects of natural and artificial lighting on the energy required for heating and cooling buildings. One computer simulation analysis made for a high-rise office building [B17] shows that for each additional watt per square foot of peak lighting demand, there is an increase of approximately (0.6 kWh/yr)/ft^2 in cooling energy required; and a decrease for heating energy required of from $(0.5 \text{ kWh/yr})/\text{ft}^2$ in cold climates to $(0.1 \text{ kWh/yr})/\text{ft}^2$ in warmer climates. These results are based on a projected occupancy of 2500 equivalent full-load hours per year.

A computer simulation of this same prototypical office structure was made to determine the effects of daylight on the energy required for artificial lighting when using fluorescent troffers for weather conditions in St. Louis, Missouri. This study showed that compared with a building with venetian blinds closed to a shading coefficient of 0.6, and in which the perimeter lighting operates constantly during working hours, turning off perimeter lights and closing venetian blinds when appropriate could save approximately 25% of the annual energy of the lighting system. The annual HVAC system cooling energy requirement decreases by 10% and the annual heating energy requirement increases by 7%. The overall net effect on the building, believed to be typical of many new high-rise office structures, by using daylight to the maximum feasible extent, is to decrease the annual energy input to the building by approximately 7% (using 3414 Btu/kWh). It must be emphasized that these results pertain only to the building studied. The data could vary appreciably for other designs and other climate conditions.

7.12 Cost analysis techniques

In either new or existing installations there are usually numerous alternatives that should be examined to determine the optimum lighting system from the energy standpoint. Evaluation techniques that rank systems on the basis of watts/square foot, watthours/square foot, Btu/hour, etc., may all be applicable, given the great variety of design techniques, legal constraints, and policies that may apply. The economics of the lighting system must also be examined and as energy costs begin to more closely reflect the actual cost, scarcity, demand, and importance of various fuels, the life cycle cost (LCC) of a lighting system can be used as a good indicator of its efficiency. The heat from the lighting system should be included as a negative cost during heating months and the energy to remove this heat during cooling months should be added as an additional energy cost.

In the annual cost model (figure 7-9), initial and operating costs are determined using actual or predicted values for hardware, installation, energy, and maintenance costs. An "owning cost" is added to the annual costs to annualize the initial cost of the installation. If the cost factors have been estimated accurately, the annual cost model provides a good estimate of actual money that will be spent.

Life cycle costing (LCC). LCC is simply the evaluation of a proposal over a reasonable time period considering all pertinent costs and the time value of the money. The evaluation can take the form of a present value analysis or uniform annual cost analysis. More sophisticated analysis would include sinking fund and rate of return on extra investment. This method of analysis is not intended to be a detailed study of various systems. It is to be used by practicing engineers as a guide for comparing the advantages of alternative design cases. A more detailed analysis taking into account other items (such as future costs) could be performed. The method given here utilizes differential costs to provide a detailed comparison of systems.

		Lighting system parameter	Base	II
Basic data	1.	Rated initial lamp lumens per luminaire		
	2.	Rated lamp life (hours) at _____ hours per start		
	3.	Group replacement interval (hours)		
	4.	Average watts per lamp		
	5.	Input watts per luminaire (including ballast losses)		
	6.	Coefficient of utilization		
	7.	Ballast factor (fluorescent)		
	8.	Lamp depreciation factor		
	9.	Dirt depreciation factor		
	10.	Effective maintained lumens per luminaire $(1 \cdot 6 \cdot 7 \cdot 8 \cdot 9)$		
	10A.	Average footcandles on work surface $(10 \div \text{ft}^2/\text{luminaire})$		
	11.	Relative number of luminaires needed for equal maintained footcandles (10 of base system ÷ 10 of system compared)		
Initial costs	12.	Net cost of one luminaire		
	13.	Wiring and distribution system cost per luminaire		
	14.	Installation labor cost per luminaire		
	15.	Net initial lamp cost per luminaire		
	16.	Total initial cost per luminaire (12 + 13 + 14 + 15)		
	17.	Annual owning cost per luminaire (15% of 12 + 13 + 14)		
	18.	Relative initial cost for equal maintained footcandles $(16 \cdot 11$ of system compared ÷ 16 of base system)		
Operating costs	19.	Burning hours per year		
	20.	Number of lamps group replaced per year $(19 \cdot = \text{lamps/unit} \div 3)$		
	21.	Number of interim spot replacements $(20 \cdot = \text{burn outs in GR interval})$		
	21A.	Number of lamps spot replaced per year—No group relamping $(19 \cdot \text{lamps/unit} \div 2)$		
	22.	Replacement lamp cost per year $(20 \text{ or } 21A \cdot \text{net lamp cost})$		
	23.	Labor cost for group replacements $(20 \cdot \text{group labor rate/lamp})$ at \$____/lamp		
	24.	Labor cost for spot replacements $(21 \cdot \text{spot labor rate/lamp})$ at \$____/lamp		
	25.	Cost of cleaning per luminaire per year		
	26.	Annual energy cost per year $(5 \cdot 19 \cdot \text{¢/kWh} \div 100\,000$ at ____ ¢ kWh		
	27.	Total annual operating cost per luminaire (22 + 23 + 24 + 25 + 26)		
	28.	Relative annual operating cost for equal maintained footcandles $(27 \cdot 11$ of system compared ÷ 27 of base system)		
Total	29.	Total annual cost—owning and operating—per luminaire (17 + 27)		
	30.	Relative total annual cost for equal maintained footcandles $(29 \cdot 11$ of system compared ÷ 29 of base system)		

Figure 7-9—Annual cost work sheet

Figure 7-10 shows an outline of one method of determining costs, called the "life cycle cost analysis."

7.13 Lighting and energy standards

In 1976, the Energy Research and Development Association (ERDA) contracted with the National Conference of States on Building Codes and Standards (NCSBCS) to codify ASHRAE/IESNA 90-75. The resulting document was called "The Model Code for Energy Conservation in New Buildings." The model code has been adopted by a number of states to satisfy the requirements of Public Laws 94-163 and 94-385.

There have been several revisions to ASHRAE/IESNA 90-75 since 1976. All were included in the lighting portion of ANSI/ASHRAE/IESNA 90A-1-1988, Energy Conservation in New Building Design, and in EMS-1981, IES Recommended Lighting Power Budget Determination Procedure.

ASHRAE/IESNA 90.1-1989, Energy Efficient Design of New Buildings Except Low-Rise Residential Buildings, is the third generation document on building energy efficiency since the first publication in 1975. It sets forth design requirements for the efficient use of energy in new buildings intended for human occupancy. The requirements apply to the building envelope, distribution of energy, system and equipment for auxiliaries, heating, ventilating, air-conditioning, service water heating, lighting, and energy management. ASHRAE/IESNA 90.1-1989 is intended to be a voluntary standard which can be adopted by building officials for state and local codes.

Since 1982, the IESNA has published the LEM series: LEM-1, Lighting Power Limit Determination, which is a further refinement by combining EMS-6 and EMS-1; LEM-2, Lighting Energy Limit Determination; LEM-3, Design Considerations for Effective Building Lighting Energy Utilization; LEM-4, Energy Analysis of Building Lighting Design and Installation; and LEM-6, IES Guidelines for Unit Power Density (UPD) for New Roadway Lighting Installations. Since 1995, the IESNA has withdrawn LEM-1, LEM-2, and LEM-4 from further publication. Only LEM-3 and LEM-6 remain as active documents.

The Energy Policy Act was signed into law on October 25, 1992 by the President. Among the many provisions, this act establishes energy efficiency standards for HVAC, lighting, and motor equipment; encourages establishment of a national window energy-efficiency rating system; and encourages state regulators to pursue demand-side management (DSM) programs.

Under this bill, lighting manufacturers will have three years to stop making F96T12 and F96T12/HO 8 ft fluorescent lamps and some types on incandescent reflectors. Standard F40 lamps except in the SP and SPX or equivalent types of high color rendering lamps, would also be phased out. General-service incandescent lamps to be eliminated would include those from 30 W to 100 W, in 115 to 130 V ratings, having medium screw bases; incandescent reflector lamps from 40 W to 205 W, in 115 V to 130 V ratings, having medium screw bases, of both reflector and PAR types, having a diameter larger than 2 3/4 inches.

Life cycle cost analysis for_____ft^2

Luminaire_____   Luminaire_____

Layout _____   Layout _____

A. Lighting and air conditioning installed costs (initial)

1. Luminaire installed costs: luminaire, lamps, material, labor $_____ $_____

2. Total kW lighting: _____kW   _____kW

3. Tons of air conditioning required for lighting: (3.41 × kW/12)   _____tons   _____tons

4. First cost of air-conditioning machinery: @ $_____/ton $_____ $_____

5. Reduction of first cost of heating equipment: $_____ $_____

6. Other differential costs: $_____ $_____
   $_____ $_____
   $_____ $_____
   $_____ $_____

7. Subtotal mechanical and electrical installed cost: $_____ $_____

8. Initial taxes: $_____ $_____

9. Total costs: $_____(A1) $_____(B1)

10. Installed costs per square foot: $_____ $_____

11. Watts per square foot of lighting: _____watts   _____watts

12. Salvage (at γ years): $_____(As) $_____(Bs)

B. Annual power and maintenance costs

1. Lamps: burning hours × kW × $/kWh $_____ $_____

2. Air conditioning: operation-hours × tons × kW/ton × $/kWh $_____ $_____

3. Air conditioning maintenance: tons × $/ton $_____ $_____

4. Reduction in heating cost fuel used: _____ $_____ $_____

5. Reduced heating maintenance: MBtu × $/MBtu $_____ $_____

6. Other differential costs: $_____ $_____
   $_____ $_____
   $_____ $_____
   $_____ $_____

7. Cost of lamps: (No. of lamps _____ @ $_____/lamp per N) (Group relamping every N years, typically every one, two, or three years, depending on burning schedule.) $_____ $_____

8. Cost of ballast replacement: (No. of ballasts _____ @ $_____/ballast per n) (n = number of years of ballast life.) $_____ $_____

9. Luminaire washing cost: No. of luminaires _____ @ $_____ each. (Cost to wash one luminaire includes cost to replace or wash lamps.) $_____ $_____

10. Annual insurance cost: $_____ $_____

11. Annual property tax cost: $_____ $_____

12. Total annual power and maintenance cost: $_____(Ap) $_____(Bp)

13. Cost per square foot: $_____ $_____

Figure 7-10—Life cycle cost (LCC) analysis

There are no immediate regulations impacting HID lamps. Within 18 months of the legislation's enactment, the DOE will determine the HID types for which standards could possibly save energy and publish testing requirements for these lamps.

As far as the general-service lamps are concerned, the legislation mandates that the FTC must, within 12 months of enactment, publish rules to govern efficiency labeling for the affected lamps. All lamp packaging will be required to be clearly labeled as to the efficiency of the lamps to be sold. The efficiency will be based on their use at 120 V.

The halogen and halogen/infrared types of reflector lamps will remain the only such type of lamps on the market. ER and BR types, those intended for rough or vibration service, and many of the types excluded under the general-service incandescent portions of the law will also be excluded here.

Table 7-13 shows the proposed efficiency standards for the fluorescent lamps, and table 7-14 shows the proposed efficiency standards for incandescent reflector lamps.

Table 7-13—Proposed efficiency standards for fluorescent lamps*

Lamp type	Nominal lamp wattage (W)	Minimum average CRI	Minimum average lamp efficacy
F40	> 35	69	75
	≤ 35	45	75
F40/U	> 35	69	68
	≤ 35	45	64
F96T12	> 65	69	80
	≤ 65	45	80
F96T12/HO	> 65	69	80
	≤ 65	45	80

*Excludes lamps designed for plant growth, cold temperature service, reflectorized/aperture, impact resistance, reprographic service, colored lighting, ultraviolet, and lamps with a CRI higher than 82.

Table 7-14—Proposed efficiency standards for incandescent reflector lamps*

Nominal lamp wattage	Minimum average lamp efficacy (LPW)
40–50	10.5
51–66	11.0
67–85	12.5
86–115	14.0
116–155	14.5

*Excludes miniature, decorative, traffic signal, marine, mine, stage/studio, railway, colored lamps, and other special application types.

Effective April 28, 1994, the Federal Trade Commission (FTC) must provide manufacturers with labeling requirements for all lamps covered: fluorescent, incandescent, and reflector incandescent. Manufacturers were required to begin applying labels by 28 April 1995.

The new Energy Policy Act is all encompassing. It promises to change the way the industries produce, distribute, and utilize the valued energy resources.

7.14 Bibliography

Additional information may be found in the following sources:

[B1] Advanced Lighting Technologies Application Guidelines, California Energy Commission, Mar. 1990.

[B2] ANSI C82.1-1985 (Reaff 1992), Specifications for Fluorescent Lamp Ballasts.

[B3] ANSI C82.2-1984 (Reaff 1989), Methods of Measurement of Fluorescent Lamp Ballasts.

[B4] ANSI C82.3-1983 (Reaff 1989), Specifications for Fluorescent Lamp Reference Ballasts.

[B5] ANSI C82.4-1992, Specifications for High-Intensity Discharge and Low-Pressure Sodium Lamp Ballasts (Multiple-Supply Type).

[B6] ANSI C82.5-1990, Specifications for High-Intensity Discharge and Low-Pressure Sodium Lamp Reference Ballasts.

[B7] ANSI C82.6-1985, Methods of Measurement of High-Intensity Discharge Lamp Reference Ballasts.

[B8] ANSI/ASHRAE/IESNA 90A-1-1988, Energy Conservation in New Building Design (Sections 1–9).

[B9] ANSI/ASHRAE/IESNA 100.3-1985, Energy Conservation in Existing Buildings—Commercial.

[B10] ANSI/ASHRAE/IESNA 100.4-1984, Energy Conservation in Existing Facilities—Industrial.

[B11] ASHRAE/IESNA 90.1-1989, Energy Efficient Design of New Buildings Except Low-Rise Residential Buildings.

[B12] Burkhardt, W. C., "High Impact Acylic for Lenses and Diffusers," *Lighting Design and Application*, vol. 7, no. 4, Apr. 1977.

[B13] Chen, K., and Guerdan, E. R., "Resource Benefits of Industrial Relighting Program," *IEEE Transactions on Industry Applications*, vol. IA-15, no. 3, May/June 1979.

[B14] Chen, K.; Unglert, M. C.; and Malafa, R. L.; "Energy Saving Lighting for Industrial Applications," *IEEE Transactions on Industry Applications*, vol. IA-14, no. 3, May/June 1978.

[B15] Chen, Kao, Energy Effective Industrial Illuminating Systems, The Fairmount Press, Lilburn, Ga., 1994.

[B16] Duffy, C. K., and Stratford, R. P., "Update of Harmonic Standard IEEE-519: IEEE Recommended Practices and Requirements for Harmonic Control in Electric Power Systems," *IEEE Transactions on Industry Applications*, vol. IA-25, no. 6, 1989.

[B17] "Energy Conservation Principles Applied to Office Lighting and Thermal Operations," Federal Energy Administration Conservation Paper No. 18, Rev. Dec. 23, 1975.

[B18] IES LEM-1-1982, Lighting Power Limit Determination.

[B19] IES LEM-2-1984, Lighting Energy Limit Determination.

[B20] IES LEM-3-1987, Design Considerations for Effective Building Lighting Energy Utilization.

[B21] IES LEM-4-1984, Energy Analysis of Building Lighting Design and Installation.

[B22] IES LEM-6, IES Guidelines for Unit Power Density (UPD) for New Roadway Lighting Installations.

[B23] IRS Lighting Handbook, vols. I and II, 1984 and 1987.

[B24] Jewell, J. E.; Selkowitz, S.; and Verderber, R.; "Solid-State Ballasts Prove to be Energy Savers," *Lighting Design and Applications*, vol. 10, no. 1, Jan. 1980.

[B25] Nuckolls, J. L., Electrical Controls, *Lighting Design and Applications*, vol. 7, no. 10, Oct. 1977.

[B26] Rensselaer Polytechnic Institute Publications, Troy, NY, Specific Reports on "Electronic Ballasts," Dec. 1991.

[B27] Rensselaer Polytechnic Institute Publications, Troy, NY, Specific Reports on "Compact Fluorescent Lamps," 1992.

[B28] Rensselaer Polytechnic Institute Publications, Troy, NY, Specific Reports on "Occupancy Sensors," 1992.

[B29] Rensselaer Polytechnic Institute Publications, Troy, NY, Specific Reports on "Specular Reflectors," July 1992.

[B30] Verderber R.; Selkowitz, S.; and Berman, S.; "Energy Efficiency and Performance of Solid-State Ballasts," *Lighting Design and Application*, vol. 9, no. 4, Apr. 1979.

[B31] Wenger, L., "Architectural Dimming Controls," *Lighting Design and Application*, vol. 7, no. 6, June 1977.

Chapter 8
Cogeneration

8.1 Introduction

Cogeneration is the simultaneous production of electric power and process heat or steam that optimizes the energy supplied as fuel to maximize the energy produced for consumption.

In a conventional electric utility power plant, considerable energy is wasted in the form of heat rejection to the atmosphere through cooling towers, ponds, lakes, rivers, the sea, and boiler stacks. In a cogeneration system, heat rejection can be minimized by systems that apply the otherwise wasted energy to process systems requiring energy in the form of steam or heat.

Generation of electricity within an industrial plant is not a new concept. Electrical generation was developed within the world's industrial plants as a source of power for electrical motors and lighting. The development of central power stations with transmission and distribution systems owned by municipalities, co-ops, private utilities, and government authorities reduced the cost of generated electrical power by taking advantage of the economics realized by large generator sizes, cheaper fuels, and advancements in machine efficiencies.

As efficiencies of central stations increased and fuel prices remained low, smaller industrial facilities gradually retired their generating systems, purchasing electrical power from a local serving utility and developing process heat with boilers. As fuel prices rose through the 1970s, however, industrial users began to consider again the economies of generating electricity as a by-product of systems producing process heat.

Introduction of a cogenerator into a commercial or industrial electrical system should be studied for its economic impact on the facility's energy requirements, and for its technical impact on the operation of the facility's electrical distribution system. This chapter will explore the concerns that should be considered in both of these areas.

8.2 Forms of cogeneration

Cogeneration systems may be grouped broadly into two types: topping cycles and bottoming cycles. Virtually all cogenerators use the topping cycle that generates electricity from high-pressure steam and that uses the exhausted steam or other hot gas for process heat.

The bottoming cycle utilizes lower working temperatures in various arrangements to produce process steam or electricity. Thermal energy is first used for the process, then the exhaust energy is used to produce electricity at the bottom of the cycle. The applications for electrical generation may be limited. This cycle is most beneficial where large amounts of heat are utilized in processing, such as in rotary kilns, furnaces, or incinerators. Table 8-1 gives a summary of cogeneration technologies.

Table 8-1—Summary of cogeneration technologies

Technology	Unit size	Fuels used (present/possible in future)	Average annual availability (%)	Full-load electric efficiency (%)	Part-load (at 50% load) electric efficiency (%)	Total heat rate (Btu/kWh)	Net heat rate (Btu/kWh)	Electricity-to-steam ratio (kWh MMBtu)
Steam turbine topping	500 kW to 100 MW	Natural gas, distillate, residual, coal, wood solid waste/ coal- or biomass-derived gases and liquids	90–95	14–28	12–25	12 000 to 24 000	4500–6000	30–75
Open-cycle gas turbine topping	100 kW to 100 MW	Natural gas, distillate, treated residual/coal- or biomass-derived, gases and liquids	90–95	24–35	19–29	9750 to 14 200	5500–6500	140–225
Closed-cycle gas turbine topping	500 kW to 100 MW	Externally fired—can use most fuels	90–95	30–35	30–35	9750 to 11 400	5400–6500	150–230
Combined gas turbine/ steam turbine topping	4 MW to 100 MW	Natural gas, distillate, residual/coal- or biomass-derived gases and liquids	77–85	34–40	25–30	8000 to 10 000	5000–6000	175–320
Diesel topping	75 kW to 30 MW	Natural gas, distillate, treated residual/coal- or biomass-derived gases and liquids, slurry or powdered coals	80–90	33–40	32–39	8300 to 10 300	6000–7500	350–700
Rankine cycle bottoming; Steam	500 kW to 10 MW	Waste heat	90	10–20	Comparable to full load	17 000 to 34 100	NA	NA
Organic	2 kW to 2 MW	Waste heat	80–90	10–20	Comparable to full load	17 000 to 34 100	NA	NA
Fuel cell topping	40 kW to 25 MW	Hydrogen, distillate/coal	90–92	37–45	37–45	7500 to 9300	4300–5500	240–300
Stirling engine topping	3 to 100 kW (expect 1.5 MW by 1990)	Externally fired—can use most fuels	Not known—expected to be similar to gas turbines and diesels	35–41	34–40	8300 to 9750	5500–6500	340–500

Source: *Industrial and Commercial Cogeneration*, Office of Technology Assessment, Washington, DC, 1983.

Figure 8-1 illustrates, in a simplified manner, a widely used topping cycle consisting of high-pressure boilers, typically 600–1500 lbf/in^2, generating steam for admission to back pressure steam turbines. The steam turbine drives an electrical generator, or serves as a mechanical driver, for such equipment as fans, pumps, compressors, etc. The advantage of this system is that only the energy content of the steam required for mechanical power and losses is utilized in the turbine. The majority of the energy content remains in the back pressure steam that will be utilized in the process system. Electrical generation in the 4500–6500 Btu/kWh range is typical as compared to the usual 10 000–12 000 Btu/kWh heat rate of the electric utility. At 100% efficiency, the conversion rate is 3412 Btu/kWh. The use of this system implies a balance between kilowatt requirements and process steam requirements. If a balance does not exist, other means shall be provided to effect a balance or the difference shall be handled by outside means, such as exchanging power with the electric utility. The system illustrated in figure 8-1 typically has an output of 30–35 kW/(1000 lb/h) of steam flow and has a thermal efficiency of approximately 80%.

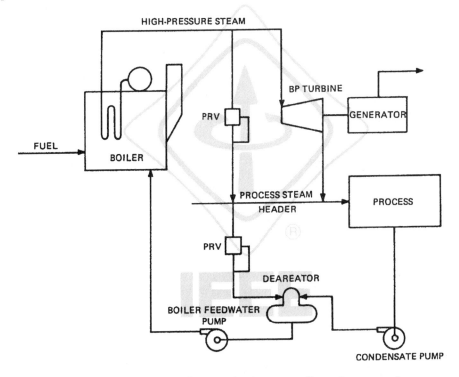

Figure 8-1—Plant topping cycle cogeneration steam system

Another highly efficient topping cycle employs the gas turbine-heat recovery boiler combination that utilizes a steam turbine in the cycle. Occasionally, the exhaust from the gas turbine can be used directly in the process, as for certain lumber drying kilns.

The combined cycle steam, as shown in figure 8-2, has a much higher kilowatt producing capability per unit of steam produced than the back pressure system in figure 8-1. Typically, the system illustrated in figure 8-2 has an output of 300–350 kW/(1000 lb/h) of steam produced at a thermal efficiency of approximately 70%.

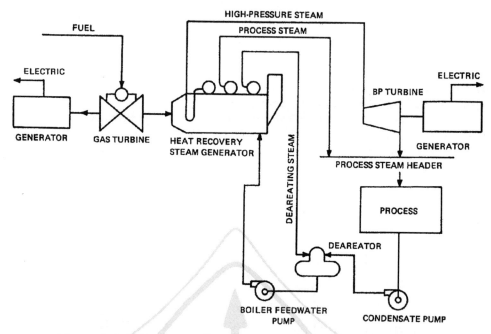

Figure 8-2—Plant combined cycle cogeneration steam system

Advanced gas turbines available for base load service have exhaust temperatures in the 900–1000 °F range with exhaust gas mass flows of typically 25–35 lb/kWh, resulting in some 5250 Btu/kWh of available exhaust energy. This high energy content can convert condensate to steam in a heat recovery boiler for use in a steam turbine or for direct use in process requirements. Generally, no additional fuel is required in the heat recovery steam generator.

The use of a steam turbine-generator results in additional incremental electrical power being generated at basically the cost of capital. There is some incremental reduction in the total quantity of waste heat steam produced because of the higher pressure level required. There is also an increase in the energy input to some small extent.

The objective in cogeneration is accomplished by utilizing the heat rejection inherent in the cycles commonly used for the production of electricity or process steam. Figure 8-3 illustrates the temperature-entropy diagram that will be used to describe the basic cycle.

In both diagrams of figure 8-3, point A represents water conditions after the boiler feed pump, and points A to B represent the energy addition in the boiler system. Point C represents the steam after going through the turbine or process and before being condensed. The area enclosed by ABCD represents the work portion of the cycle. The cross-hatched area under CD represents the rejected heat loss of the system.

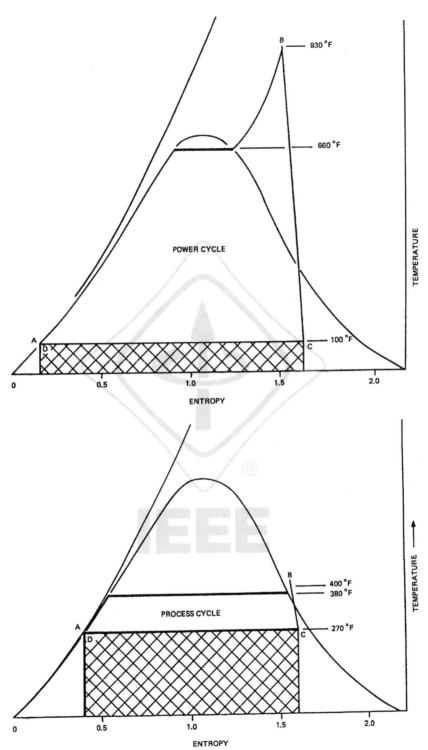

**Figure 8-3—Entropy diagrams for generation:
(a) electric output, (b) steam output**

The typical power cycle usually has a thermal efficiency in the 35% range. The condenser losses (heat rejection) are approximately 48%, and stack and miscellaneous boiler losses are approximately 17%.

The typical process steam generating cycle may operate at efficiencies in the 85% range with stack and miscellaneous boiler losses of approximately 15%.

The cogeneration approach attempts to minimize these heat rejection losses by combining the production of electricity and steam into a common facility. In a sense, the process load replaced the condenser so that useful energy is extracted from the exhaust steam. The overall efficiency is approximately 70%. However, it must be noted that each potential cogeneration facility will have unique requirements in the amount of electricity and steam required. Each facility will have varying degrees of energy savings.

8.2.1 Noncogeneration interconnections

Industrial and commercial generators may operate in parallel with a utility without recovering heat for use in a process. The Public Utilities Regulatory Policies Act of 1978, Section 210 (PURPA 210) [B10][1] recognizes generating systems that use renewable fuels such as wind, solar, and water as their primary energy source. Conventional steam generators using trash as a fuel, without heat recovery, qualify under PURPA 210, mandating utility interconnection.

Normal fossil fuel generating systems that do not recover heat and operate in parallel with a utility for peak shaving, or to provide energy during contracted interruptible load curtailment periods, are not recognized by PURPA 210 as "qualified." Utility participation to interconnect is not mandatory. The financial success of such undertakings should be evaluated carefully. Since energy efficiencies are much less than those in cogeneration systems, the savings realized are solely dependent on the generators' effect on the billing demand registered by the facility's serving utility. Redundancy requirements for the generation should be studied. Electrical requirements for interconnection would be identical to a cogeneration facility operating in parallel.

8.2.2 Types of prime movers

Selection of a prime mover must be based on the advantages and disadvantages of a particular prime mover type in a specific application. The primary considerations for the selection of a prime mover for a cogeneration application are the electrical and thermal capacity profiles of an engine generator type and how the profiles meet the load needs of the host facility. Selection is usually based upon an evaluation of the thermal and electrical load requirements, prime mover characteristics (i.e., electrical power output, recoverable thermal energy, available fuel types, etc.), and on an overall system life cycle economic analysis. Table 8-2 compares the advantages of both reciprocating and combustion turbine prime movers. Steam turbine prime movers are usually selected under a separate evaluation that is primarily contingent on the available steam power supply and facility steam needs. A brief description of each type of prime mover follows in 8.2.2.1 through 8.2.2.3.

---

[1]The numbers in brackets correspond to those of the bibliography in 8.6.

Table 8-2—Reciprocating engine and combustion turbine comparison

Reciprocating engine advantages	Combustion turbine advantages
High-speed engines have lower equipment costs. Many low-speed engines exceed combustion turbine costs.	Lower maintenance costs
Higher thermal efficiency	More efficient for high thermal loads
Greater variety of sizes and manufacturers	Lower installation costs: — Lighter weight — Smaller size — Less vibration — Engine air-cooled
Gas pressure requirements are satisfied by available utility gas distribution systems.	Superior in full-load transient frequency response
Full loading capabilities in 10 s	

8.2.2.1 Reciprocating engines

Reciprocating engines typically operate in the 720–1800 r/min range and are of either two- or four-cycle design. Ignition can be either spark or diesel, depending on the fuel selection and engine type. Electrical power output capability ranges up to 8 MW with custom designed engine generators capable of gaseous fuels. Gaseous fuel engines can be equipped with multiple carburetors, thus allowing them to operate with different gas fuels. Gaseous fuel engines may either require diesel oil primer or permit greater leakage of lube oil into the cylinders to assist combustion. Liquid-fueled engines may be spark ignited, may burn gasoline, or a diesel engine may burn light or heavy oils. For cogeneration, heat is recovered from the jacket water cooling and engine exhaust.

8.2.2.2 Combustion turbines

Combustion turbines typically operate at 3600 r/min or at higher base load speeds. Units operating above 3600 r/min use gear reducers or other types of speed reducers to drive generators at 3600 r/min or 1800 r/min. Combustion turbine generator output capacities are approaching 200 MW, with small units available below 5 MW. Combustion turbines generally require less maintenance than high- or medium-speed reciprocating engines and will generally have a lower installation cost, since they are smaller, lighter, and do not require vibration-absorbing foundations. In addition, the combustion turbine is air-cooled and does not require an elaborate cooling system. Combustion turbines use both liquid and gaseous fuels. These fuels included light and heavy oils, natural gas, and other gaseous fuels. For cogeneration, heat is recovered from the engine exhaust system.

8.2.2.3 Steam turbines

Steam turbines usually operate at either 1800 r/min or 3600 r/min with a wide variety of power output capabilities. They range in size from single-stage to heavy duty multistage machines. They may be condensing or noncondensing, with or without steam extraction, and may be of the direct drive or reduction gear drive type. In typical cogeneration applications, they are used in combined cycle facilities with combustion turbines where combustion turbine exhaust heat is recovered for steam generation.

8.3 Determining the feasibility of cogeneration

Decisions made to install a cogeneration facility are based on economic and process requirements that will vary depending on the system's operating parameters, availability of fuels, and electrical energy costs of the serving utility. It is important to compare the cost of the cogeneration system to the cost of purchasing electrical power from the utility and installing a separate boiler to satisfy the process heat requirements. This study should extend over the expected life of the plant, and be evaluated at the present worth values of the total costs. Major items for consideration are as follows:

a) Amount of process heat required (lb/h for steam)

b) Operating parameters of the heat system (steam pressure, lb/in^2)

c) Electrical plant size (kW)

d) Cost of generation equipment

    1) Heat plant

    2) Prime mover

    3) Generator and associated electrical equipment

e) Cost of capital financing—business plan

f) Hours of operation

g) Operation and maintenance costs

h) Fuel cost

i) Purchased electrical energy cost

    1) Firm contract

    2) Backup and/or maintenance electrical power

j) Expected revenues from excess electrical energy sold

8.3.1 Sizing the system

The first step in determining costs is to size the cogeneration system. The amount of heat that must be delivered to the process must be determined, and assumptions must be made on the losses in the system and the expected heat rate of the generator. With this information, the total energy requirement for the system can be determined.

Comparing the heat and electrical generation inputs and outputs of a cogeneration facility can be simplified if all energy is converted into an equivalent thermal unit. Kilowatts, barrels of oil, or British thermal units are commonly used. Figures 8-4 and 8-5 compare a typical cogeneration system supplying heat and electrical energy versus the traditional arrangement of generating heat with a boiler and purchasing electrical power generated by a utility. The cogeneration facility can produce the heat and electrical requirements of the facility with a 30% reduction in the total energy input to the system. The energy savings are attributed to the cogenerator's reduced system losses. Utilities condense the exhaust steam of a turbine and reject this heat to the atmosphere.

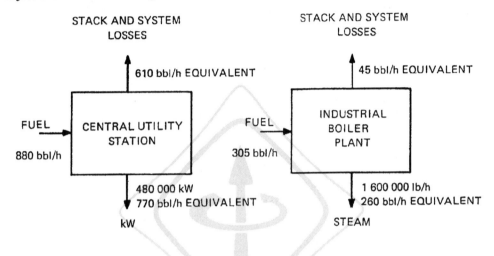

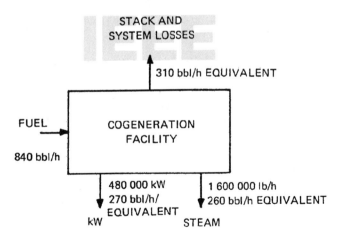

Figure 8-4—Cogeneration fuel saving potential

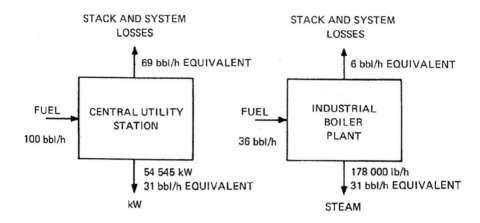

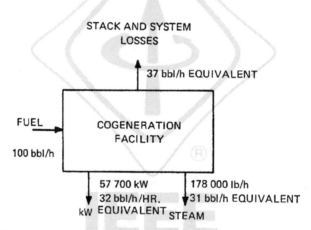

**Figure 8-5—Cogeneration fuel-saving potential
(unit comparison)**

Although the total energy consumed by the system is less, the facility's boiler capacity and fuel consumed is almost three times greater than would be necessary to supply heat for the process with a separate boiler. The cost for the increased capacity, fuel consumed, and maintenance should be offset by the savings realized in the facility's reduced electrical consumption from the utility.

8.3.2 Utility energy costs

The impact of the electrical energy supplied by the cogeneration facility should be evaluated. The facility's electrical energy output and hours of operation should be estimated for a full year, including shutdown periods for scheduled maintenance, and forced outages.

The serving utility is an excellent resource in determining the estimated annual savings that the cogeneration facility can realize. Most industrial rate structures are complex. They are sensitive to demand and time of use, which will vary during the time of day as well as from season to season. The cogeneration operation profile should be studied hour by hour, month by month on a yearly basis to ensure that the study accounts for all of the utility's rate clauses. Using a facility's yearly average cost per kilowatthour in the analysis of the generator's impact will not predict an accurate analysis of the savings involved.

The cost of maintenance and backup energy should be considered in the utility cost impact study. Most utilities offer cogenerators a special rate for either or both of these items. Maintenance energy purchases usually require notification of the shutdown in advance. The utility usually has the right to force the cogeneration shutdown to a time of day or month of year when it will have the least impact on the utility system. Backup energy, on the other hand, supplies electrical energy to the facility when the cogeneration unit experiences a forced outage from electrical or mechanical failure. The utility may have limitations on the number of hours that this service can be used during a given month or year, and will generally charge more for this energy than maintenance energy. If these tariffs are not offered or attached to the agreement between the utility and the cogenerator, costs of electrical energy during shutdowns will be charged according to the normal rates applied to the type of service supplied by the utility.

Another major item for consideration is the value of electricity sold to the utility, or wheeled to another utility over the serving utility's transmission system. This value is generally negotiable and may include credits for generation capacity. Each utility will have its own tariffs concerning energy purchases and wheeling rates on its transmission system. These energy purchase agreements are usually approved by the state public utility or service commission that regulates the serving utility.

8.3.3 Business plan

One of the early requirements in developing a cogeneration system, especially when more than one party is involved, is the development of a business plan. This plan is essential for the generation of equitable and workable systems in the design, construction, operation, and management of a multi-owned cogeneration facility. A business plan provides for the initial management of the project, provides the vehicle for financial requirements, and provides a forum for the joint development of the project by all participants. One major consideration is the state or local public utility regulations. The joint venture will probably come under the scrutiny of various federal and state agencies. Should the economics appear favorable, legal analysis of regulation is then performed.

Proper ownership shall be established. In many single-party projects, the ownership alternatives are limited. In a multi-party project, the ownership shall be decided on the basis of an equitable return to all, with consideration for the favorable tax and credit incentives available. The final decision will vary for each case and will seldom be the same for any two cases.

The economic analysis of the project should consider the ever-changing legislative environment that includes some of the following factors:

a) Environmental regulations
b) Fuel use taxes or credits
c) Tax life and depreciation schedules

8.4 Electrical interconnection

Once a feasibility study has determined that cogeneration has an economic advantage, a decision can be made concerning connection of the generation to the facility's electrical system. Most industrial or commercial facility cogeneration studies will require the size of the generator to be determined by the usable heat required for the facility's process. Operating the generator in parallel with a utility source permits the cogeneration unit to optimize the facility's requirement for heat and electrical output of its generation. The utility provides a "sink" that can accept any excess electrical generation, or that provides electrical energy if the generator operates at an output below the facility's electrical demand. Parallel operation also provides an increase in the facility's electrical system reliability. With proper relay protection and selection of interconnection equipment, the utility can provide backup energy for internal generation failures or maintenance, and the internal generation can provide backup energy during utility curtailment periods or interruptions.

Operating a cogeneration system as an isolated system is another option to the facility. This decision will require the heat source, prime mover, and generator to be sized for the facility's full electrical requirements. The process heat requirements become secondary. Efficiencies may be compromised as a result of excess heat being rejected.

8.4.1 Utility interconnection

PURPA 210 [B10] was passed to encourage cogeneration and small power production systems. It requires utilities to cooperate with these projects, to permit interconnection with their generators, and to purchase the power supplied. Interconnecting the cogeneration unit with the facility's utility source provides the greatest flexibility for optimizing generation output, heat production, and electrical reliability.

Most utilities have specific requirements governing connection of generators operating in parallel with their systems. It is important to communicate with the utility early in the project's feasibility study. Utility interconnection requirements may have a major impact on the initial costs of the facility. Each utility will have its own standards affecting the interconnection equipment and relaying selected. IEEE Std 1001-1988 is an excellent source of technical information concerning cogeneration interconnections with utilities. [2]

---

[2]Information on references may be found in 8.5.

The purpose of the interconnection requirements is to permit the safe and effective operation of independently owned generation in parallel with the utility's electrical system. The interconnection requirements for generator configurations, transformers, and relaying and circuit breaker equipment are intended to protect the utility's system and its customers from problems caused by the connection of private generation to its transmission and distribution facilities. The major concerns for these systems are as follows:

a) Energizing or back-feeding lines that are downed or open and/or equipment that would otherwise be de-energized

b) Connecting private generation to a utility system to which it is not synchronized

c) Adding to the magnitude, and/or diverting or redistributing the flow of fault current

d) Causing undervoltage and/or overvoltage on isolated systems

e) Protecting the public, utility maintenance personnel, and equipment, and maintaining the integrity of the unfaulted system

Interconnection requirements will also vary with the size and type of generator being connected. Synchronous machines will affect the system differently than induction machines. The larger the interconnected unit, the more impact the generator will have on the above mentioned concerns. Utilities may require periodic testing of the interconnection protection equipment to ensure proper operation and protection of their systems.

8.4.2 System design considerations

A major decision involves determining where the generator will be connected to the facility's electrical system. It is best to consider the generator as another source, interconnected at the service entrance disconnect. Connection to the system at buses further into the electrical system may cause difficulties with relay coordination. Avoid basing the interconnection point strictly on geographical convenience. The decision should also consider the system configuration in the event that loads will be isolated with the generation during utility electrical interruptions. This may require grouping critical loads with the generator or adding a load-shed scheme, especially if the generation capacity is not sufficient to supply the facility's entire electrical load. This decision will also have a major impact on the generator voltage rating.

8.4.3 System grounding considerations

The generator's winding connections and grounding considerations will be determined by several important factors as follows:

a) The configuration of the loads being supplied
b) The utility's system configuration
c) Ground-fault current protection coordination for the facility's and the utility's systems
d) The limit of damage from ground faults within the generator
e) The generator's ability to limit harmonic currents

A generator operating in parallel with a facility's distribution system operating at 600 V or less must also consider the single point grounding requirement as dictated by the National Electrical Code® (NEC)®(NFPA 70-1996).

8.4.4 System studies

Addition of a generator to an existing industrial or commercial power system will require studies of the resulting system. Effects on power flow, var flow, and fault current should be analyzed. The designer should also consider how the generator will react to system disturbances initiated within the facility or from the utility system.

8.4.4.1 Power flow

The output power of a cogeneration unit is optimized when the heat required for the process is satisfied. Interconnection to a utility allows this optimization. Electrical power requirements that are not fulfilled by the generator output will be supplied by the utility. Likewise, if the generation output level exceeds the facility's electrical energy needs, the excess electrical output can be sold to the interconnected utility or wheeled to another utility offering a better price. The design of the machine's governor will determine if the machine will react to changes in the process heat requirement, regulate the utility interchange energy, or be capable of following the plant electrical load if operated independently of the utility, such as during a utility interruption.

The facility's electrical system has to be capable of delivering the energy to the utility interconnection, or of distributing the energy within the facility without overloading the distribution system components, and of keeping voltages and frequencies within load tolerances.

8.4.4.2 Var flow

An analysis should be made of the generator's impact on the facility's power factor at the interconnection. Some utilities have strict power factor requirements (0.95–1.0 lagging), measure the power factor continuously, and automatically apply a surcharge if the power factor is measured outside set limits. This is of great concern when induction generators are installed. Synchronous generators can be equipped with automatic exciter controls to assist in generating vars for the facility load and to help maintain the power factor within the utility's requirements.

Whether induction or synchronous machines are installed, capacitors or synchronous condensers may be connected to help satisfy the facility var requirement. Care should be taken, however, to disconnect capacitors in the event of a sudden loss of load. The capacitors should also be switched independently of the generator's main circuit breaker or contactor.

8.4.4.3 Fault current

An analysis of the facility's electrical system for fault current flow is essential when operating a generator in parallel with the utility. Adding parallel generation will often raise distribution system fault currents to a level exceeding the interrupting and momentary current ratings of existing switchgear. The addition of current limiting reactors, dedicated unit transformers, or high-impedance grounding may avoid replacing or rebuilding of existing switchgear. The effect of fault current contributed to the utility's distribution system should also be evaluated.

8.4.4.4 System disturbances

The stability of generators operating in parallel with a utility is an important consideration. Since most cogenerators will be connected to facilities supplied by utility distribution systems, the machine will be subjected to the normal operating conditions that the distribution system must tolerate. Slow speed fault clearing, single phasing, unbalanced loads, storm damage, and vehicles striking poles are just a few. Since the utility interconnection protection is designed to isolate the machine during these problems, the utility interconnected cogenerator may be tripped more often and for longer intervals than a cogenerator that operates independently of a utility system.

The designer needs to determine if the generator and its prime mover will be tripped or if the generator will be isolated with the facility's electrical load during these utility system disturbances. A decision to isolate the machine with the facility's load will require a study to determine the stability response of the machine to the disconnection from the utility. The governor, prime mover, and exciter should be capable of responding to the suddenly increasing or decreasing load condition. During the disturbance, voltage and frequency should remain within the tolerances of the facility's loads, or a total or partial shutdown of the facility's process may result.

An excellent source for information on generator and load stability during transient conditions is IEEE Std 399-1990.

Cogenerators interconnected to utilities through dedicated distribution facilities or transmission facilities have less exposure to utility system disturbances that cause problems. Greater flexibility is also realized in designing protective systems that will respond quickly enough to permit a stable response to disturbances. This type of interconnection has a greater initial cost that is usually paid by the cogenerator.

8.4.5 Generator protection

8.4.5.1 General

Generators must be protected against mechanical, electrical, and thermal damage that may occur as a result of abnormal conditions in the plant or in the utility system to which the plant is electrically connected.

The abnormal operating conditions that may arise must be detected automatically and timely corrective action must be taken to minimize the impact. Some of these actions will be provided by the control system. The remaining actions should be provided by protective relays.

The type and extent of the protection will depend upon many considerations, some of which are as follows:

a) Capacity, number, and type of units in the plant
b) Type of power system
c) Interconnecting utility requirements
d) Facility's dependence on the plant for power
e) Manufacturer's recommendations
f) Equipment capabilities
g) Control location and extent of monitoring

The design must detect abnormal conditions quickly and isolate the affected equipment as rapidly as possible so as to minimize the extent of damage and yet retain the maximum amount of equipment in service. IEEE Std 1020-1988 and IEEE Std 242-1986 are excellent resources for information concerning protection of generators. For additional information, see also [B8].

Interconnecting and operating a generator in parallel with the utility will create some special concerns for protection of the machine. Schemes protecting the machine against unbalanced load currents, phase faults, unbalanced fault currents, out-of-step protection, and synchronizing need additional study because the utility's distribution system is constantly changing and beyond the control of the facility.

8.4.5.2 Unbalanced loads

Unbalanced loads can be detected by negative sequence voltage and/or current relays. A study must be made to determine if the typical 5% negative sequence voltage setting would allow sufficient unbalanced currents to damage a machine. Negative sequence overcurrent relays may be more effective and necessary for larger units.

8.4.5.3 Utility system faults

A machine operating in parallel with a utility will experience more disturbances than a machine operating as an isolated system. Faults on the utility's distribution circuit serving the facility will impact the generator. Coordination studies are required to determine selectivity of circuit breakers at various levels between the generator and the utility interconnection point. Decisions should be made as to how much of the facility's system will remain connected to the generator if a utility system disturbance causes automatic disconnection. The amount of load remaining with the generator will determine the impact of these disturbances to the facility's use of its backup power agreement with the serving utility.

8.4.5.4 Synchronizing

Synchronizing of the generator with the utility is vital to machine operation. Synchronizing equipment should be provided at all circuit breakers that may act as an isolation point between the generator and the serving utility. This equipment includes circuit-breaker control, generator control, synchronizing equipment, and supervising equipment. Obviously, this includes the generator's main circuit breaker. Other circuit breakers in the facility should be equipped to synchronize if all or some of the facility's load will be operated with the generator as an island when the generator is disconnected during utility distribution system disturbances.

The utility's use of automatic reclosures must be studied when designing the generator protection. If autoreclosing of the distribution system is used, the utility should be requested to block closing onto a circuit that has a voltage. The utility will object to this request if it hinders its ability to restore service to customers.

8.4.5.5 Isolated generation and utility load

Most utility parallel generation interconnection and generator protective relaying scheme requirements are not infallible in their ability to determine when the utility distribution source has been disconnected. Some utility disturbances resulting in substation circuit-breaker tripping are not caused by faults on the line serving the interconnected generation. Typical examples of these occurrences are distribution bus relay operations, operator errors, line conductors breaking and grounding on the utility side of the break, and circuit-breaker misoperations.

If the utility circuit breaker opens and load conditions in the facility and on the distribution circuit are within the ability of the generator to maintain voltage and frequency, the generator can remain connected. The only absolute protection to prevent utility reclosing on the generator is to serve the facility with a dedicated circuit and employ a transfer trip scheme between the utility's circuit breaker and the facility. Dedicated lines are usually not offered unless the facility pays for all expenses not justified by the load being served. Use of transfer trip schemes on standard distribution systems may provide little protection. Utilities may switch the facility from one line to another without notice, especially during emergency conditions. The facility might not be guaranteed a trip signal from the utility circuit breaker serving the circuit to the facility.

8.5 References

This chapter shall be used in conjunction with the following publications:

IEEE Std 242-1986 (Reaff 1991), IEEE Recommended Practice for Protection and Coordination of Industrial and Commercial Power Systems (IEEE Buff Book) (ANSI).[3]

IEEE Std 399-1990, IEEE Recommended Practice for Industrial and Commercial Power Systems Analysis (IEEE Brown Book) (ANSI).

IEEE Std 1001-1988, IEEE Guide for Interfacing Dispersed Storage and Generation Facilities with Electric Utility Systems (ANSI).

---

[3]IEEE publications are available from the Institute of Electrical and Electronics Engineers, 445 Hoes Lane, P.O. Box 1331, Piscataway, NJ 08855-1331.

IEEE Std 1020-1988 (Reaff 1994), IEEE Guide for Control of Small Hydroelectric Power Plants (ANSI).

IEEE Std 1109-1990, IEEE Guide for the Interconnection of User-Owned Substations to Electric Utilities (ANSI).

IEEE Std C37.95-1989 (Reaff 1994), IEEE Guide for Protective Relaying of Utility-Consumer Interconnections (ANSI).

NFPA 70-1996, National Electrical Code® (NEC®).[4]

8.6 Bibliography

Additional information may be found in the following sources:

[B1] California PUC Staff Report on California Cogeneration Activities: Resolution on Cogeneration, Jan. 10, 1978. "Cogeneration," *Power Engineering*, vol. 82, no. 3, Mar. 1978.

[B2] Comtois, Wilfred H., "What Is the True Cost of Electric Power from a Cogeneration Plant?" *Combustion*, pp. 8–14, Sept. 1978.

[B3] Harkins, H. L., "Cogeneration for the 1980s," Conference Record, *IEEE Industry Applications Society Annual Meeting,* 1978, no. 78CH1346-61A, pp. 1161–1168, 1978.

[B4] "Industrial and Commercial Cogeneration," Office of Technology Assessment, Congress of the United States, Washington, DC, 1983.

[B5] Javetski, John, "Cogeneration," *Power,* vol. 122, no. 4, Apr. 1978.

[B6] McFadden, R. H., "Grouping of Generators Connected to Industrial Plant Distribution Buses," *IEEE Transactions on Industry Applications,* Nov./Dec. 1981.

[B7] Nichols, Neil, "The Electrical Considerations in Cogeneration," *IEEE Transactions on Industry Applications,* vol. IA–21, no. 4, May/Jun. 1985.

[B8] *On-Site Power Generation,* Electrical Generating Systems Association.

[B9] Pope J. W., "Parallel Operation of Customer Generation," *IEEE Transactions on Industry Applications,* Jan./Feb. 1983.

[B10] Public Utility Regulatory Policies Act of 1978, 16 USC, Section 210 (PURPA 210), U.S. Consolidated Annotated Federal Statutes, Section 824A-3.

[B11] Smith, A. J., "Remarks on Cogeneration," presented at the American Society of Mechanical Engineers Conference, Oct. 25, 1977.

---

[4]The NEC® is available from Publications Sales, National Fire Protection Association, 1 Batterymarch Park, P.O. Box 9101, Quincy, MA 02269-9101.

INDEX

A

G

Gas turbine topping, 324–326
Gear drives, 170–171
General lighting, definition of, 279
Generator protection, cogeneration systems, 337–339
  isolated generation and utility load, 339
  synchronizing, 338–339
  unbalanced loads, 338
  utility system faults, 338
Geothermal heating, 132
Glare, 305
  control, luminaires, 303–304
  definition of, 279
Glass products industry, electricity consumption by, 226
Governmental regulatory agencies, 9–10
Grinders, pneumatic, air consumption by, 117
Grounding of cogeneration systems, 335
Guaranteed efficiency, definition of, 140

H

Harmonic analyzer, definition of, 193
Harmonic factor, definition of, 193
Harmonic order, definition of, 194
Harmonics, 313–314
  definition of, 193
  filters, 191
  limits, 193–196
  motors and, 172–175
Heat exchangers, 130–133
Heating
  commercial facilities, electrical energy use pattern, 92
  dielectric materials, radio frequency heating of, 105–106
  electron beam (EB), 135
  induction, 107–108
  materials
    infrared (IR), 100–101
    resistance heating. See Resistance heating of materials
    ultraviolet (UV), 99–100

motor drive consumption, 137
space heating, 101–102, 122
tungsten halogen, 106–107
Heating, ventilating, and air conditioning (HVAC) systems. See HVAC and energy management
Heat leaks, eliminating, 16
Heat pipes (gas-to-gas heat recovery), 130–132
Heat pumps
  electric industrial, 133–134
  solar-assisted, 113–114
Heat-recovery heat pumps, 133
Heat-recovery systems, 129–134
  efficiency of heat recovery, 130
  electric industrial heat pumps, 133–134
  equipment, 130–133
  heat pipe, 131–132
  liquid-to-liquid plate heat exchangers, 132–133
  run-around coil (liquid-coupled indirect heat exchanger), 131
  thermal wheel, 131
  indirect evaporative cooling, 130
Heat tracing, 102–103
Helical gears, 170–171
Helical rotary compressors, 128–129
  economizers, 128
  variable-volume ratio, 129
High head dams, 4
High-intensity discharge (HID) lighting, 291–293. See also Lighting
  ballasts, 294, 296, 299–303
  electronic, 301–303
  high-pressure sodium (HPS) lamps, 294, 301
  metal halide, 294
  characteristics, 287–288
  dimming, 309
  efficacy, 286
High-low ballasts, 299
High-pressure sodium (HPS) lighting, 292–293, 301
  ballasts, 294, 301
  efficacy, 285, 315
  new luminaires, 307
Hoists, 99

NOTES

NOTES

NOTES

NOTES

NOTES

NOTES

NOTES

NOTES

NOTES

NOTES

NOTES